Roofing

Level Two

Trainee Guide
Second Edition

Pearson

NCCER
President and Chief Executive Officer: Boyd Worsham
Senior Director of Innovation and Advancement: Jennifer Wilkerson
Chief Learning Officer: Lisa Strite
Roofing Project Manager: Lauren Corley
Senior Manager of Projects: Chris Wilson
Senior Production Manager: Erin O'Nora
Testing/Assessment Project Manager: Elizabeth Schlaupitz
Testing and Assessment Project Coordinator: Chelsi Santana
Technical Writers: Karyn Payne, Natalie Hasty
Art Manager: Carrie Pelzer
Production Artist: Judd Ivines
Permissions Specialists: Kay Lewis, Amanda Smith
Desktop Publishing Manager: James McKay
Production Specialists: Gene Page, Eric Caraballoso, Daphney Milian
Digital Content Coordinators: Rachael Downs, Yesenia Tejas
Managing Editor: Graham Hack
Editors: Jordan Hutchinson, Karina Kuchta, Hannah Murray, Zi Meng

Pearson
Manager of Project Management: Vanessa Price
Senior Content Producer: Alexandrina B. Wolf
Employability Solutions Coordinator: Monica Perez
Content Producers: Alma Dabral, Aurodeep Mukherjee
Composition: NCCER

Text Fonts: Palatino and Univers
Content Technologies: Gnostyx

Cover Image
Cover image courtesy of F.J.A. Christiansen Roofing Co., Inc., A Tecta America Company

Credits and acknowledgments for content borrowed from other sources and reproduced, with permission, in this textbook appear at the end of each module.

ISBN-13: 978-0-13-749131-5

Pearson

ScoutAutomatedPrintCode

Preface

To the Trainee

Roofs are the first line of defense against the elements. A roof means more than shelter; it means security. There is a wealth of career potential and opportunity in roofing for people with various skill sets and levels of education. The modules contained in this level address foundational knowledge essential for all roofing professionals.

NCCER wishes you success as you embark on your second year of training in the roofing trade, and we hope that you'll continue your training outside of this series. By taking advantage of training opportunities as they arise, you'll demonstrate initiative and a desire to learn—qualities that are present in the industry's best professionals.

New with *Roofing Level Two*

NCCER and Pearson are pleased to present *Roofing Level Two* in partnership with the National Roofing Contractors Association (NRCA).

Each module in this book provides an in-depth review of a different roof system, familiarizing trainees with their tools and components and preparing them to install each system safely and effectively. NCCER module 16212, *Roofing Service and Repair*, introduces the repair careers in the industry and overviews basic repair procedures.

This program offers two interim credentials as potential career paths: Steep-Slope Roofer and Low-Slope Roofer. Interim credentials provide training pathways and credentials for trainees who are seeking specialized careers. These paths allow trainees to develop and prove competency in their particular area of expertise—whether steep- or low-slope roofing. Trainees may select to pursue one or both interim credentials.

For more details on requirements for interim credentials and other information about this curriculum, visit the Roofing craft details page in the Program Resources section of NCCER's website.

For more details on ProCertification®, visit NRCA's website at **www.nrca.net/procertification**.

We wish you success as you progress through this training program. If you have any comments on how NCCER might improve upon this textbook, please complete the User Update form using the QR code at the back of each module or by using the form on our website. NCCER appreciates and welcomes its customers' feedback. You may submit yours by emailing **support@nccer.org**. When doing so, please identify feedback on this title by listing *#RoofingL2* in the subject line.

Our website, **www.nccer.org**, has information on the latest product releases and training.

NCCER Standardized Curricula

NCCER is a not-for-profit 501(c)(3) education foundation established in 1996 by the world's largest and most progressive construction companies and national construction associations. It was founded to address the severe workforce shortage facing the industry and to develop a standardized training process and curricula. Today, NCCER is supported by hundreds of leading construction and maintenance companies, manufacturers, and national associations. The NCCER Standardized Curricula was developed by NCCER in partnership with Pearson, the world's largest educational publisher.

Some features of the NCCER Standardized Curricula are as follows:

- An industry-proven record of success
- Curricula developed by the industry, for the industry
- National standardization providing portability of learned job skills and educational credits
- Compliance with the Office of Apprenticeship requirements for related classroom training (*CFR 29:29*)
- Well-illustrated, up-to-date, and practical information

NCCER also maintains the NCCER Registry, which provides transcripts, certificates, and wallet cards to individuals who have successfully completed a level of training within a craft in NCCER's Curricula. *Training programs must be delivered by an NCCER Accredited Training Sponsor in order to receive these credentials.*

For information on NCCER's credentials and the NCCER Registry, contact NCCER Customer Service at 1-888-622-3720 or visit **www.nccer.org**.

Special Features

In an effort to provide a comprehensive and user-friendly training resource, this curriculum showcases several informative features. Whether you are a visual or hands-on learner, these features are intended to enhance your knowledge of the construction industry as you progress in your training. Some of the features you may find in the curriculum are explained below.

Introduction

This introductory page, found at the beginning of each module, lists the module Objectives, Performance Tasks, and Trade Terms. The Objectives list the knowledge you will acquire after successfully completing the module. The Performance Tasks give you an opportunity to apply your knowledge to real-world tasks. The Trade Terms are industry-specific vocabulary that you will learn as you study the module.

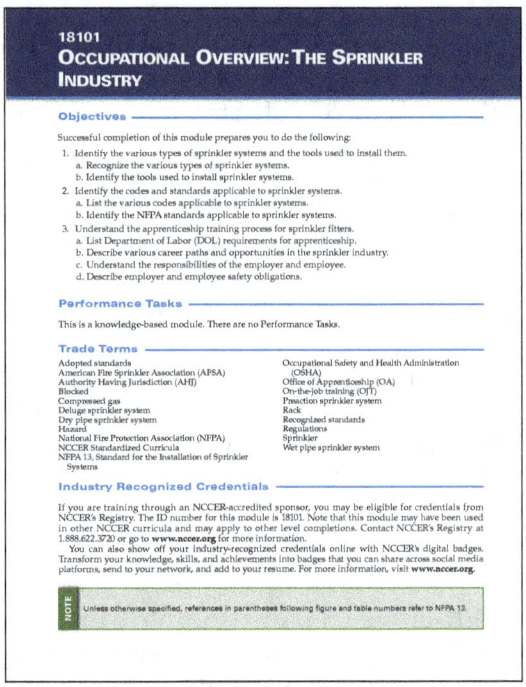

18101
OCCUPATIONAL OVERVIEW: THE SPRINKLER INDUSTRY

Objectives

Successful completion of this module prepares you to do the following:

1. Identify the various types of sprinkler systems and the tools used to install them.
   a. Recognize the various types of sprinkler systems.
   b. Identify the tools used to install sprinkler systems.
2. Identify the codes and standards applicable to sprinkler systems.
   a. List the various codes applicable to sprinkler systems.
   b. Identify the NFPA standards applicable to sprinkler systems.
3. Understand the apprenticeship training process for sprinkler fitters.
   a. List Department of Labor (DOL) requirements for apprenticeship.
   b. Describe various career paths and opportunities in the sprinkler industry.
   c. Understand the responsibilities of the employer and employee.
   d. Describe employer and employee safety obligations.

Performance Tasks

This is a knowledge-based module. There are no Performance Tasks.

Trade Terms

Adopted standards
American Fire Sprinkler Association (AFSA)
Authority Having Jurisdiction (AHJ)
Blocked
Compressed gas
Deluge sprinkler system
Dry pipe sprinkler system
Hazard
National Fire Protection Association (NFPA)
NCCER Standardized Curricula
NFPA 13, Standard for the Installation of Sprinkler Systems
Occupational Safety and Health Administration (OSHA)
Office of Apprenticeship (OA)
On-the-job training (OJT)
Preaction sprinkler system
Rack
Recognized standards
Regulations
Sprinkler
Wet pipe sprinkler system

Industry Recognized Credentials

If you are training through an NCCER-accredited sponsor, you may be eligible for credentials from NCCER's Registry. The ID number for this module is 18101. Note that this module may have been used in other NCCER curricula and may apply to other level completions. Contact NCCER's Registry at 1.888.622.3720 or go to **www.nccer.org** for more information.

You can also show off your industry-recognized credentials online with NCCER's digital badges. Transform your knowledge, skills, and achievements into badges that you can share across social media platforms, send to your network, and add to your resume. For more information, visit **www.nccer.org**.

NOTE: Unless otherwise specified, references in parentheses following figure and table numbers refer to NFPA 13.

Trade Features

Trade features present technical tips and professional practices based on real-life scenarios similar to those you might encounter on the jobsite.

Bowline Trivia

Some people use this saying to help them remember how to tie a bowline: "The rabbit comes out of his hole, around a tree, and back into the hole."

Figures and Tables

Photographs, drawings, diagrams, and tables are used throughout each module to illustrate important concepts and provide clarity for complex instructions. Text references to figures and tables are emphasized with *italic* type.

Figure 31 Marking a cutting line.

Notes, Cautions, and Warnings

Safety features are set off from the main text in highlighted boxes and categorized according to the potential danger involved. Notes simply provide additional information. Cautions flag a hazardous issue that could cause damage to materials or equipment. Warnings stress a potentially dangerous situation that could result in injury or death to workers.

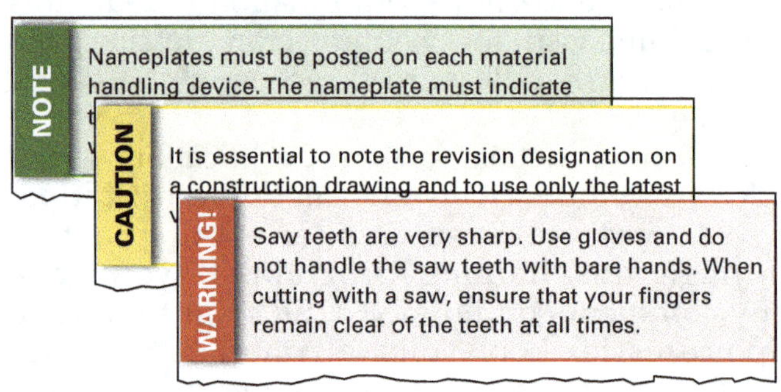

Going Green

Going Green looks at ways to preserve the environment, save energy, and make good choices regarding the health of the planet. Through the introduction of new construction practices and products, you will see how the greening of America has already taken root.

> **Going Green**
>
> **Big Horn Wind Farm**
>
> The Big Horn wind farm uses 133 GE turbines, each rising 119 meters (389 feet) from the ground to the top of the blade tips. The wind farm uses only a

Did You Know?

The *Did You Know?* features offer hints, tips, and other helpful bits of information.

> Did You Know?
>
> **How Blueprints Started**
>
> The process for making blueprints was developed in 1842 by an English astronomer named Sir John F. Herschel. The method involved coating a paper with a special chemical. After the coating dried, an original hand drawing was placed on top of the paper. Both papers were then covered with a piece of glass and set in the sunlight for about an hour. The coated paper was developed much like a photograph. After a cold-water wash, the coated paper turned blue, and the lines of the drawing remained white.

Step-by-Step Instructions

Step-by-step instructions are used throughout to guide you through technical procedures and tasks from start to finish. These steps show you not only how to perform a task but also how to do it safely and efficiently.

> Perform the following steps to erect this system area scaffold:
>
> *Step 1* Gather and inspect all scaffold equipment for the scaffold arrangement.
>
> *Step 2* Place appropriate mudsills in their approximate locations.
>
> *Step 3* Attach the screw jacks to the mudsills.
>
> *Step 4* Adjust the screw jacks to near their lowest position.
>
> *Step 5* Determine the location of the highest

Trade Terms

Each module presents a list of Trade Terms that are discussed within the text and defined in the Glossary at the end of the module. These terms are denoted in the text with **bold, blue type** upon their first occurence. To make searches for key information easier, a comprehensive Glossary of Trade Terms from all modules is located at the back of this book.

> During a rigging operation, the **load** being lifted or moved must be connected to the apparatus, such as a crane, that will provide the power for movement. The connector—the link between the load and the apparatus—is often a sling made of synthetic, chain, or **wire rope** materials. This section focuses on three types of slings:
>
> - Synthetic slings
> - Alloy steel chain slings
> - Wire rope slings

Section Review

Section Review questions can be found at the end of each section to test your knowledge of the content.

> **1.0.0 Section Review**
>
> 1. The type of sprinkler system that might be found in an aircraft hangar is a _____.
>   a. dry pipe sprinkler system
>   b. wet pipe sprinkler system
>   c. preaction sprinkler system
>   d. deluge sprinkler system
> 2. A type of ladder that provides a platform for standing is a(n) _____.
>   a. A-frame
>   b. extension trestle ladder
>   c. podium ladder
>   d. movable leg ladder

Review Questions

Review Questions are provided to reinforce the knowledge you have gained. This makes them a useful tool for measuring what you have learned.

> **Review Questions**
>
> 1. Identification tags for slings must include the _____.
>   a. type of protective pads to use
>   b. type of damage sustained during use
>   c. color of the tattle-tail
>   d. manufacturer's name or trademark
> 2. The type of wire rope core that is susceptible to heat damage at relatively low temperatures is the _____.
>   a. fiber core
>   b. strand core
>   c. independent wire rope core
>   d. metallic link supporting core
> 3. Synthetic slings must be inspected _____.
>   a. once every month
>   b. visually at the start of each work week
>   c. before every use
>   d. once wear or damage becomes apparent
> 4. An alloy steel chain sling must be removed from service if there is evidence that _____.
> 7. Chain hoists are able to lift heavy loads by utilizing a _____.
>   a. rope and pulley system
>   b. rigger's strength
>   c. stationary counterweight
>   d. gear system
> 8. Before attempting to lift a load with a chain hoist, make sure that the _____.
>   a. hoist is secured to a come-along
>   b. load is properly balanced
>   c. tag lines are properly anchored
>   d. tackle is connected to its power source
> 9. A hitch configuration that allows slings to be connected to the same load without using a spreader beam is a _____.
>   a. double-wrap hitch
>   b. choker hitch
>   c. bridle hitch
>   d. basket hitch
> 10. To make the emergency stop signal that is

NCCER Standardized Curricula

NCCER's training programs comprise more than 80 construction, maintenance, pipeline, and utility areas and include skills assessments, safety training, and management education.

Construction and Maintenance

Boilermaking
Cabinetmaking
Carpentry
Concrete Construction
Construction Craft Laborer
Construction Technology
Core: Introduction to Basic Construction Skills
Drywall
Electrical
Electronic Systems Technician
Heating, Ventilating, and Air Conditioning
Heavy Equipment Operations
Heavy Highway Construction
Hydroblasting
Industrial Coating and Lining Application Specialist
Industrial Maintenance Electrical and Instrumentation Technician
Industrial Maintenance Mechanic
Instrumentation
Ironworking
Manufactured Construction Technology
Masonry
Mechanical Insulating
Millwright
Mobile Crane Operations
Painting
Painting, Industrial
Pipefitting
Pipelayer
Plumbing
Reinforcing Ironwork
Rigging
Roofing
Scaffolding
Sheet Metal
Signal Person
Site Layout
Sprinkler Fitting
Tower Crane Operator
Welding

Maritime

Maritime Industry Fundamentals
Maritime Electrical
Maritime Pipefitting
Maritime Structural Fitter
Maritime Welding
Maritime Aluminum Welding

Green/Sustainable Construction

Building Auditor
Fundamentals of Weatherization
Introduction to Weatherization
Sustainable Construction Supervisor
Weatherization Crew Chief
Weatherization Technician
Your Role in the Green Environment

Energy

Alternative Energy
Introduction to the Power Industry
Introduction to Solar Photovoltaics
Power Generation Maintenance Electrician
Power Generation I&C Maintenance Technician
Power Generation Maintenance Mechanic
Power Line Worker
Power Line Worker: Distribution
Power Line Worker: Substation
Power Line Worker: Transmission
Solar Photovoltaic Systems Installer
Wind Energy
Wind Turbine Maintenance Technician

Pipeline

Abnormal Operating Conditions, Control Center
Abnormal Operating Conditions, Field and Gas
Corrosion Control
Electrical and Instrumentation
Field and Control Center Operations
Introduction to the Pipeline Industry
Maintenance
Mechanical

Safety

Fall Protection Orientation
Field Safety
Safety Orientation
Safety Technology

Supplemental Titles

Applied Construction Math
Tools for Success

Management

Construction Workforce Development Professional
Fundamentals of Crew Leadership
Mentoring for Craft Professionals
Project Management
Project Supervision

Spanish Titles

Acabado de concreto: nivel uno (*Concrete Finishing Level One*)
Aislamiento: nivel uno (*Insulating Level One*)
Albañilería: nivel uno (*Masonry Level One*)
Andamios (*Scaffolding*)
Carpintería: Formas para carpintería, nivel tres (*Carpentry: Carpentry Forms, Level Three*)
Currículo básico: habilidades introductorias del oficio (*Core Curriculum: Introductory Craft Skills*)
Electricidad: nivel uno (*Electrical Level One*)
Herrería: nivel uno (*Ironworking Level One*)
Herrería de refuerzo: nivel uno (*Reinforcing Ironwork Level One*)
Instalación de rociadores: nivel uno (*Sprinkler Fitting Level One*)
Instalación de tuberías: nivel uno (*Pipefitting Level One*)
Instrumentación: nivel uno, nivel dos, nivel tres, nivel cuatro (*Instrumentation Levels One through Four*)
Orientación de seguridad (*Safety Orientation*)
Paneles de yeso: nivel uno (*Drywall Level One*)
Seguridad de campo (*Field Safety*)

Acknowledgments

This curriculum was revised as a result of the vision and leadership of the following sponsors:

CMR Construction & Roofing
Deer Park Roofing
Don Kennedy Roofing Co.
Eagle Roofing
Florida Roofing and Sheet Metal Contractors Association
GAF
Garlock-French Roofing
National Roofing Contractors Assocation
Nations Roof
PDM Architects
The Arizona Roofer
University of Florida

This curriculum would not exist were it not for the dedication and unselfish energy of those volunteers who served on the Authoring Team. A sincere thanks is extended to the following:

Alison LaValley
Amy Staska
Brian Davis
Bryan Karel
Clay Thomas
Doug Duncan
Gary Howes
Glenn Watson
Grant Brock
Henry Staggs
Jean-Paul Grivas
John Schehl
Jon Goodman
Josey Parks
Mark Graham
Mike Silvers
Nick Sabino
Rich Trewyn
Tom Shanahan
Tyler Allwood

NCCER Partners

NCCER partnering organizations are national associations and organizations that share a common interest in the goals and objectives of NCCER. To learn more about NCCER business partners, go to **www.nccer.org/about-us/partners**.

You can also scan this code using the camera on your phone or mobile device to view these partnering organizations.

NRCA
NATIONAL ROOFING CONTRACTORS ASSOCIATION

Figure Credit: Courtesy of the National Roofing Contractors Assoc.

The National Roofing Contractors Association (NRCA) is one of the construction industry's most respected trade associations. It is the voice of roofing professionals and the leading authority in the roofing industry for information, education, technology, and advocacy.

Through NRCA's ProCertification®, experienced and skilled roofing workers can advance their careers by becoming certified in specific roof system installations. ProCertifications currently offered by NRCA include the following:

- Architectural Metal Flashings and Accessories Installer
- Asphalt Shingles Installer
- Clay and Concrete Tile Systems Installer
- EPDM Systems Installer
- Roofing Foreman
- Thermoplastic Systems Installer (e.g., TPO, PVC)

For more information, visit NRCA's website at **www.nrca.net**.

Contents

Module One

Asphalt Shingle Roof Systems

Discusses asphalt shingles as well as the tools and procedures used in the application of these systems. Describes appropriate attachment techniques and safety precautions involved in asphalt shingles installation and repair. (Module ID 16201; 15 Hours)

Module Two

Clay and Concrete Tile Roof Systems

Introduces tile roof systems as well as the components and tools unique to their installation. Describes tile roof application procedures, safety considerations, and flashing techniques. Discusses tile repair methods. (Module ID 16202; 12.5 Hours)

Module Three

Wood Roof Systems

Outlines the components, tools, and installation methods used for wood roof systems. Distinguishes between wood shingles and wood shakes. Identifies common mistakes in the installation of wood roof systems and introduces basic repair methods. (Module ID 16203; 12.5 Hours)

Module Four

Slate Roof Systems

Describes slate roof system components, styles, tools, and safety considerations. Details appropriate procedures for the installation and repair of these systems. (Module ID 16204; 12.5 Hours)

Module Five

Metal Roof Systems

Describes the various types of metal roofs, including architectural, structural, and metal shingle systems. Discusses the seam types, methods, and tools used in metal roof installation. Details the proper tools and techniques used for the installation of these features. (Module ID 16205; 15 Hours)

Module Six

Thermoplastic Roof Systems

Describes types of thermoplastic roof systems and outlines their installation and repair. Identifies types of field membranes, pre-manufactured flashing accessories, attachment components, and related tools. (Module ID 16206; 10 Hours)

Module Seven

EPDM Roof Systems

Describes types of EPDM roof systems and their components. Identifies field membrane types, pre-manufactured flashing accessories, attachment methods, and components of EPDM roof systems. Describes installation and repair of these systems. (Module ID 16207; 10 Hours)

Module Eight

Built-Up Roof Systems

Introduces types of built-up roof (BUR) systems and their components. Outlines basic installation and repair methods associated with BUR systems, as well as related tools. (Module ID 16208; 15 Hours)

Module Nine

Modified Bitumen Roof Systems

Describes types and characteristics of modified bitumen (mod-bit) roof systems. Discusses the components, installation procedure, basic repair methods, and tools used for mod-bit systems. (Module ID 16209; 15 Hours)

Module Ten

Liquid-Applied Roofing

Describes liquid-applied roofing products and explains the difference between roof coatings and liquid-applied roof membranes. Outlines procedures and tools used to install and repair liquid-applied roofing. (Module ID 16210; 12.5 Hours)

Module Eleven

Roofing Service and Repair

Introduces service and repair in the roofing industry and related career opportunities for roofers. Describes the general process of repair as well as specific repairs for specific steep- and low-slope roof systems. (Module ID 16212; 15 Hours)

Glossary

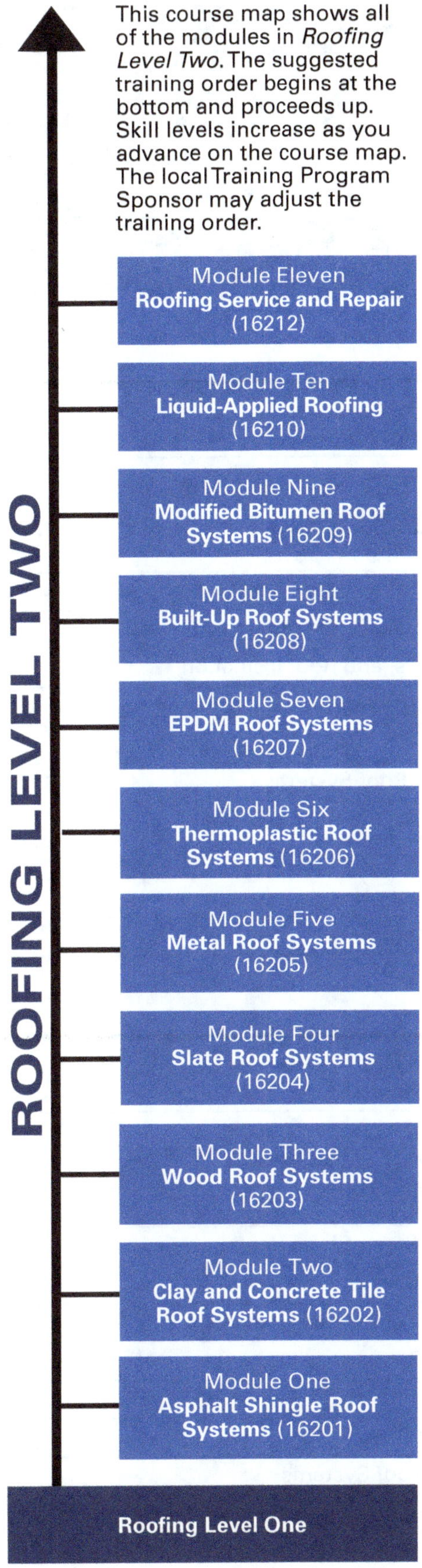

This course map shows all of the modules in *Roofing Level Two*. The suggested training order begins at the bottom and proceeds up. Skill levels increase as you advance on the course map. The local Training Program Sponsor may adjust the training order.

Roofing

The Roofing curriculum has two possible career paths: Steep-Slope Roofer and Low-Slope Roofer. The tables below indicate the modules and registry ID numbers required for each credential. The tables also indicate the level of the craft where each module can be found.

Steep-Slope Roofer	Module ID	Level
Introduction to Roofing	16101	Level One
Roofing Safety	16102	Level One
Fall Protection Orientation	75901	Level One
Drawings in Roofing	16103	Level One
Introduction to Steep-Slope Roofing	16105	Level One
Introduction to Low-Slope Roofing	16106	Level One
Substrates, Decks, and Roof Insulation	16104	Level One
Sheet Metal in Roofing	16107	Level One
Rigging Practices	38102	Level One
Asphalt Shingle Roof Systems	16201	Level Two
Clay and Concrete Tile Roof Systems	16202	Level Two
Wood Roof Systems	16203	Level Two
Slate Roof Systems	16204	Level Two
Metal Roof Systems	16205	Level Two

Low-Slope Roofer	Module ID	Level
Introduction to Roofing	16101	Level One
Roofing Safety	16102	Level One
Fall Protection Orientation	75901	Level One
Drawings in Roofing	16103	Level One
Introduction to Steep-Slope Roofing	16105	Level One
Introduction to Low-Slope Roofing	16106	Level One
Substrates, Decks, and Roof Insulation	16104	Level One
Sheet Metal in Roofing	16107	Level One
Rigging Practices	38102	Level One
Metal Roof Systems	16205	Level Two
Thermoplastic Roof Systems	16206	Level Two
EPDM Roof Systems	16207	Level Two
Built-Up Roof Systems	16208	Level Two
Modified Bitumen Roof Systems	16209	Level Two
Liquid-Applied Roofing	16210	Level Two

Asphalt Shingle Roof Systems

Overview

Since the introduction of asphalt shingles at the start of the twentieth century, this roof system has established itself as the market choice for an affordable, more fire-resistant alternative to traditional wood shakes and shingles. Asphalt shingles come in a variety of colors, shapes, and thicknesses designed to suit a wide variety of residential roofs. Roofing craft professionals need to be familiar with features of asphalt shingle roofs so they are prepared to safely install this popular steep-slope system.

Module 16201

Trainees with successful module completions may be eligible for credentialing through the NCCER Registry. To learn more, go to **www.nccer.org** or contact us at 1.888.622.3720. Our website, **www.nccer.org**, has information on the latest product releases and training.

Your feedback is welcome. You may email your comments to **curriculum@nccer.org**, send general comments and inquiries to **info@nccer.org**, or fill in the User Update form at the back of this module.

This information is general in nature and intended for training purposes only. Actual performance of activities described in this manual requires compliance with all applicable operating, service, maintenance, and safety procedures under the direction of qualified personnel. References in this manual to patented or proprietary devices do not constitute a recommendation of their use.

16201 V1.0

16201
Asphalt Shingle Roof Systems

Objectives

Successful completion of this module prepares you to do the following:

1. Identify the background of asphalt shingles and the tools commonly used to install this system.
   a. Describe the types and styles of asphalt shingles.
   b. Identify the hand tools and power tools used in asphalt shingle installation.
2. Explain asphalt shingle installation.
   a. Learn the procedures used to safely install asphalt shingles.
   b. Identify the flashings and water-shedding details involved with asphalt shingle installation.
   c. Overview how to replace a damaged shingle.

Performance Tasks

Under supervision, you should be able to do the following:

1. Don all appropriate PPE for asphalt shingle installation.
2. Install drip edge metal and manufacturer-recommended underlayment to a wood deck.
3. Properly install starter strip shingles and courses of field shingles according to manufacturer instructions. Be sure to use the correct exposure and offset measurements for your system.
4. Install valley underlayment, valley metal, and valley shingles according to manufacturer instructions.
5. Add appropriate flashings to a penetration or vertical-surface protrusion.
6. Identify a shingle that needs replacement and perform the repair.

Trade Terms

Backnailing
Fiberglass base
Hook blade
Magnet sweeper
Roofing hammer
Starter strip shingles
Taper
Tin snips
Ultraviolet (UV) radiation resistance
Valley centerline

Industry Recognized Credentials

If you are training through an NCCER-accredited sponsor, you may be eligible for credentials from NCCER's Registry. The ID number for this module is 16201. Note that this module may have been used in other NCCER curricula and may apply to other level completions. Contact NCCER's Registry at 1.888.622.3720 or go to **www.nccer.org** for more information.

You can also show off your industry-recognized credentials online with NCCER's digital badges. Transform your knowledge, skills, and achievements into badges that you can share across social media platforms, send to your network, and add to your resume. For more information, visit **www.nccer.org**.

Videos and Digital Resources

How to Access Resources

This craft has additional videos and resources to enhance your learning experience. To view these resources, scan the QR below. The videos and resources are separated by module.

You can scan this code using the camera on your phone or mobile device to view these videos and resources.

Contents

1.0.0 Background and Tools 1
1.1.0 Types and Styles 2
1.1.1 Base 2
1.1.2 Styles 3
1.2.0 Tools 3
1.2.1 Utility Knife 4
1.2.2 Tin Snips 4
1.2.3 Roofing Hammer 4
1.2.4 Tear-off Tools 4
1.2.5 Magnet Sweeper 4
1.2.6 Nail Gun 5
1.2.7 Chalk Lines 5
1.2.8 Roofing Nails 5
2.0.0 Installation and Repair 7
2.1.0 Installation Procedures and Components 7
2.1.1 Underlayments 8
2.1.2 Starter Strip Shingles 9
2.1.3 Field Installation 9
2.1.4 Valleys 10
2.1.5 Hips and Ridges 15
2.2.0 Flashings 15
2.2.1 Eaves and Rakes 15
2.2.2 Vertical-Surface Protrusions 17
2.2.3 Penetrations 19
2.3.0 Replacing a Damaged Shingle 19

Figures

Figure 1 Asphalt shingles on a home 1
Figure 2 Wood-shake roof 2
Figure 3 Anatomy of a shingle 2
Figure 4 Fiberglass roll 3
Figure 5 Architectural shingles 3
Figure 6 Specialty shingles 3
Figure 7 Three-tab shingles 3
Figure 8 Utility knife with straight-edge blade inserted and separate hook blade 4
Figure 9 Set of color-coded tin snips 5
Figure 10 Roofing hammers 5
Figure 11 Magnet sweeper 5
Figure 12 Nail gun 6
Figure 13 Roofing nail installation 6
Figure 14 Steep-slope roof anatomy 7
Figure 15 Underlayment installation 8
Figure 16 Starter strip roll 9
Figure 17 Aligning starter strip shingles 10
Figure 18 Three-tab shingle offset pattern 10
Figure 19 Example of stair-step offset 11
Figure 20 Open versus closed-cut valleys 12
Figure 21 Closed-cut valley 13
Figure 22 Open valley 14
Figure 23 No-cut valley 16
Figure 24 Hip and ridge shingles 17
Figure 25 How to cut hip and ridge caps from three-tab shingles 17
Figure 26 Roof area flashing 17
Figure 27 Drip edge flashing types 17
Figure 28 Drip edge installation 17
Figure 29 Skylight with vertical-protrusion flashings 18
Figure 30 Chimney protrusion with cricket, step, and apron flashings 20
Figure 31 Kickout flashing installed where lower-level roof intersects wall 20
Figure 32 Vent boot 20
Figure 33 Vent boot alignment 21
Figure 34 Sealant usage around a vent pipe penetration 21
Figure 35 Torn shingle in need of replacement 22

Section One

1.0.0 Background and Tools

Objective

Identify the background of asphalt shingles and the tools commonly used to install this system.

a. Describe the types and styles of asphalt shingles.
b. Identify the hand tools and power tools used in asphalt shingle installation.

Trade Terms

Fiberglass base: An asphalt shingle base made with glass spun into fibers and combined using plastic polymers.

Hook blade: A blade that has sharp, rounded edges used to cut materials, such as underlayment and asphalt shingles, with great precision.

Magnet sweeper: A push-broom tool that uses magnets to attract and collect metal waste at a jobsite.

Roofing hammer: A hammer that allows users to orient and nail shingles using just one tool.

Tin snips: Useful tools for cutting and trimming thin but tough materials, such as sheet metal and shingles.

Ultraviolet (UV) radiation resistance: The ability of shingles to resist ultraviolet radiation from the sun degrading their quality over time.

Asphalt shingles (*Figure 1*) are the most popular steep-slope roof covering. Since its introduction at the turn of the twentieth century, this roof system has been the top choice in residential roofing due to its aesthetics, affordability, and simplicity of installation.

Studying the history of this roof system will help you understand asphalt shingles' current popularity. Before installing asphalt shingle systems, roofing professionals must also be familiar with the common types and styles of this system, as well as the tools used during installation.

WARNING!

It is crucial to consider safety during asphalt shingle installation. For more information on roofing safety, review NCCER Module 16102, *Roofing Safety*.

NOTE

Throughout NCCER's Roofing program, the term *steep-slope* refers to a roof slope of 3:12 or greater. This differs from OSHA's definition of steep-slope roofs (slope of 4:12 or greater).

When installing or repairing roof systems, always check company, local, state, and federal codes. If any of the applicable codes contradict each other, follow the strictest one.

Figure 1 Asphalt shingles on a home.

Before the introduction of asphalt shingles, roof systems were often made of wood, clay, and slate. While these systems are still used today, asphalt shingles offer a far more popular, practical, and affordable alternative. Additionally, asphalt shingles are more fire resistant than wood systems (*Figure 2*), and they weigh less than clay and slate systems.

Once asphalt shingles were adopted into the roofing market, attempts to improve them quickly followed. The first adaptation cut individual shingles out of rolls that had been covered with asphalt and surfaced with granules. While shingles were first manufactured in individual pieces, strip shingles were invented and gained popularity in the mid-1900s. All shingles were originally made with an organic base, and a fiberglass base was introduced in the 1960s.

Various versions of asphalt shingles over the years have increased their longevity, fire and hail resistance, **Ultraviolet (UV) radiation resistance**, and wind uplift prevention. As a result, asphalt shingles have become the top-selling residential roof system in the market.

Today, these water-shedding shingles come in a wide variety of colors and styles. Modern asphalt shingles also offer customizable options, such as algae-resistant shingles that prevent algae from staining or deteriorating the roof system.

1.1.0 Types and Styles

Asphalt shingle systems are *hydrokinetic*, or water-shedding. These shingles are constructed with a fiberglass base that is coated with asphalt and surfaced with a granule layer (*Figure 3*). Asphalt shingles commonly contain adhesive strips to help seal the shingles to the shingle course below and prevent wind uplift. Asphalt shingles come in a variety of styles. Styles of asphalt shingles include popular market designs, such as architectural, specialty, and three-tab shingles.

1.1.1 Base

Fiberglass is most often used as the base for asphalt shingles. Fiberglass-based shingles are made using rolls of glass spun into fibers and combined using plastic polymers (*Figure 4*). To manufacture asphalt shingles, the base material is covered on both sides with asphalt or modified asphalt. Granules are also adhered on the top surface, and the bottom of the shingles may be treated with a substance such as sand or covered with plastic film to prevent the shingles from sticking to one another before installation.

Figure 2 Wood-shake roof.

Organic Base

Asphalt shingles were originally made using an organic base. Organic bases include materials such as cellulose, wood fiber, and paper. These bases are more prone to fire damage than shingles made with fiberglass bases. Fiberglass-based shingles are now the standard asphalt shingle type due to their lighter weight and greater fire resistance.

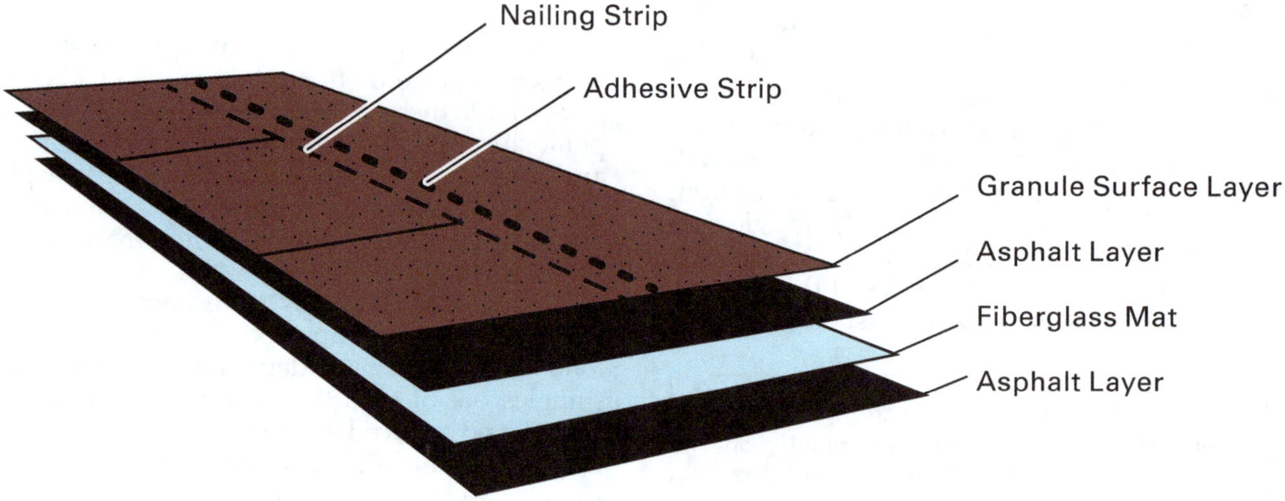

Figure 3 Anatomy of a shingle.

Figure 4 Fiberglass roll.

1.1.2 Styles

Asphalt shingles come in a wide array of shapes, colors, and styles. The most common of these styles are:

- *Architectural* — Also known as *dimensional shingles* and *laminated shingles,* this style features a traditional look and is one of the most common (*Figure 5*). Architectural shingles offer better quality and fewer necessary repairs and replacements.
- *Specialty* — These shingles have historically included designer shapes, such as scalloped, fish scale, diagonal, diamond, and octagon. *Figure 6* shows an example of specialty shingles. Current specialty shingles are available in a wide variety of designs.
- *Three-Tab* — Three-tab shingles are typically thinner, and they may have to be repaired and replaced more often than architectural shingles (*Figure 7*).

> **NOTE**
> This module only covers some common asphalt shingle styles. Roofers may encounter additional styles on the job and new products are regularly introduced. Always consult manufacturer instructions when installing shingles and talk to your supervisor if you have installation questions.

1.2.0 Tools

It is crucial for roofers to know how to safely use the hand and power tools necessary to install asphalt shingles. Along with understanding tool usage, roofing professionals are also responsible for ensuring their tools are clean and in good condition before using them to install asphalt shingle roofs.

Figure 5 Architectural shingles.

Figure 6 Specialty shingles.

Figure 7 Three-tab shingles.

Continuing Your Education as a Craft Professional

This curriculum cannot possibly teach you all the things you need to know about the many roof materials and applications you will encounter. At the end of a long and successful career, you will still be learning.

Throughout this level, trainees are directed to read and follow the manufacturer's instructions for a roofing product before installation begins. Some of the best training you can hope to receive are found in these documents. As you encounter new products, study all the available information, including marketing information. Understand the materials' strengths and weaknesses and ask questions when you need deeper understanding. This is one way to get ahead, become a resource to those around you, and move into a position of leadership in your chosen craft.

NOTE

This section highlights some of the tools you will encounter when installing asphalt shingle roof systems. Other common tools used in asphalt shingle installation, such as hammers and circular saws, are covered in greater detail in NCCER Module 00103, *Introduction to Hand Tools*, and NCCER Module 00104, *Introduction to Power Tools*.

1.2.1 Utility Knife

Utility knives, practical tools used for general purposes, have retractable blades that make them safe to carry on a tool belt. A utility knife may have a traditional, straight-edge blade or a **hook blade** (*Figure 8*). Hook blades have sharp, rounded edges that allow roofers to cut materials, such as underlayment and asphalt shingles, with great precision. Due to its hook shape, this blade grips a layer and separates it from any layer underneath to prevent the user from accidentally slicing through lower layers.

1.2.2 Tin Snips

Tin snips are useful tools for cutting and trimming thin but tough materials, such as sheet metal and shingles. These tools are sometimes colored red, yellow, or green to signify the different directions the snips are made to cut, as shown in *Figure 9*. Yellow snips cut straight. Red snips usually cut to the left. Green snips typically cut to the right.

WARNING!

To use tin snips safely, be sure that the grip shown on the handle fits closely against your thumb.

1.2.3 Roofing Hammer

A **roofing hammer** (*Figure 10*) allows roofing professionals to orient and nail shingles using just one tool. The claw, or back end of the hammer, may be used to help lift shingles. In some hammers, the gauge, or the topmost surface of the hammer, can assist with measuring correct shingle placement. The hammer's face is used to drive nails into a roof substrate.

1.2.4 Tear-off Tools

Tearing off shingles may require the use of hand tools, such as roofer's pitchforks, shovels, and spades, that allow roofers to gain leverage with the handle of each tool and more effectively remove any unwanted shingles.

WARNING!

During tear-off projects, be sure that the ground below is protected from the possibility of any falling debris. Be sure to tape off the construction area to keep pedestrians away from potential hazards until the job is finished and the site is cleaned.

1.2.5 Magnet Sweeper

After any roofing project, leftover metal debris can pose a serious safety risk—for example, a pedestrian can harm himself on a sharp nail that was not properly cleaned up after a new roof installation.

Figure 8 Utility knife with straight-edge blade inserted and separate hook blade.

To assist with cleanup and promote a safer jobsite, a **magnet sweeper** (*Figure 11*) is another essential tool to have in stock. Magnet sweepers operate like push-broom tools that use magnets to attract and collect metal waste at a jobsite. These sweepers efficiently pick up nails, pins, and other metal debris that would otherwise be difficult to collect by hand.

Magnet sweepers typically include a release button to help release metal waste after it is collected. These tools help maintain jobsite safety by removing hazards that may otherwise be easy to overlook.

1.2.6 Nail Gun

An alternative to roofing hammers is the nail gun (*Figure 12*). This tool speeds up the nailing process by allowing its user to simply click a button to drive nails rather than having to drive nails manually.

1.2.7 Chalk Lines

Chalk lines help roofing professionals align shingles during installation before they are attached with nails. Chalk lines are used by anchoring the end of the line to a fixed point, and then extending the chalk line across an area before snapping it onto the surface. Once a chalk line is snapped to a surface, it will temporarily imprint a line of chalk that roofers can use as a guiding point for installation.

When using chalk to mark up a roof, be sure *not* to use vibrant colors, like red, that risk leaving permanent marks on the roof.

1.2.8 Roofing Nails

Roofing nails are designed with a strong shank and a large head to better grip the deck and hold the shingle. Most asphalt shingle roofs are installed using roofing nails. Make sure you do not under-drive or over-drive nails by installing them with either too little or too much force. Nails should also be driven straight into the roof system so that they are not installed crookedly (*Figure 13*).

> **CAUTION**
>
> When working in cold weather, it is best to drive nails by hand to help prevent shingle fracturing.

Figure 9 Set of color-coded tin snips.

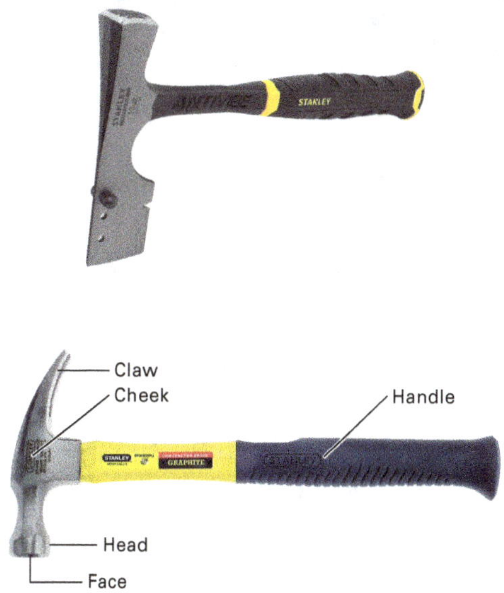

Figure 10 Roofing hammers.

Figure 11 Magnet sweeper.

Figure 12 Nail gun.

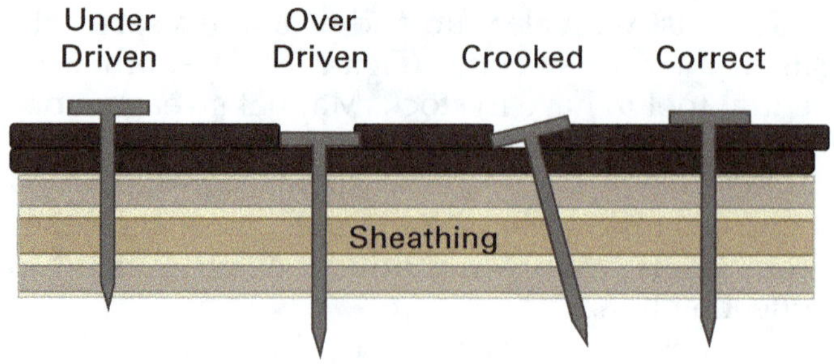

Figure 13 Roofing nail installation.

1.0.0 Section Review

1. Which shingle style includes popular shape designs, such as diamond-shaped shingles?
   a. Dimensional
   b. Architectural
   c. Specialty
   d. Three-Tab

2. Tin snips with yellow handles usually cut straight.
   a. True
   b. False

SECTION TWO

2.0.0 INSTALLATION AND REPAIR

Objective

Explain asphalt shingle installation.

a. Learn the procedures used to safely install asphalt shingles.
b. Identify the flashings and water-shedding details involved with asphalt shingle installation.
c. Overview how to replace a damaged shingle.

Performance Tasks

1. Don all appropriate PPE for asphalt shingle installation.
2. Install drip edge metal and manufacturer-recommended underlayment to a wood deck.
3. Properly install starter strip shingles and courses of field shingles according to manufacturer instructions. Be sure to use the correct exposure and offset measurements for your system.
4. Install valley underlayment, valley metal, and valley shingles according to manufacturer instructions.
5. Add appropriate flashings to a penetration or vertical-surface protrusion.
6. Identify a shingle that needs replacement and perform the repair.

Trade Terms

Backnailing: Nailing the concealed portion of a roofing component so the fasteners are covered by the next ply, or course, and are not exposed to the weather in the finished roof system.

Starter strip shingles: Roll roofing or shingle strips applied along the downslope eave line before the first course of roof covering. These are intended to fill spaces between cutouts and joints of the first course.

Taper: Increasing width uniformly to gradually widen or narrow.

Valley centerline: The center of the valley; no fasteners should be used within 6" (15 cm) of this point.

In addition to understanding how to safely install asphalt shingles, roofing professionals must know how to flash a roof to ensure it is correctly shedding water. Also, it is important to review some basics about asphalt shingle repair so that you can identify and fix potential problem areas during installation. *Figure 14* highlights some sections of a steep-slope roof that must be carefully considered during installation.

2.1.0 Installation Procedures and Components

Installing an asphalt shingle system involves attaching drip edge metal, underlayments, starter strip shingles, and various types of shingles and flashings. Once the drip edge, underlayment, and starter strip shingles have been added to a roof deck, roofers must know how to safely install the rest of the shingles and flashings.

Since there are a variety of asphalt shingle styles, it is crucial to abide by the manufacturer recommendations of each system during installation. Each installation procedure has the potential to vary from any other shingle installation a roofing professional has encountered in the past.

Before learning more specifics about asphalt shingles, it is important to have a big-picture outline of how these systems are typically installed. The remainder of this module will provide more information about the system components mentioned in this outline.

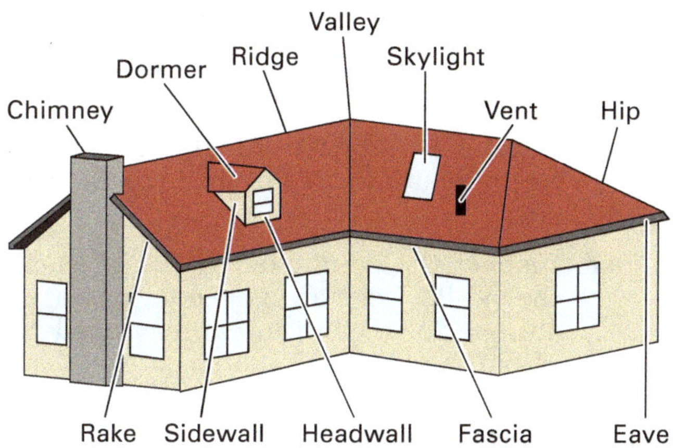

Figure 14 Steep-slope roof anatomy.

The following provides a general outline for asphalt shingle system installation:

> **CAUTION**
>
> This outline is intended only to overview a general installation sequence. It cannot be used in place of manufacturer instructions or building-code requirements.

Step 1 Don all proper PPE for the job. Ensure that the appropriate fall protection equipment is used or in place. Depending on the project and environment, this may be a personal fall arrest system (PFAS) with approved anchorage, or appropriate guardrails or scaffolding.

Step 2 Ensure that the roof deck is clean, dry, and smooth before you install the roof's drip edge metal, underlayment, and valley metal.

Step 3 Starting at the roof's eave, install a starter course.

Step 4 On top of your starter course, measure and install the first shingles course from one direction (either left-to-right or right-to-left). Manufacturer instructions must be consulted for the exposure and offset requirements of each specific system.

Step 5 Install shingles at the roof's valleys following manufacturer instructions and local building codes.

Step 6 Install secondary flashings, such as penetration, wall, and counter flashings.

Step 7 Install hip and ridge shingles. These may be premanufactured or made on the jobsite.

Step 8 Inspect the completed roof and clean the jobsite. Keep an eye out for damaged shingles or flashings that may need replacing.

2.1.1 Underlayments

Underlayments (*Figure 15*) are among first components a roofing professional will install on a steep-slope roof.

Underlayments provide some weather protection to the roof deck before a roof system is installed. They also serve as a secondary barrier for any water that penetrates past the roof system after it is installed, and they separate the roof covering from the deck substrate below. Asphalt shingle systems may incorporate different kinds of underlayments depending on the project design. It is crucial to consult the manufacturer's instructions for more specific details about any given underlayment.

The types of underlayments most often used with asphalt shingle systems are as follows:

- *Asphalt felts* — One of the most commonly used underlayments, this type is made with organic, inorganic, or combination reinforcing mats coated with asphaltic bitumen. Organic mats are made of substances such as wood and cellulose fiber. Inorganic mats are often made from glass or polyester fibers. This type of underlayment is typically mechanically attached.
- *Synthetic sheets* — Underlayment products that have gained widespread acceptance in the market. Synthetic underlayments, made of polymer reinforcements, are strong, light weight, and have low moisture absorption rates. They are often preferred when projects need longer periods of temporary protection. This type of underlayment is also typically mechanically attached.

> **NOTE**
>
> ASTM standards are reviewed in NCCER Module 16105, *Introduction to Steep-Slope Roofing.*

- *Leak Barrier* — Also called *self-adhering sheets* and, in colder climates, *ice dam protection*. These underlayment sheets have adhesive bitumen applied at the factory and provide labor savings with installation on the job.

Figure 15 Underlayment installation.

Installing Mechanically Attached Underlayments

Before installing underlayment, ensure that your roof deck is clean, dry, and smooth and that you have consulted the underlayment's manufacturer instructions for any specific directions.

If you are mechanically attaching an underlayment, your fastener type (for example, roofing nails) must follow manufacturer guidelines and relevant building codes. Be sure not to install underlayment in weather conditions that are not recommended. Always consult manufacturer instructions about the minimum and maximum temperatures required for installation.

To mechanically attach an underlayment, apply the material horizontally and in laps as per the manufacturer's recommendations. Underlayments are often installed over the eave metal and under the rake metal.

Be sure that the vertical end laps are offset, usually by a minimum of 4" (10 cm). All underlayments should be folded up vertical walls and curbs by at least 4" (10 cm).

Installing Leak Barrier

Leak barriers provide protection to a roof deck by sealing around nails driven through it. Typically, manufacturers recommend that self-adhering underlayment is installed with its end pieces overlapping at least 6" (15 cm) and in an offset pattern. All underlayments should be folded up vertical walls and curbs by at least 4" (10 cm).

Applying leak barrier vertically, rather than horizontally, is called *strapping* and can be used in some steep-slope systems when recommended. This leak-barrier application may require backnailing to prevent slippage. After installing underlayment, consult your manufacturer instructions for next steps.

2.1.2 Starter Strip Shingles

Starter strip shingles should be applied directly over the underlayment along the perimeter, or eaves and rake edges, of a roof. *Figure 16* shows an example of a starter strip roll commonly sold by manufacturers. Starter strip shingles are used under the first course of shingles to help shed water that may pass through that course.

While some manufacturers sell starter strip shingles, you may also create your own by cutting off and using the tab portion of a shingle strip. However, using manufacturer-supplied starter strip shingles ensures the adhesives on the strips are properly placed for the upper layers of shingles installed.

Figure 16 Starter strip roll.

Drip edge is flashing installed at rakes and eaves that will be further discussed later in this module. Starter strip shingles should extend less than an inch beyond the perimeter eave and rake edges if the drip edge metal is L-type. If the flashing drip edge metal is T-type, the starter strip shingles should be installed evenly with the outer edge of the drip edge.

> **NOTE**
> Drip edge metal, along with other steep-slope flashings, were introduced in NCCER Module 16105, *Introduction to Steep-Slope Roofing*.

2.1.3 Field Installation

After your starter strip shingles have been installed (*Figure 17*), you will begin aligning your first course of shingles. Snapping a chalk line across the field may help you align your first course so it appears straight and even. If necessary, cut the ends of your shingles as you install them so they are not hanging off the end of the roof where the course terminates.

Most asphalt shingles are applied in laps that extend across and diagonally up the roof. End joints of shingles must be offset a minimum of 4" (10 cm) for each course.

In normal conditions, a minimum of four roofing nails are typically used to install each shingle strip. In regions with high-wind zones, a minimum of six nails are typically required to install each strip. Always consult manufacturer instructions and local building codes to make sure you are using the correct minimum number of nails per shingle strip.

Figure 17 Aligning starter strip shingles.

Three-Tab Installation

As with all asphalt shingle installation, three-tab shingles must be installed in an overlapping, offset pattern. This installation method ensures that water does not pass through any openings in the shingles.

Figure 18 pictures the most common offset pattern for three-tab installation, the 6" (15 cm) method. This method aligns cutouts every other course by offsetting consecutive shingle courses by 6" (15 cm).

A minimum of four roofing nails should be used per shingle when attaching them to a roof deck that has been covered with any necessary underlayment and starter courses. Consult local building codes to ensure that you are not working in a windier region that typically requires a minimum of six roofing nails per shingle.

Laminate Installation

Different shingle installations may require different offset patterns. Always consult manufacturer instructions to make sure you are installing your system correctly.

Figure 19 illustrates one example of an offset pattern for laminate shingles. Install nails at the thickest point near a shingle strip's nailing line to ensure that the courses most effectively attach to the roof system.

Laminate installation also requires roofing professionals to use a minimum of four roofing nails per each shingle. Consult local building codes to ensure that using four roofing nails is enough to prevent wind uplift in your region.

CAUTION

When working with very steep slopes, refer to the manufacturer's specifications for positioning additional nails and supplemental sealants. These promote further protection against wind uplift and slippage.

Hand-sealing is another method that involves using sealant between shingles' courses to further seal the laps together and prevent wind uplift and water leakage.

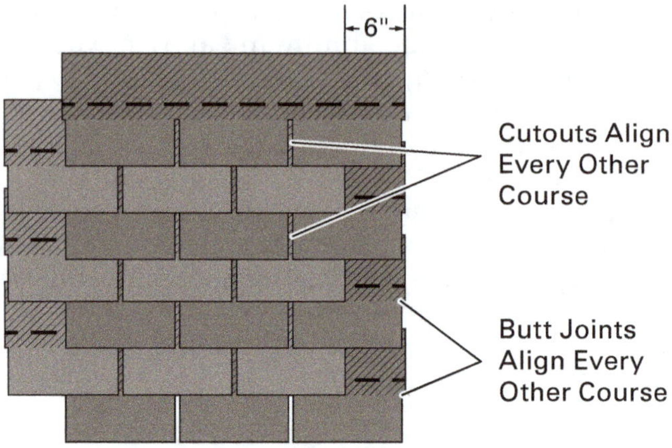

Figure 18 Three-tab shingle offset pattern.

2.1.4 Valleys

Valleys are the V-shaped angle that forms where the downslopes of two roof fields intersect. Since a valley is at the center of two sloping planes, water will shed in its direction and make the area particularly vulnerable to water leakage.

Leak barrier should be installed under the valley flashings. Always consult manufacturer instructions for leak barrier requirements and installation recommendations.

Several different methods of valley installation can be used in shingle systems. These methods include closed-cut, open, and no-cut valleys. Open and closed-cut valleys (*Figure 20*) are defined by whether shingles are coursed across the valley's centerline (closed-cut valleys) or trimmed back from that area on either side to reveal valley flashing (open valleys). No-cut valleys can be used in both open and closed-cut valley installations.

In all valley methods, the courses of underlayment from either side of the field should overlap the valley's underlayment by 6" (15 cm).

The valley metal should extend slightly past the drip edge metal installed at the roof's eave to facilitate water runoff. Use snips to trim the valley metal if necessary.

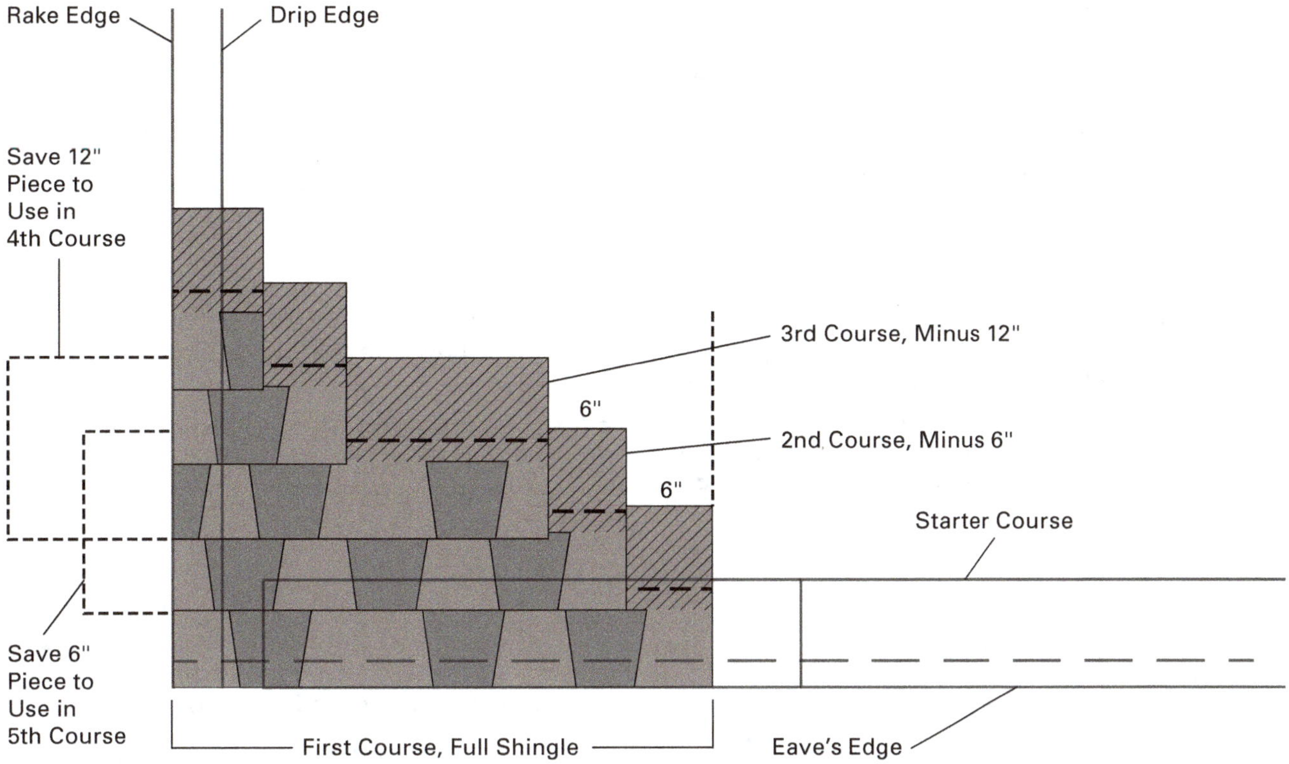

Figure 19 Example of stair-step offset.

Valley metals are formed into either a V or a W shape. The rib in the center of the W-shape valley metal is a splash diverter that helps guide rainfall off the roof.

> **NOTE**
> A fourth type of valley method, woven valleys, interlaces shingles from either plane to close the valley. This method of installation is not recommended because it increases the likelihood that water will migrate underneath the woven shingles.

Closed-Cut Valley

Closed-cut valley installations attach shingles across the valley from one direction. The shingles that intersect at the valley centerline from the other direction must be trimmed back about 2" (5 cm) short of the valley's center so that there is just one course and direction of shingles crossing through the centerline.

This method can be used with most types of shingles. Since the valley centerline must not have any penetrations that water could get through, the closed-cut valley is only recommended for shingle systems that are installed in strips and do not require any nails to be used near the valley centerline.

Closed-cut valleys are installed using the following general steps:

Step 1 After the valley and field underlayment is installed, lay the first course of shingles starting from the lower-sloped side and continuing through the valley. If the slopes are the same size, lay the first course on the plane with the shortest distance to the ridge. Do not use nails within 6" (15 cm) of the valley's centerline.

Step 2 Two nails should be used at the end of each shingle before the strip reaches within 6" (15 cm) of the valley's centerline. The shingle strip that passes through the valley should extend 12" (30 cm) past the centerline.

> **CAUTION**
> To ensure that no nails or screws are installed within 6" (15 cm) of the valley's centerline, you may need to cut a shingle strip short and attach a new strip that will pass through the valley without needing to be nailed close to the centerline.

Step 3 Using a chalk line, mark your valley centerline on the first course of shingles you have installed through the valley center.

Step 4 When installing the second course of shingles on the intersecting slope of the valley, use a utility knife to trim back shingle strips approximately 2" (5 cm) from the point of the chalk line that you marked on the first course of shingles installed through the valley. Use sealant underneath each of the top-layer shingles near the valley centerline to secure the strips.

Step 5 Trim the upper corner of the top layer of shingle strips at the valley's centerline. This will help prevent any water that may leak underneath the shingles from passing through the membrane, as depicted in *Figure 21* (see the black circle lettered *A*).

Open Valley

Open Valley with a Metal Flashing: The valley should be 4" to 6" wide at the top and should widen 1/8" per foot as you move down the valley.

Closed-Cut Valley

Closed-cut valley still requires a leak barrier underneath.

Figure 20 Open versus closed-cut valleys.

Open Valley

Unlike closed-cut valleys, open valleys trim shingles back on either side of the valley centerline to reveal flashing below. Open valleys promote increased drainage that is particularly useful when used on a roof with shallow slopes or surrounding foliage. This valley flashing is typically made from corrosion-resistant metal—usually steel, aluminum, or copper. Valley metal is most effective when it is at least 24" (61 cm) wide.

Open valleys are installed using the following general steps:

Step 1 Underneath the field underlayment, install a minimum 36" wide (91 cm) layer of valley underlayment that is oriented to cover the valley centerline.

Step 2 Before adding shingles to the perimeter of the open valley, install the valley metal. To guarantee correct alignment, start installation from the eave and work your way up to the top of the valley.

Step 3 Overlap conjoining valley metal pieces at least 6" (15 cm) with each upward extension. Use sealant to secure overlapping pieces and use nails only on the perimeter of the valley metal.

Step 4 Using chalk lines, mark the metal flashing to indicate the point where shingles should overlap on either side. If possible, taper the open valley so its width uniformly widens as it runs downslope. Tapering allows for increased drainage and helps melt and clear ice buildup. In most climates, it is recommended for a valley taper to be 1/8" (0.3 cm) per 1' (0.3 m) of valley run.

Step 5 As you install shingles up the valley, trim the upper corner of the top layer of shingle strips closest to the valley's centerline. This will help prevent any water that may leak underneath the shingles from passing through the membrane, as depicted in *Figure 22* (see the black circle lettered *A*).

Step 6 Snip and close off the end of the valley metal to prevent leakage.

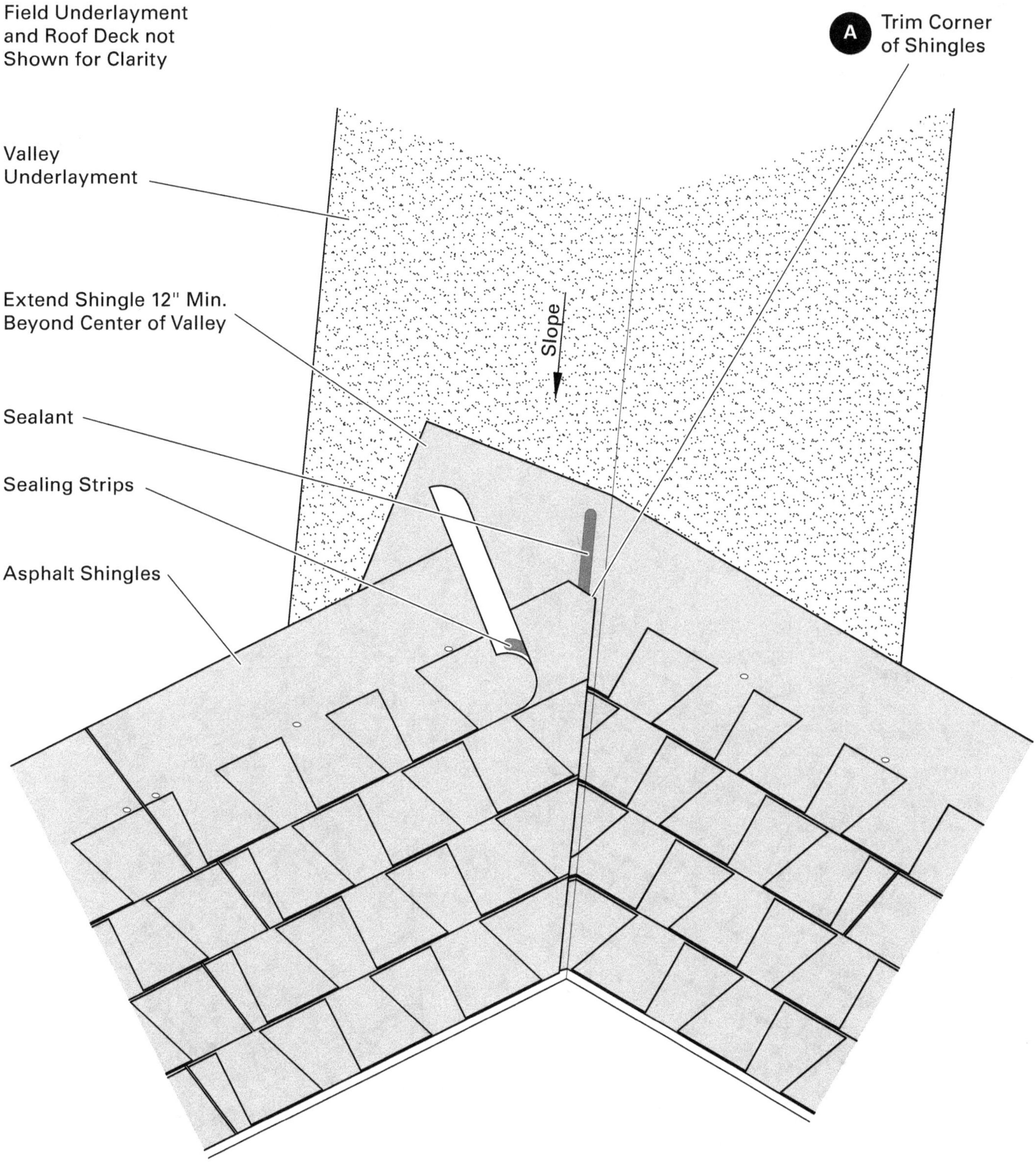

Figure 21 Closed-cut valley.

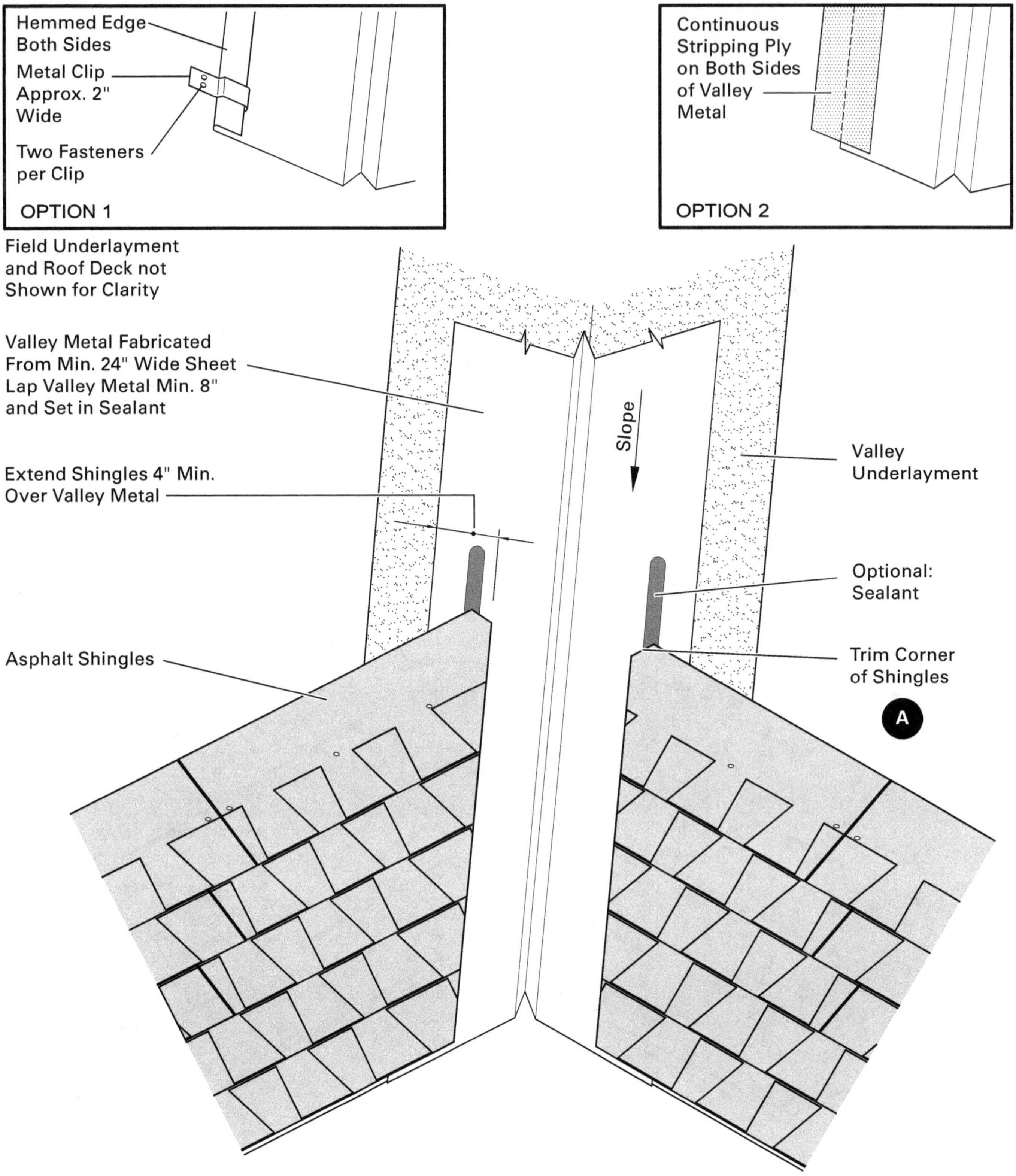

Figure 22 Open valley.

No-Cut Valley

No-cut valleys are sometimes called *California valleys* (*Figure 23*). To install a no-cut valley, apply a row of shingles vertically, flush against the valley flashing (if working with an open valley) or centerline (if working with a closed-cut valley). The horizontal shingle courses should then be installed with the tip of each course covering the vertical shingle.

> **NOTE**
> No-cut valleys only work with architectural shingles. This valley installation method should not be used for three-tab shingles.

2.1.5 Hips and Ridges

Hips, or the outer angle formed by the meeting of two sloping sides of a roof, should facilitate water running down and away from them. Similarly, ridges, or the horizontal junction of two roof slopes, should also help water run down and away from them. To accomplish this, most manufacturers sell shingles that are specifically made to cover and water-shed hips and ridges (*Figure 24*). However, factory-produced ridge and hip shingles may not be provided for three-tab shingle strips.

If you are installing a three-tab roof system, you can manufacture your own hip and ridge caps. Using a hook blade, cut a small angle into the nailable part of a shingle strip. Continue this pattern on each individual strip until each strip is separate and tailored to cover hips and ridges. *Figure 25* illustrates where and how the shingles should be cut, demonstrated by thin, dotted lines.

The following provides a general installation overview for hip and ridge shingles:

Step 1 Snap chalk lines to ensure that hip and ridge shingles are correctly aligned for installation. Chalk lines help improve the appearance of the system.

> **CAUTION**
> Hip and ridge shingles must be flexible enough to bend over the hip or ridge line without cracking. When installing them during cold weather, allow a warm-up period after removing them from the packaging and before installing them onto the roof.

Step 2 Begin application at the lower end of the hip. Each piece should be applied using the tailored, nailable portion of the shingle. Since these shingles are either pre-designed or tailored for this part of the roof, the nailed-in portion of the shingle will be concealed by the butt end of the next shingle applied.

Step 3 Using roofing nails that penetrate a minimum of $^{3}/_{4}$" (2 cm) through the roof deck, install each piece in overlapping succession. Be sure that the nailable portion of the previous shingle is covered by the butt end of the next shingle installed.

Step 4 Cover any exposed fasteners with the appropriate sealant or roof cement on a run's final ridge cap.

2.2.0 Flashings

Roof flashing is an integral part of any roof system because it helps protect the building against water infiltration. So, what exactly is roof flashing, and how does it work?

Roof flashing helps provide a seal where there is a roof joint or penetration (like a chimney or vent), where there is an intersection of two roof surfaces, and where the roof meets an exterior wall. Flashing protects these areas by acting both as a barrier and as a way to direct water down and off a roof.

Most roof flashing is made of metal, including galvanized steel, copper, aluminum, and stainless steel. Metal flashing can be pre-formed to fit general specifications or hand-cut to fit your unique roof design. Any seams in flashing should be soldered or sealed for secure weatherproofing.

Flashing can be installed using adhesives, roofing nails, or roofing screws. Flashing must be installed whenever the roof plane is interrupted, including on valleys, hips, penetrations, headwalls, and sidewalls. *Figure 26* identifies areas of the roof where some flashing and weatherproofing is essential.

2.2.1 Eaves and Rakes

Eaves and rakes comprise the perimeter of a roof where the field terminates on all sides. These areas require drip edge flashings that come in both T- and L-type profiles (*Figure 27*). They can come premanufactured or can be made on the jobsite. Always follow manufacturer instructions for guidance on how to fasten drip edge in each particular system.

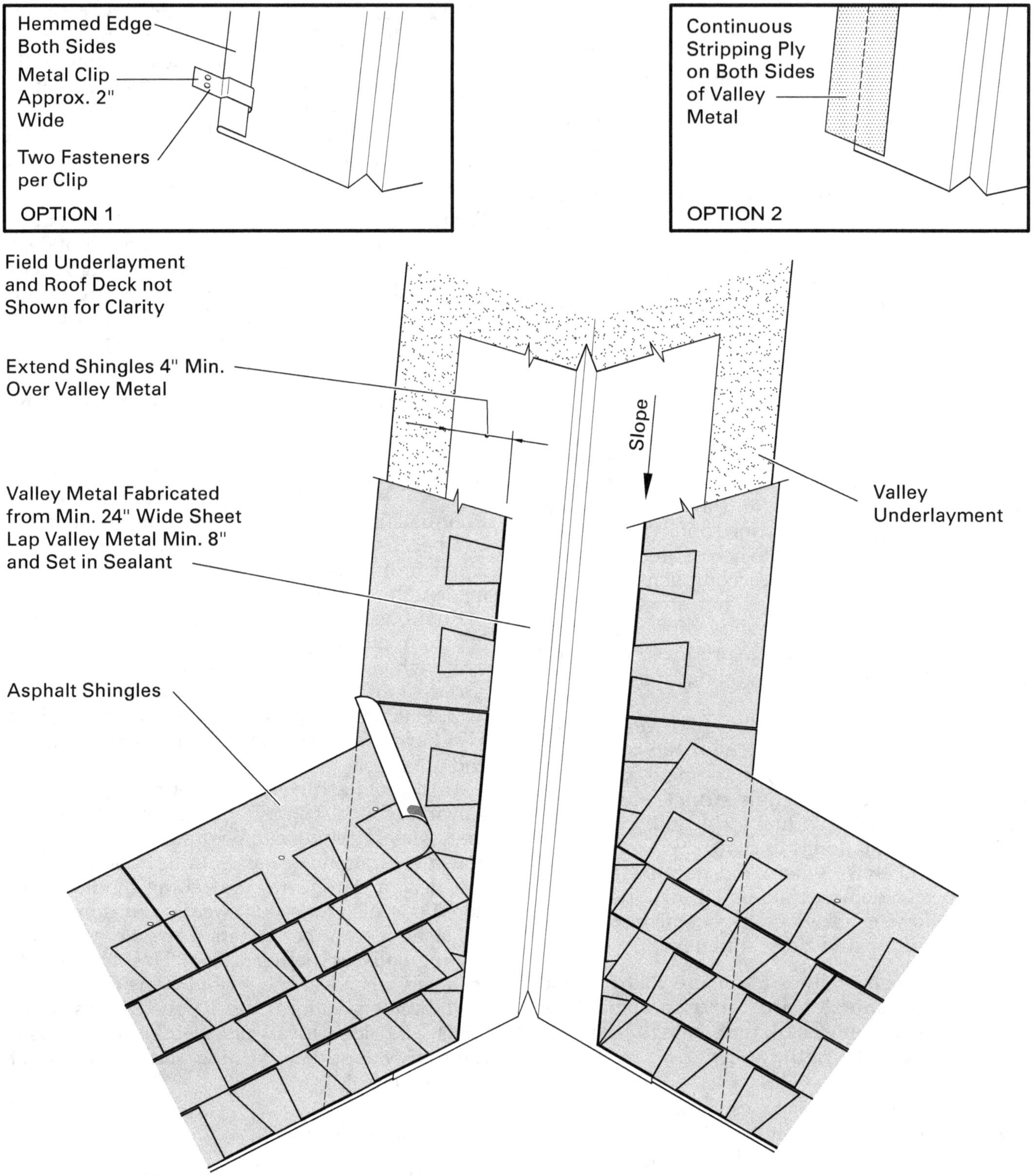

Figure 23 No-cut valley.

Drip edge is typically installed flush with the roof's decking and fascia and usually fastened in overlapping succession along the entire roof's perimeter. *Figure 28* shows drip edge metal installation. Sometimes, prefabricated drip edges do not properly cover the area flush with the roof's fascia and will need to be modified on-site so that the area is weatherproofed.

> **NOTE**
> Consult manufacturer instructions for whether underlayment should be installed under or over drip edge. Depending on what region you are in, you should install perimeter metal with a leak barrier.

2.2.2 Vertical-Surface Protrusions

Vertical-surface protrusions, such as walls, dormers, and chimneys, interrupt the roof's field and require flashing to finish the area. *Figure 29* shows a skylight with an overview of the required vertical-surface flashings.

Figure 24 Hip and ridge shingles.

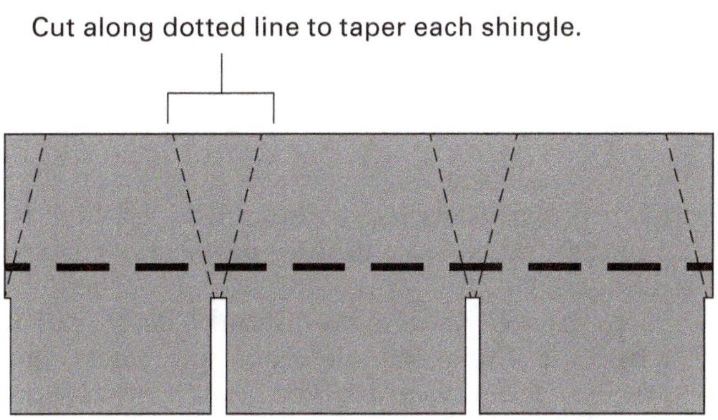

Figure 25 How to cut hip and ridge caps from three-tab shingles.

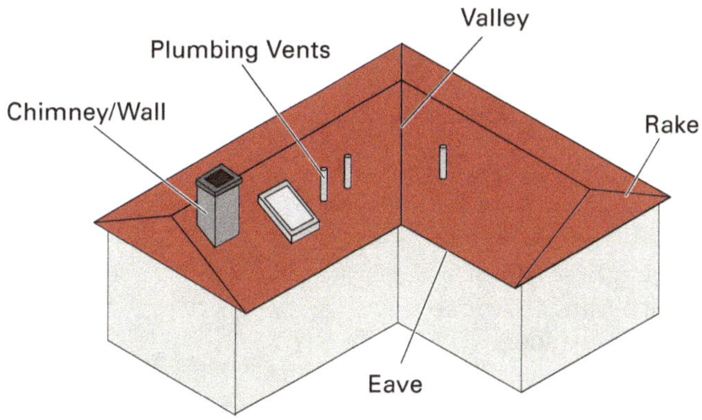

Figure 26 Roof area flashing.

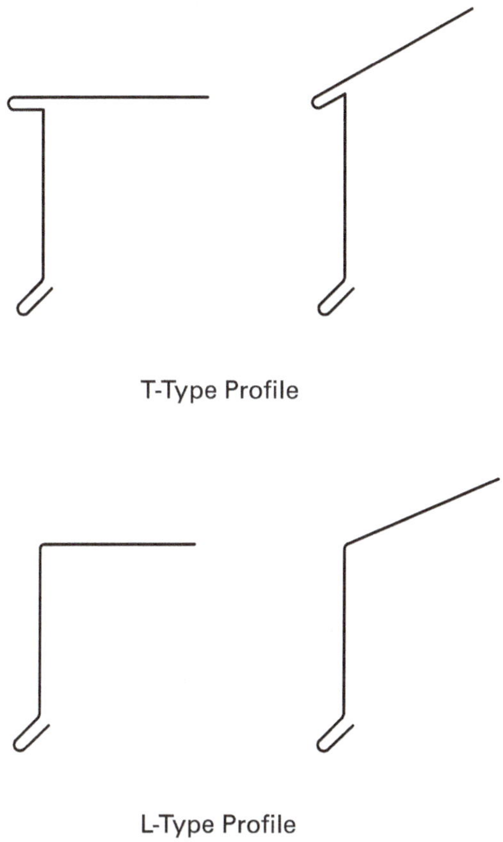

Figure 27 Drip edge flashing types.

Figure 28 Drip edge installation.

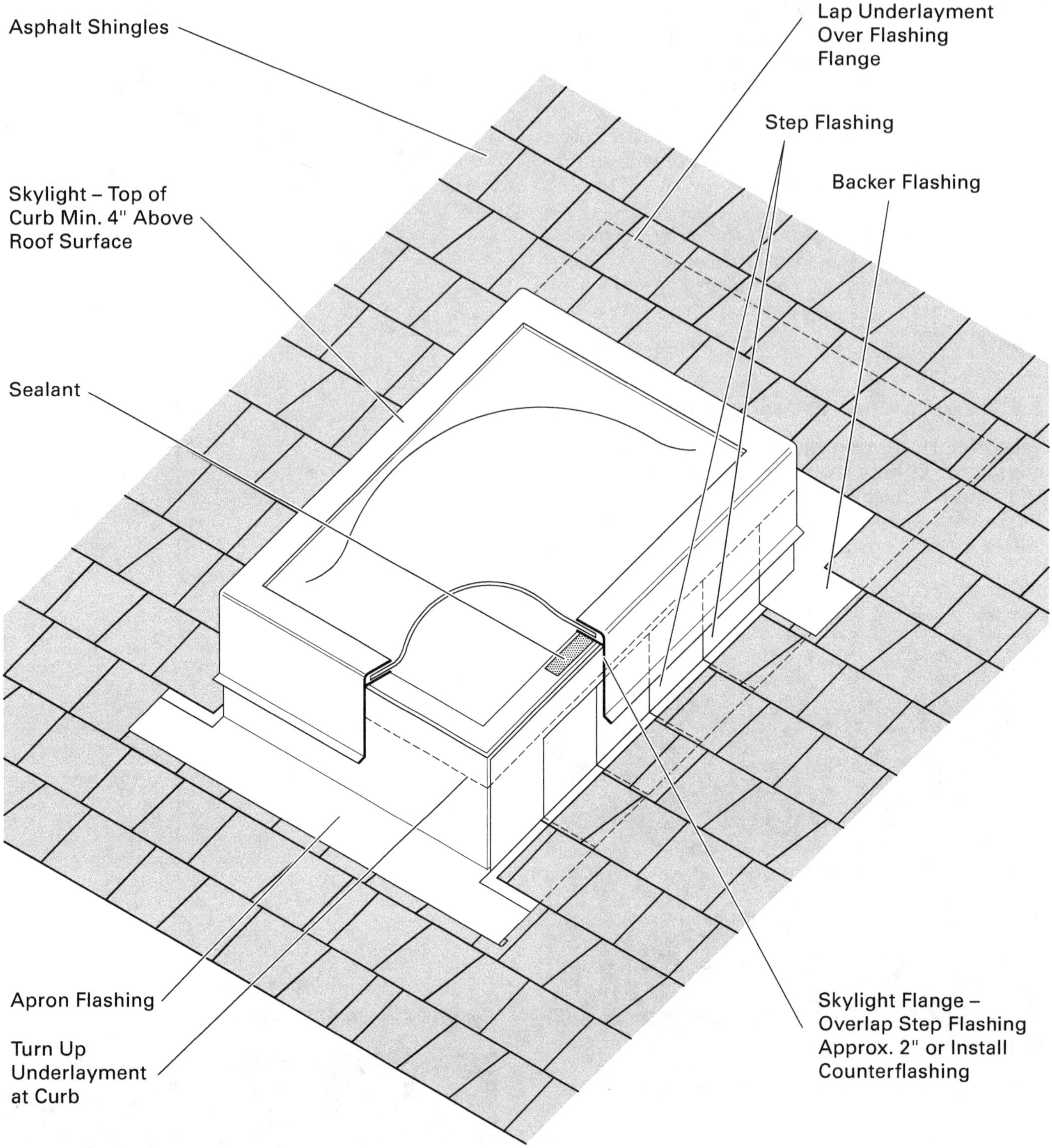

Figure 29 Skylight with vertical-protrusion flashings.

The following are the four types of flashings used at locations where metal shingles intersect a vertical surface:

- *Apron flashings* — Weatherproof the area where the roof meets a headwall.
- *Channel/step flashings* — Should be applied on metal shingle systems that intersect a vertical sidewall.
- *Backer/cricket flashings* — Should be installed where a metal-shingle roof intersects the upslope side of a vertical roof penetration.

Backer flashings are used on penetrations that are 24" (61 cm) wide or less, and cricket flashings are used on penetrations greater than 24".

- *Counterflashing* — Should be applied to base flashings, such as apron, step, channel, cricket, and backer flashings to protect the base flashings' top edges from water intrusion.

NOTE

In other systems that only have one layer of roof covering, such as metal-panel systems, flashing at sidewalls is called *pan flashing*.

The term *step flashing* only applies to systems with more than one layer of roof covering, such as asphalt shingle systems.

Figure 30 illustrates the location of apron flashing, step flashing, cricket flashing, and counterflashing on a chimney protrusion.

CAUTION

Higher-end shingles may require copper flashings. Always consult manufacturer instructions to be sure you are installing the correct type of flashings for a given system.

Kickout Flashing

In some cases, a lower-level roof edge may intersect with a vertical wall if a building has a multilayered roof system. *Kickout flashing* (*Figure 31*) is a metal flashing piece installed at the eave end of a roof-to-wall transition and designed to direct runoff away from the wall or wall cladding.

Kickout flashing may come premanufactured, or it may be necessary to create on the job. You can make a kickout flashing by hand by cutting and shaping a piece of step flashing. Fold the piece so that the material can reach under the shingle course and up the intersecting wall. Depending on the placement of the intersecting wall, the material should curve left or right to divert water away from that wall. You can use rivets, screws, and adhesives to seal the joints of your hand-made kickout flashing.

2.2.3 Penetrations

Penetrations interrupt the roof's field and require flashing to weatherproof the opening they create in the roof's deck. Penetrations include features such as plumbing vent pipes, exhaust vents, and exhaust fans. To weatherproof these areas, preformed boots or collars can be applied around the penetration and secured with roofing nails and sealants (*Figure 32*).

A patch of leak barrier should be used around the penetration to seal the sides from potential leakage. Boots and collars must be installed under the asphalt shingles on the upslope side of a penetration and over the shingles on the downslope side, as pictured in *Figure 33*.

WARNING!

OSHA considers a hole any opening that is 2" by 2" (5 cm by 5 cm) or larger. Be sure that all necessary safety precautions are taken to protect against any hazards presented by holes.

Prefabricated boots and collars are customized to fit around commonly found roof penetrations with a flange that attaches to the roof membrane and a cylindrical or rectangular section that extends above the flange to fit around pipe- or box-shaped penetrations. Using additional sealant under the shingles in the area of the penetration (*Figure 34*) increases that area's weatherproofing.

2.3.0 Replacing a Damaged Shingle

Along with understanding how to install a roof system, from underlayments to flashings, you should notice and fix mistakes that may occur during installation (*Figure 35*). Roofing professionals should pay particular attention to the quality and alignment of the asphalt shingles, drip edge metal, base flashings, and counterflashings. If a shingle or flashing is damaged, it should be replaced.

To replace a damaged shingle, use a flat-bar or flat-bladed tool to break the weather sealant of the damaged shingle and the two shingles directly above it. Insert the flat-bar tool under the damaged shingle and locate all four nails (or however many nails were used to install the shingle). With your tool, grasp each nail's shank and apply upward pressure on the shingle and nail head until it pulls out of the deck. Lift each nail in the same fashion. Remove the nails. Repeat this process for the two shingles directly above the damaged shingle. Install new shingles. Drive nails within approximately 1" (2.5 cm) of the original nails' positions. Use sealant to fill all nail holes and be sure the new pieces you have installed are fully adhered to the system.

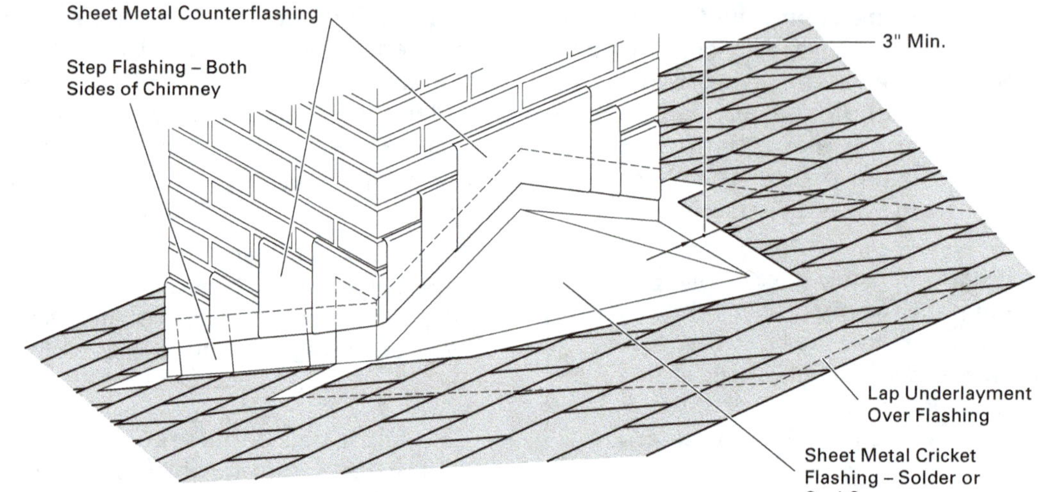

(A) Cricket Flashing at Masonry Chimney

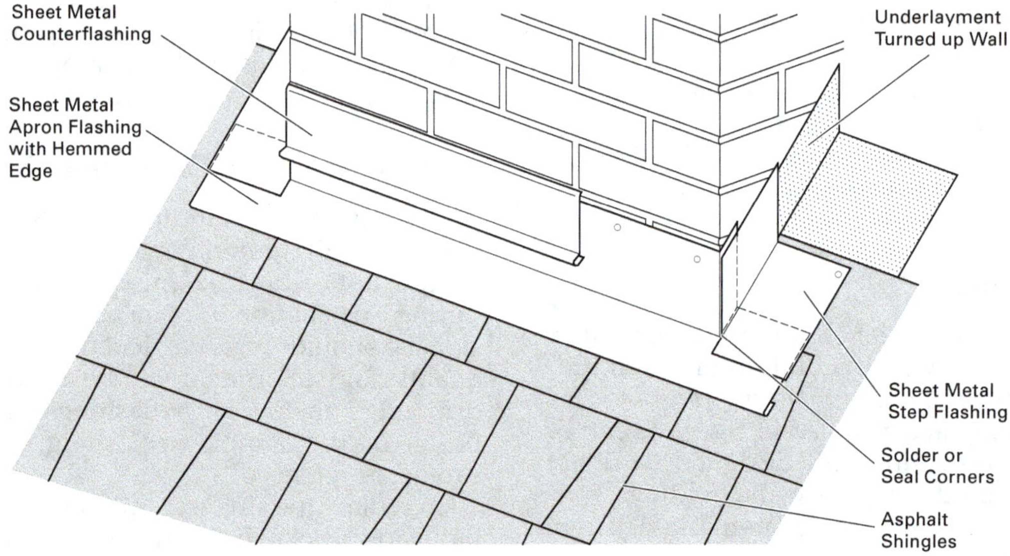

(B) Apron Flashing at Masonry Chimney

Figure 30 Chimney protrusion with cricket, step, and apron flashings.

Figure 31 Kickout flashing installed where lower-level roof intersects wall.

Figure 32 Vent boot.

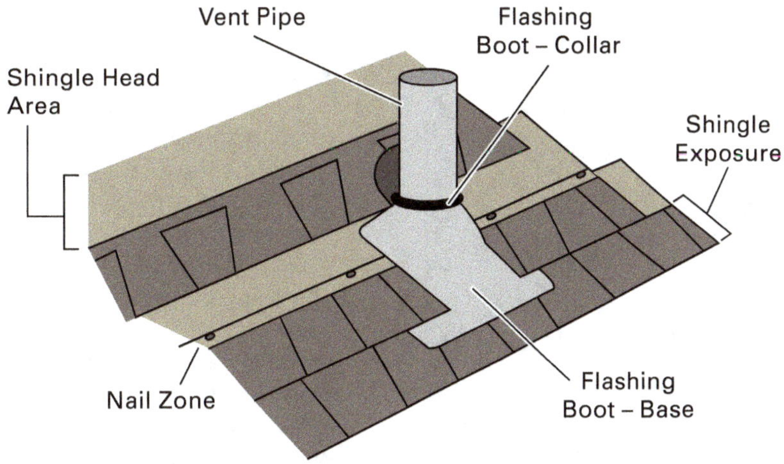

Figure 33 Vent boot alignment.

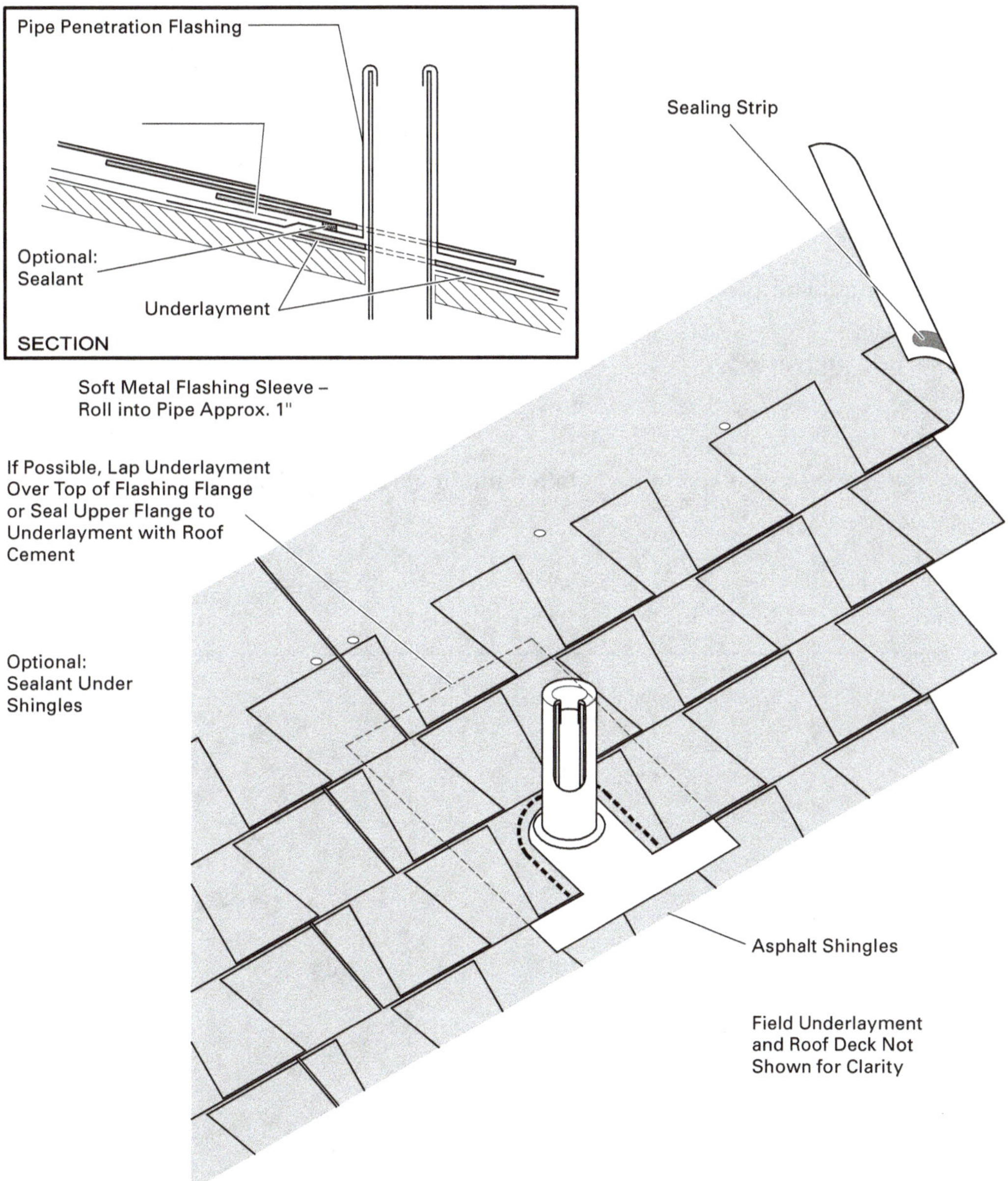

Figure 34 Sealant usage around a vent pipe penetration.

Figure 35 Torn shingle in need of replacement.

Recycling Shingles

Roofing shingles, nails, and underlayment that cannot be used on a project due to damage or new roofing can still be recycled. Instead of throwing away unusable shingles, find a recycling facility near you that will repurpose the material. The shingle manufacturer you are working with may have a recycling program already available. Earth 911, an online site, can also be used to find your closest shingle-recycling facility. For more information, visit the website at **www.earth911.com**.

2.0.0 Section Review

1. You should install starter strip shingles before you install your first course of shingles.
   a. True
   b. False
2. Drip edge metal should be applied to the roof's _____.
   a. eaves and rakes
   b. vertical-surface protrusions
   c. headwalls
   d. penetrations
3. Which of the following tools is *best* to use when removing a damaged shingle?
   a. Magnet sweepers
   b. A hammer
   c. A flat-bar
   d. Tin snips

Review Questions

1. Which part of asphalt shingles is *most* helpful in sealing courses and preventing wind uplift?
   a. Adhesive strip
   b. Nailing strip
   c. Asphalt coating
   d. Fiberglass base

2. Why are the undersides of asphalt shingles often covered with a plastic strip?
   a. To help the shingles adhere together
   b. To improve the way the product looks
   c. To ensure the asphalt does not melt during transportation
   d. To keep shingles from sticking together during transportation

3. Which type of asphalt shingle style includes diagonal shapes?
   a. Locking
   b. Architectural
   c. Specialty
   d. Three-tab

4. _____ are sometimes colored red, yellow, and green to signify different cutting directions.
   a. Utility knives
   b. Compressors
   c. Magnet sweepers
   d. Snips

5. Leak barriers are sometimes called *self-adhering sheets.*
   a. True
   b. False

6. What is *strapping*?
   a. Fastening underlayment in vertical runs
   b. A type of underlayment brand
   c. Using starter strip shingles before installing a first course of shingles
   d. A mechanical attachment device used in shingle systems

7. On what style of valley are shingles attached across the valley from one direction?
   a. Valley centerlines
   b. Open valleys
   c. Closed-cut valleys
   d. Metal flashings

8. Premanufactured ridge and hip shingles are *less* likely to be provided with _____.
   a. laminate shingles
   b. three-tab shingles
   c. interlocking shingles
   d. dimensional shingles

9. _____ should be installed at the backwall of a vertical-surface protrusion.
   a. Drip edge flashings
   b. Kickout flashings
   c. Step flashings
   d. Backer flashings

10. _____ are installed at the end of every shingle course that intersects a vertical wall.
    a. Step flashings
    b. Drip edge flashings
    c. Cricket flashings
    d. Apron flashings

Trade Terms Introduced in This Module

Backnailing: Nailing the concealed portion of a roofing component so the fasteners are covered by the next ply, or course, and are not exposed to the weather in the finished roof system.

Fiberglass base: An asphalt shingle base made with glass spun into fibers and combined using plastic polymers.

Hook blade: A blade that has sharp, rounded edges used to cut materials, such as underlayment and asphalt shingles, with great precision.

Magnet sweeper: A push-broom tool that uses magnets to attract and collect metal waste at a jobsite.

Roofing hammer: A hammer that allows users to orient and nail shingles using just one tool.

Starter strip shingles: Roll roofing or shingle strips applied along the downslope eave line before the first course of roof covering. These are intended to fill spaces between cutouts and joints of the first course.

Taper: Increasing width uniformly to gradually widen or narrow.

Tin snips: Useful tools for cutting and trimming thin but tough materials, such as sheet metal and shingles.

Ultraviolet (UV) radiation resistance: The ability of shingles to resist ultraviolet radiation from the sun degrading their quality over time.

Valley centerline: The center of the valley; no fasteners should be used within 6" (15 cm) of this point.

Additional Resources

This module presents thorough resources for task training. The following reference material is suggested for further study.

Earth911, **www.earth911.com**.

National Roofing Contractors Association (NRCA), **www.nrca.net**.

NCCER Module 16102, *Roofing Safety*.

NCCER Module 16105, *Introduction to Steep-Slope Roofing*.

NCCER Module 75901, *Fall Protection Orientation*.

Occupational Outlook Handbook, Roofers, Bureau of Labor Statistics, US Department of Labor, **www.bls.gov/ooh/construction-and-extraction/roofers.htm**.

Occupational Safety and Health Administration (OSHA), US Department of Labor, **www.osha.gov**.

OSHA 3755-05, *Protecting Roofing Workers*. US Department of Labor. **www.osha.gov/Publications/OSHA3755.pdf**.

Figure Credits

iStock@SMJoness, Module Opener

iStock@rickmartinez, Figure 1

Courtesy of the National Roofing Contractors Assoc., Figures 2, 13, 17, 21–23, 26, 28–30, 32–34

iStock@PragasitLalao, Figure 4

Courtesy of GAF, Figures 5–7, 15, 16, 24

Image property of Stanley Black & Decker. Used with permission., Figures 8–10, 12

Courtesy of DryFlekt, Inc., Figure 31

Courtesy of Prugar Consulting, Inc., Figure 35

Section Review Answer Key

Section 1.0.0

Answer	Section Reference	Objective
1. c	1.1.2	1a
2. a	1.2.2	1b

Section 2.0.0

Answer	Section Reference	Objective
1. a	2.1.3	2a
2. a	2.2.1	2b
3. c	2.3.0	2c

User Update

NCCER makes every effort to keep its textbooks up-to-date and free of technical errors. We appreciate your help in this process. If you find an error, a typographical mistake, or an inaccuracy in NCCER's curricula, please submit a User Update form by visiting **https://www.nccer.org/olf**. You can also scan the QR code using the camera on your phone or mobile device to access the form.

Clay and Concrete Tile Roof Systems

Overview

Clay tile systems have deep roots in our history and culture. Concrete tile systems became a popular and more affordable alternative to clay systems in the 1800s. Today, tile roofs remain a popular aesthetic in roof design and offer extended roof life due to the enduring nature of these systems. Roofing professionals need to be familiar with features of tile roofs so that they are prepared to safely install these steep-slope systems.

Module 16202

Trainees with successful module completions may be eligible for credentialing through the NCCER Registry. To learn more, go to **www.nccer.org** or contact us at 1.888.622.3720. Our website, **www.nccer.org**, has information on the latest product releases and training.

Your feedback is welcome. You may email your comments to **curriculum@nccer.org**, send general comments and inquiries to **info@nccer.org**, or fill in the User Update form at the back of this module.

This information is general in nature and intended for training purposes only. Actual performance of activities described in this manual requires compliance with all applicable operating, service, maintenance, and safety procedures under the direction of qualified personnel. References in this manual to patented or proprietary devices do not constitute a recommendation of their use.

16202 V1.0

16202
Clay and Concrete Tile Roof Systems

Objectives

Successful completion of this module prepares you to do the following:

1. Explain the details of tile systems and the tools used to install these systems.
   a. Describe types, styles, and accessory pieces of tile roofs.
   b. Identify the hand tools and power tools used in tile installation.
2. Describe tile installation.
   a. Describe the safe procedures to install tile.
   b. Explain the flashings and water-shedding details used in tile installation.
   c. Describe basic repair methods of tile.

Performance Tasks

Under supervision, you should be able to do the following:

1. Don all appropriate PPE for tile installation.
2. Apply drip edge, manufacturer-recommended underlayment, and base flashings to a wood deck.
3. Install valley underlayment, valley metal, and valley tile according to manufacturer instructions.
4. Attach clay or concrete tile over underlayment according to manufacturer instructions.
5. Complete any secondary flashings.
6. Identify a tile that needs repair and perform the repair.

Trade Terms

Apex tile
Batten-hung
Eave closures
Exposure
Interlocking
Mechanical fasteners
Profiles
Tapered edge boards
Tile adhesives
Trim tile

Industry Recognized Credentials

If you are training through an NCCER-accredited sponsor, you may be eligible for credentials from NCCER's Registry. The ID number for this module is 16202. Note that this module may have been used in other NCCER curricula and may apply to other level completions. Contact NCCER's Registry at 1.888.622.3720 or go to **www.nccer.org** for more information.

You can also show off your industry-recognized credentials online with NCCER's digital badges. Transform your knowledge, skills, and achievements into badges that you can share across social media platforms, send to your network, and add to your resume. For more information, visit **www.nccer.org**.

Videos and Digital Resources

How to Access Resources

This craft has additional videos and resources to enhance your learning experience. To view these resources, scan the QR below. The videos and resources are separated by module.

You can scan this code using the camera on your phone or mobile device to view these videos and resources.

Contents

1.0.0 Background and Tools 1
1.1.0 Styles and Profiles 2
1.1.1 Styles 2
1.1.2 Profiles 3
1.1.3 Accessories 4
1.2.0 Tools 5
1.2.1 Trowels 5
1.2.2 Chalk Lines 6
1.2.3 Manual Tile Chippers 6
1.2.4 Screw Gun 6
1.2.5 Saws 7
1.2.6 Mixers 7
2.0.0 Installation and Repair 8
2.1.0 Installation Procedures and Components 8
2.1.1 Underlayments 8
2.1.2 Attachment Methods 9
2.1.3 Installation Outline 9
2.1.4 Field Installation 10
2.1.5 Valleys 10
2.1.6 Hips, Ridges, and Rakes 13
2.2.0 Flashings 13
2.2.1 Perimeter Flashings 13
2.2.2 Vertical-Surface Protrusions 13
2.2.3 Penetrations 14
2.3.0 Basic Repair 19

Figures

Figure 1 Tile installation..... 1
Figure 2 Antique clay tile roofs in China..... 2
Figure 3 Clay tile..... 2
Figure 4 Concrete tile..... 2
Figure 5 Interlocking tile..... 3
Figure 6 Batten lugs on the underside of tile..... 3
Figure 7 Tile classifications..... 3
Figure 8 Low-profile tile..... 4
Figure 9 Medium-profile tile..... 4
Figure 10 High-profile tile..... 4
Figure 11 Accessory pieces for a tile roof..... 5
Figure 12 A pointing trowel..... 5
Figure 13 Chalk line..... 6
Figure 14 Manual tile chipper..... 6
Figure 15 Screw gun..... 6
Figure 16 Masonry bit..... 7
Figure 17 Portable mixer..... 7
Figure 18 Tile installed on a batten and counter-batten system..... 10
Figure 19 Raised fascia board..... 11
Figure 20 Eave closure..... 11
Figure 21 Open valley..... 12
Figure 22 Closed-valley installation..... 13
Figure 23 Ridge-tile installation..... 14
Figure 24 Rake tile..... 15
Figure 25 Drip edge metal..... 16
Figure 26 Vertical-surface flashings..... 17
Figure 27 Penetration flashing..... 18
Figure 28 Damaged tile..... 19

Section One

1.0.0 Background and Tools

Objective

Explain the details of tile systems and the tools used to install these systems.

a. Describe types, styles, and accessory pieces of tile roofs.
b. Identify the hand tools and power tools used in tile installation.

Trade Terms

Apex tile: An accessory piece that helps waterproof systems by covering the area where hips and/or ridges converge from different directions.

Batten-hung: A style of clay and concrete tile that has a batten lug, typically on the backs of the tile pieces to help hang the tile off batten systems.

Eave closures: Accessory pieces used to close off and flash the open space at a roof's perimeter edge. Also called *eave risers* for low-profile tile and *bird stops* for medium- to high-profile tile.

Interlocking: A style of clay and concrete tile that has integrated connecting mechanisms on each tile piece that help link tile together during installation.

Profiles: The ratio of a tile's height to its width; the lower the profile, the flatter the tile.

Trim tile: An accessory piece provided by the manufacturer used to attractively weatherproof various areas of a roof, such as at ridges, hips, and rakes.

Tile systems (*Figure 1*) combine valuable weather protection with aesthetically pleasing designs. Before installing tile, roofing professionals must be familiar with the common styles, profiles, and accessory pieces of these systems as well as the tools used during installation.

WARNING! Safety is a crucial consideration in tile installation. For more information on roofing safety, review NCCER Module 16102, *Roofing Safety*.

NOTE Throughout NCCER's Roofing program, the term *steep-slope* refers to a roof slope of 3:12 or greater. This differs from OSHA's definition of steep-slope roofs (slope of 4:12 or greater). When installing or repairing roof systems, always check company, local, state, and federal codes. If any of the applicable codes contradict each other, follow the strictest one.

Clay tile roofs have deep roots in culture and history. The origin of this roof system can be traced as far back as 3000 BC, when clay tile installation began in China (*Figure 2*) and the Middle East. Clay tile systems spread across the world, and clay tile became an iconic steep-slope system known for its durability and pleasant design. Since its early establishment as a roof system, clay tile has lost and gained popularity throughout the course of roofing history according to the changing needs of society and the introduction of alternate roof systems.

Concrete tile roofs gained popularity as an affordable alternative to clay tile roofs in the 1900s. Concrete tile could also be manufactured to look like clay tile, and it offered similar fire-protection properties.

Today, both clay and concrete tile systems are popular choices for steep-slope roofing. While these roof systems used to be more challenging to craft, the introduction of assembly-line technology has allowed the market to provide tile at a lower price and greater quantity.

Figure 1 Tile installation.

Figure 2 Antique clay tile roofs in China.

Forbidden City

The Forbidden City, a Chinese palace constructed from 1406 AD to 1420 AD, is the largest still-existing palace in the world. This palace has 980 surviving buildings and measures approximately 178 acres (720,000 m^2). The color of the glazed-clay roof tiles used throughout the palace indicated the political significance of the people who were housed inside a building.

Figure Credit: iStock@Vincent_St_Thomas

1.1.0 Styles and Profiles

Tile systems are hydrokinetic, or *water-shedding*. Clay tile (*Figure 3*) is made by molding a tile shape and then firing the clay tile in kilns at temperatures that reach up to 2,000°F (1,093°C). Concrete tile (*Figure 4*) is made by mixing cement, water, and fine aggregate or sand before setting or extruding the mixture into a tile shape and allowing it to dry.

Roofing tile comes in a variety of styles and profiles. Manufacturers of roof tile also provide accessory pieces designed to cover parts of the roof other than the field, such as tile designed for hips and ridges.

1.1.1 Styles

Since tile systems have been used for centuries, there are many styles that roofing professionals will encounter on the job. Some major differences between style types are whether the tile is designed to be **interlocking** or noninterlocking; another important distinction is between **batten-hung** and non-batten-hung tile. These style features determine the way that tile should be attached to a roof.

Figure 3 Clay tile.

Figure 4 Concrete tile.

Interlocking tile (*Figure 5*) has connecting mechanisms made with individual pieces that link the tile together during roof installation. Note that not all tiles have interlocking pieces. Similarly, batten-hung tile has a hooking feature, or *batten lug*, on the back of each tile piece. This allows a more seamless installation on batten systems (*Figure 6*); non-batten-hung tile does not have this feature.

> **NOTE**
> This module only covers common clay and concrete tile styles. Roofers may encounter additional style variations on the job. If you have questions about how to install a particular style, consult the manufacturer's instructions and talk to your supervisor.

Figure 5 Interlocking tile.

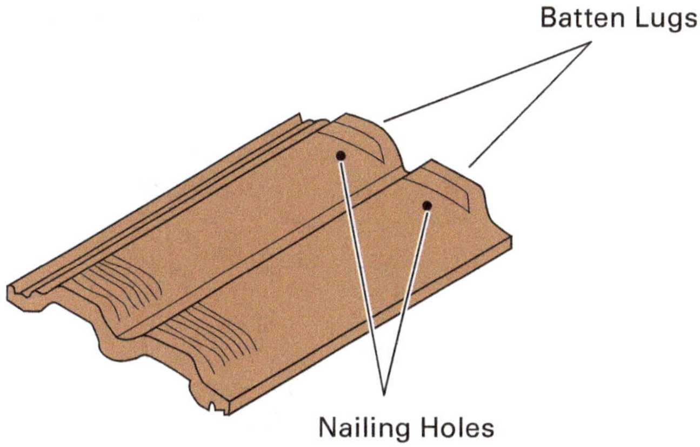

Figure 6 Batten lugs on the underside of tile.

1.1.2 Profiles

In addition to style variety, roofing tile comes in several profiles. Profiles describe the ratio of the tile's height to its width. The lower the profile, the flatter the tile. *Figure 7* illustrates the differences between tile classifications.

Profiles for both clay and concrete tile include:

- *Low* — This profile type has a top surface rise of $^1/_2$" (1 cm) or less. An example of low-profile clay and concrete tile is *flat tile*. *Figure 8* shows low-profile tile.

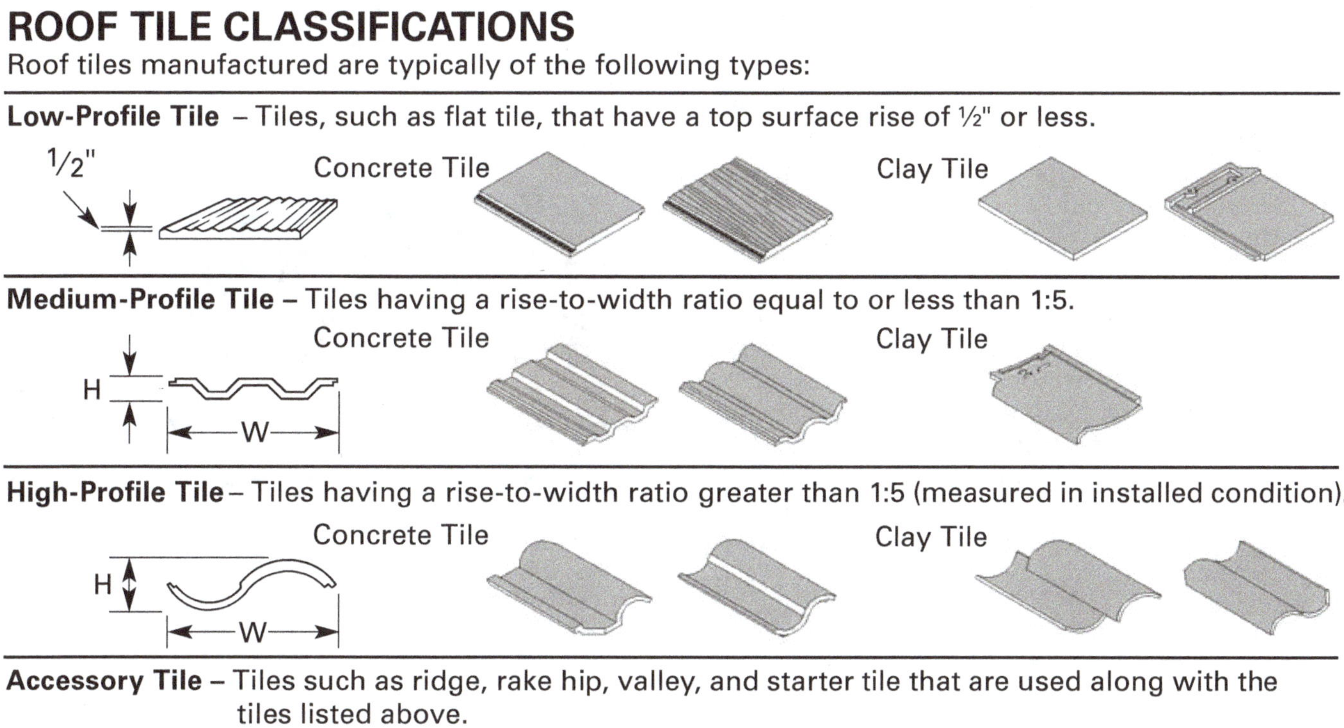

Figure 7 Tile classifications.

Figure 8 Low-profile tile.

- *Medium* — This profile type has a rise-to-width ratio of 1:5 or less. An example of medium-profile tile is *French tile*. Medium-profile tile is shown in *Figure 9*.
- *High* — This profile type has a rise-to-width ratio greater than 1:5 when measured in installed condition. An example of high-profile tile is *S-tile*. *Figure 10* shows high-profile tile.

1.1.3 Accessories

Tile systems contain accessory pieces created by the manufacturer to aid installation. While the individual pieces included may vary from one system to another, there are some common accessory pieces roofing professionals should be able to recognize. *Figure 11* depicts some examples of **trim tile**, which is an accessory piece designed for a tile roof.

Overall, accessory pieces promote positive water drainage and structural integrity by providing custom-fit items that protect areas of the roof not covered with field tile.

Some common examples of accessory pieces are:

- **Apex tile** helps waterproof systems by covering the area where hips and/or ridges converge from different directions.
- Hip and ridge tile are pieces made specifically for those areas of the roof.
- **Eave closures**—also called *eave risers* for low-profile tile and *bird stops* for medium- to high-profile tile—close off the open space at a roof's perimeter edge.

Figure 9 Medium-profile tile.

CAUTION

To ensure that you are properly informed about the specific accessory pieces included in a given roof system, consult the manufacturer instructions before beginning tile installation.

Figure 10 High-profile tile.

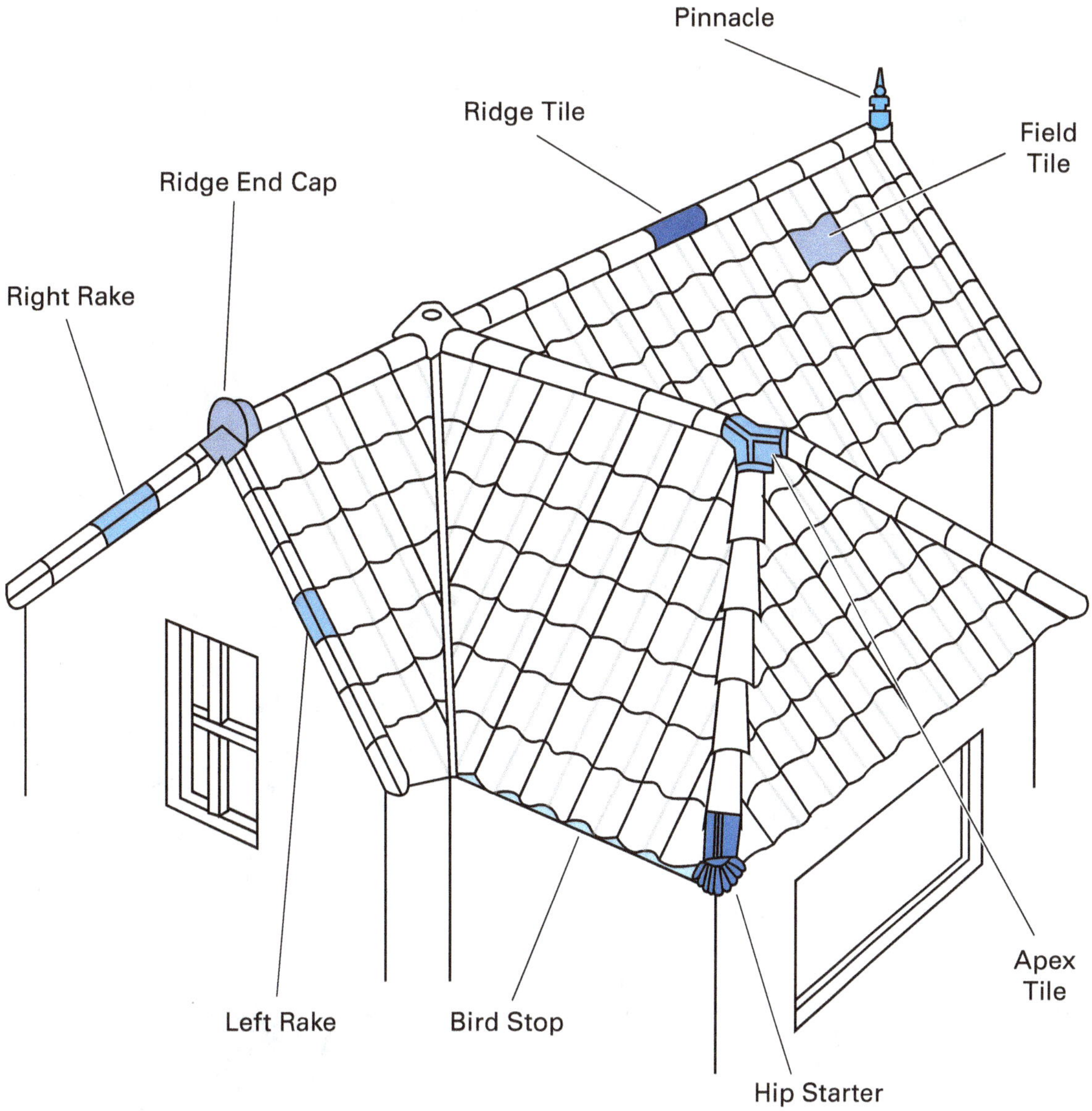

Figure 11 Accessory pieces for a tile roof.

1.2.0 Tools

Roofing professionals must know how to safely use the common hand and power tools involved in clay and concrete tile installation. Along with understanding tool usage, roofing professionals are also responsible for ensuring their tools are clean and in good condition before using them to install tile roofs.

> **NOTE**
> This section highlights some of the tools you will encounter when installing tile roof systems. Other common tools used in tile installation, such as hammers and circular saws, are covered in greater detail in NCCER Module 00103, *Introduction to Hand Tools*, and NCCER Module 00104, *Introduction to Power Tools*.

1.2.1 Trowels

Trowels are hand tools with flat metal blades held by short handles. Pointing trowels (*Figure 12*) and margin trowels are varieties that may be useful during underlayment and tile installation. Trowels can assist with stirring and applying mixes such as mortar. The fine point at the end of a trowel adds precision to the application process.

Figure 12 A pointing trowel.

1.2.2 Chalk Lines

Chalk lines (*Figure 13*) are hand tools used by anchoring one end of the line at a certain point on the roof and reeling the line away from its setting point. Once a chalk line is hovered over its desired position, it should be snapped onto the roof deck or underlayment. This leaves a chalk imprint that can help orient and align underlayment and tile.

1.2.3 Manual Tile Chippers

Manual tile chippers (*Figure 14*), or *manual cutters*, are hand tools that help limit exposure to silica dust by carefully cutting a clay or concrete tile with a precise, handle-powered blade. To use a hand-powered tile chipper, start by marking your tile where it should be cut. Then, align the tile in the chipper, pumping the handle as you move the tile to cut along your marked line. Once the tile is cut, the pieces will collect in the removeable bin at the bottom of the tool.

1.2.4 Screw Gun

Power equipment used in tile installation will vary by region. Since different geographical areas experience different weather patterns, the way that a particular roof system is installed can vary greatly from one region to another. In some areas, such as Florida, wind uplift presents a greater risk; therefore, tile installed in regions with this weather pattern may be both screwed into place and adhered with foam. A screw gun (*Figure 15*) with a modified adapter for installing screws into tile is a useful power tool for these high-wind regions.

Figure 13 Chalk line.

Masonry bits (*Figure 16*) can be added to a drill to create a power tool that allows roofing professionals to easily drill any additional holes needed into a clay or concrete tile. As with saws, be sure to use safety precautions, such as gloves and a mask, to avoid breathing silica particles released into the air while drilling.

Figure 14 Manual tile chipper.

Figure 15 Screw gun.

Figure 16 Masonry bit.

1.2.5 Saws

Saws with diamond blades, such as cordless saws, skill saws, and grinders, are power tools that will prove useful in any region as they assist with cutting tile. Some saw types are dustless, meaning they have dust collection systems that minimize the dust typically emitted through tool usage.

> **WARNING!**
> Roofing tiles often contain silica and traces of other hazardous substances. Dust produced while cutting, drilling, or grinding tile can irritate or harm the lungs and skin. Always follow necessary safety precautions, such as wearing gloves or a mask, when working with these power tools. For more information, refer to *29 CFR 1926.1153(c)(1), Table 1.*

1.2.6 Mixers

Mixers are power tools that thoroughly combine liquids such as cement and mortar so that they are fresh before roof application. Mixers may be handheld or come in the form of a portable container (*Figure 17*).

Figure 17 Portable mixer.

1.0.0 Section Review

1. Which of the following is a type of tile classification?
   a. Apex tile
   b. Interlocking
   c. Batten-hung
   d. Low

2. Masonry bits can be used to drill additional holes into tile.
   a. True
   b. False

Section Two

2.0.0 Installation and Repair

Objective

Describe tile installation.

a. Describe the safe procedures to install tile.
b. Explain the flashings and water-shedding details used in tile installation.
c. Describe basic repair methods of tile.

Performance Tasks

1. Don all appropriate PPE for tile installation.
2. Apply drip edge, manufacturer-recommended underlayment, and base flashings to a wood deck.
3. Install valley underlayment, valley metal, and valley tile according to manufacturer instructions.
4. Attach clay or concrete tile over underlayment according to manufacturer instructions.
5. Complete any secondary flashings.
6. Identify a tile that needs repair and perform the repair.

Trade Terms

Exposure: The amount of tile surface exposed to the weather and not overlapped by the next course or surrounding tiles.

Mechanical fasteners: Attachment methods that secure objects together through fastening. Examples include nails and screws.

Tapered edge boards: Strips used to modify the angle where the roofing membrane meets an eave; uniformly props up tile courses starting from the eave.

Tile adhesives: Attachment methods that adhere, or stick, objects together. For example, foam or mortar.

Since tile roofing comes in many different styles and profiles, installation methods also vary between different projects and regions. While roofing professionals should always consult manufacturer instructions before and during installation, they should also be familiar with general installation procedures and accessories so they can safely and knowledgably begin installing tile systems. It is also important to familiarize yourself with the types of flashings installed on tile roofs as well as some basics of tile repair that may come in handy on the jobsite.

2.1.0 Installation Procedures and Components

Tile systems are known for being long-lasting. Roofing professionals must know how to install each component of these systems safely and correctly—from underlayments to flashings—to make sure they live up to their potential lifespans.

2.1.1 Underlayments

Tile systems are compatible with several underlayment types. Always consult manufacturer instructions to ensure you are using an underlayment that works well with the specific tile system you will install. Regional weather patterns and slope also influence the type of underlayment that should be used in a given project. Roofing professionals must make sure a deck is clean, dry, and in good condition before applying underlayment.

The following underlayment options may be used during tile roof installation depending on manufacturer recommendations, regional codes, and roof slope:

- *Double-layer asphalt felts* — Install underlayments in two layers. This method may combine a variety of underlayment types and installation methods, such as mechanically attaching a first layer of #30 felt and adhering a cap sheet as the second layer. When applying double-layer underlayment, install the sheets horizontally in shingle fashion.
- *Single-layer asphalt felts* — Install underlayment, such as #30 felt, #40 felt, or #43 felt, in a single layer.
- *Single-ply membrane* — In some unique cases, single-ply membrane may be used as underlayment for tile systems.
- *Leak barrier* — Also called *self-adhering sheets* and, in colder climates, *ice dam protection.* These underlayment sheets have adhesive bitumen applied at the factory and provide labor savings with installation on the job.

NOTE

Some building codes require certain fastener types to be used with certain underlayment-installation methods. Always abide by local building-code requirements when installing roof systems.

#30 Felt

When #30 felt entered the market, the product weighed 30 lb (13.6 kg). Due to advancements in production, #30 felt tends to weigh less today than it originally did. The same is true for different types of felt, such as #15 and #40.

Even though the weights of these felt products have changed, their naming conventions have remained. Remember that if you are dealing with #15 felt, you may be using felt that does not weigh 15 lb (6.8 kg), despite its name.

2.1.2 Attachment Methods

Before beginning installation, roofing professionals should be familiar with the methods of attaching tile roofs. Tile can be attached to a roof system with **tile adhesives**, such as foam and mortar, and **mechanical fasteners**, such as nails and screws. The best attachment method for a given project will depend on factors such as regional weather, deck type, slope, building codes, and manufacturer recommendations.

The following attachment methods are commonly used in tile application:

- *Adhesive* — Examples include mortar and foam adhesive; this attachment method may involve laying a tile in a bed of foam adhesive, and steep-slope applications usually require the use of additional securement methods such as nails.
- *Clips* — Sometimes referred to as *wind clips* or *storm anchors*. They are useful to use in combination with other attachment methods, particularly in high-wind regions.
- *Lug-hung* — Batten-hung tile may be designed with a lug on their bottom surfaces that hang the tile over a batten. In steep-slope applications, lug hanging is most often used in combination with other securement methods.
- *Roofing nails* — Nails are the most common type of tile fastener. When attaching tile with roofing nails, be sure that the nails are long enough to penetrate all layers of roofing materials and are securely anchored to underlying battens or roof deck.
- *Screws* — Some tile roof systems require screws to be used. Screw fasteners should be long enough to penetrate all layers of roofing materials and securely anchor to underlying battens or roof deck.
- *Wire-tied and strapping* — In cases of non-nailable decks, hanging tile with wire can be an effective securement method. Wire-tied application is commonly combined with nails and screws to secure the wiring to a roof substrate.

These attachment methods are often used in combination when installing a tile roof. For example, roofing professionals may encounter tile systems that require the use of nails, screws, and adhesives. As always, be sure to consult your local building codes and manufacturer recommendations to understand which attachment method(s) would be best.

2.1.3 Installation Outline

Tile systems may be installed directly onto a roof deck or batten system depending on weather patterns, building codes, and manufacturer recommendations. Steep-slope decking, battens, and counter-battens (*Figure 18*) are further detailed in NCCER Module 16105, *Introduction to Steep-Slope Roofing*.

Since there is a variety of styles, profiles, underlayment types, and attachment methods used in tile installation, it is crucial to understand and apply manufacturer recommendations during installation. Each tile system installation has the potential to vary from any other tile system a roofing professional has previously encountered.

The following provides a general outline for clay and concrete tile installation:

WARNING! This outline is intended only to overview general installation sequence. It cannot be used in place of manufacturer instructions or building-code requirements.

Step 1 Don all proper PPE for the job. Ensure that the appropriate fall protection equipment is used or in place. Depending on the project and environment, this may be a personal fall arrest system (PFAS) with approved anchorage, or appropriate guardrails or scaffolding.

Step 2 Begin with a roof deck that is clean, dry, and in good condition.

Step 3 Install the roof's underlayment and drip edge metal.

Step 4 If applicable, install battens. Manufacturer instructions will provide guidance on whether tile should be installed on a batten system as well as how a given batten system should be spaced and installed.

Step 5 Install base flashings, eave closures (also called *eave risers* or *bird stops*), trim boards, and hip and ridge channels.

Step 6 Measure and install the first course of field tile. Be sure to abide by the manufacturer-recommended **exposure** on each tile course.

Step 7 Install vents.

> **NOTE**
> It is recommended to loose lay field tile to ensure any penetrations in the roof are accounted for before field installation.

Step 8 Finish installing field tile.

Step 9 Cut valley tile and field tile located at the roof's hips. If necessary, cut tile to fit into the valley.

Step 10 Install tile.

Step 11 Vacuum or sweep the roof. In addition to promoting a safer work environment, this will get rid of the dust spread during installation that can ruin the look of a roof.

Step 12 Install secondary flashings, such as penetration, wall, and counter flashings. Depending on the region, you may need to further weather block flashings and trim boards with mortar or flexible flashing.

Step 13 Install trim tile.

Figure 18 Tile installed on a batten and counter-batten system.

Step 14 If applicable, install rake tile.

Step 15 If applicable, apply finishing mortar wherever needed.

Step 16 Inspect the completed roof and clean up any remaining dust. Identify broken tile that may need replacing and perform any necessary replacements.

2.1.4 Field Installation

Field installation method is determined by the clay and concrete tile style: for example, whether tile will be batten-hung or not and whether tile will be interlocking or not depend on the design of the tile. The best way to know how to approach clay and concrete tile systems is to consult manufacturer recommendations and local building codes for any given system.

Accessories you are likely to encounter during various tile field installation are:

- Starter courses may be provided for flat-tile systems. After laying a starter course, succeeding courses are typically laid in offset patterns to head lap the tile.
- Interlocking and plain-tile roof systems commonly require elevation to be installed on the roof deck to create a uniformed slope angle for the system's tile. Pieces sometimes used to achieve elevation with interlocking and plain-tile roof systems include raised fascia boards and **tapered edge boards**. These pieces also close off any open space at a roof's perimeter edge (*Figure 19*).
- Eave closures—also called *eave risers* for low-profile tile and *bird stops* (*Figure 20*) for medium- to high-profile tile—are used with some tile systems to close off and flash the open space at a roof's perimeter edge.

2.1.5 Valleys

Valleys are the V-shaped angle that forms where the downslopes of two roof fields intersect. This area is particularly vulnerable to leakage since water will shed in its direction. The two methods of valley installation used in tile systems are open valley and closed-cut valley.

> **CAUTION**
> All valley construction must place fasteners away from the centerline. This is to ensure there is a clear pathway for water to shed off the roof after it collects in valleys.

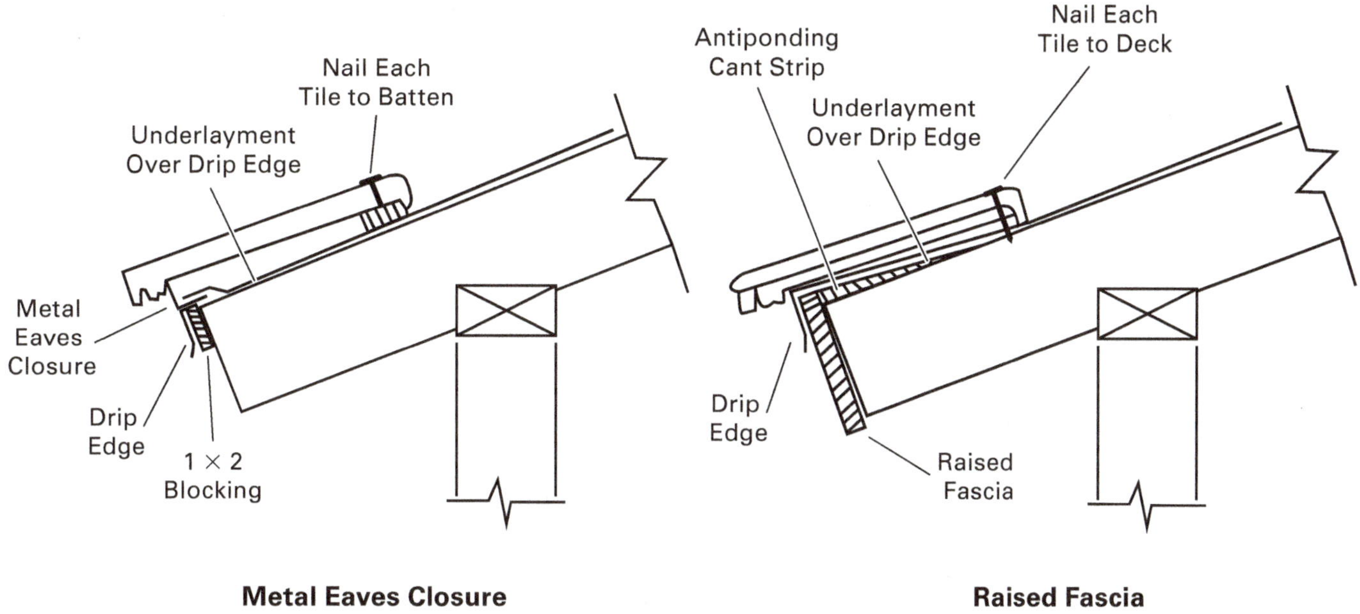

Figure 19 Raised fascia board.

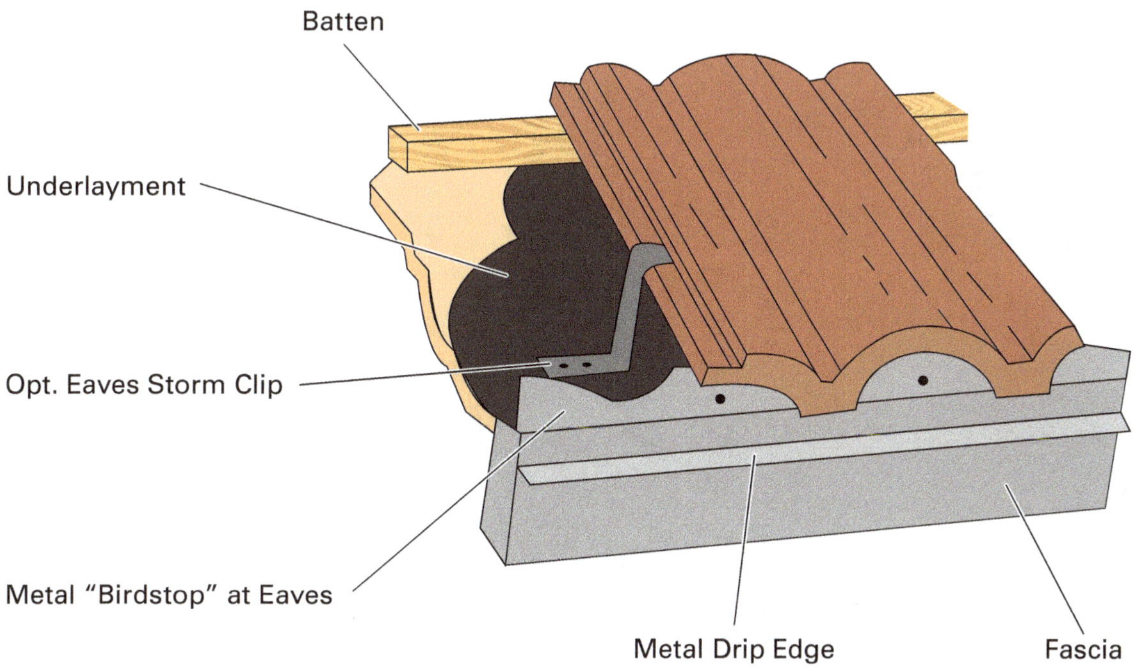

Figure 20 Eave closure.

Open Valley

Open valleys require flashing metal to be installed starting from the low point of the valley up to its high point. The most common valley metal for tile systems is a W-shaped piece with a splash diverter in its center.

Install tile on both sides so that it overlaps at least 4" (10 cm) over the valley metal. In some installations, underlayment or stripping ply may also overlap open-valley metal (*Figure 21*).

Tapering the valley so that it uniformly widens as it transitions downslope is recommended to allow more space for water runoff at the end of the flashing. In colder areas, tapering an open valley also helps melting ice slide down and off the valley rather than getting lodged and refreezing.

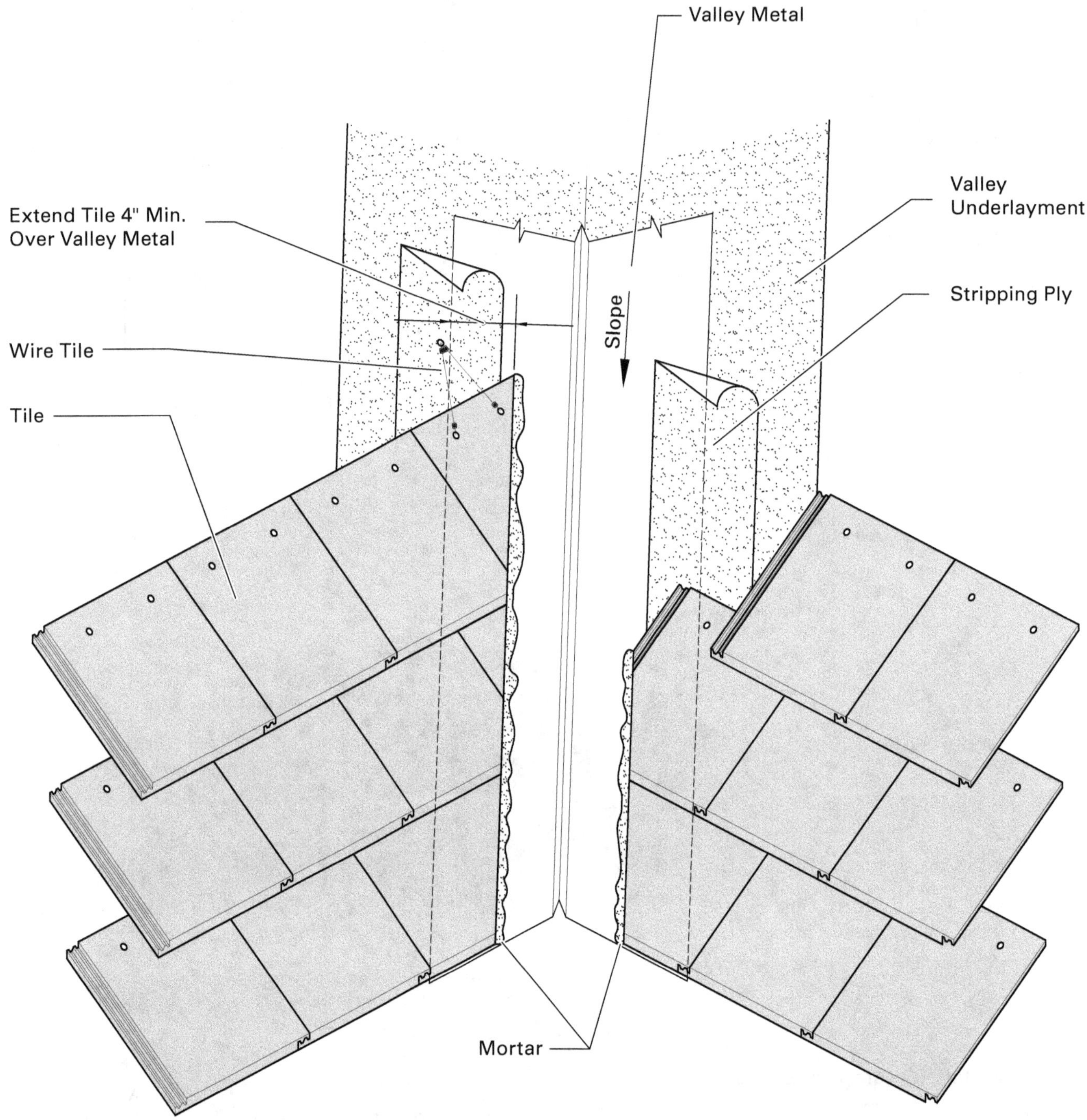

Figure 21 Open valley.

Closed-Cut Valley

Tile systems can create a closed-cut valley, also called a *closed valley* (*Figure 22*), for plain and flat interlocking styles. This is done by cutting tile on both sides at an angle parallel to the valley centerline and butting those sides together to form a miter joint. It is possible to further weatherproof these joints by placing individual pieces of flashing under each tile course that runs along the centerline.

Another closed-valley option requires using valley metal to divert water away from tile joints in that area. Wire-attachment methods and adhesives help roofing professionals install tile close to the valley centerline without using attachment devices, such as nails and screws, that might obstruct that centerline.

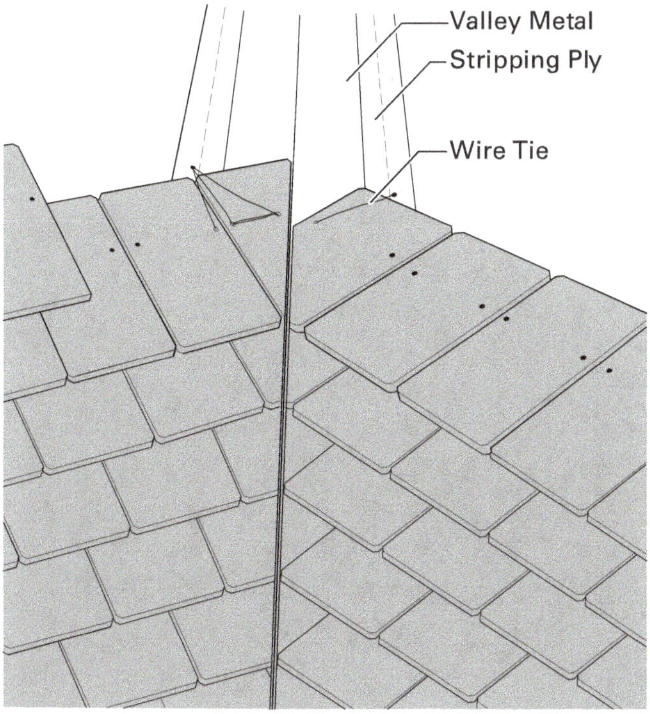

Figure 22 Closed-valley installation.

CAUTION

In areas with heavy vegetation, closed valleys may not be recommended. Leaves, moss, and other organics are more likely to collect in closed valleys and obstruct water runoff.

2.1.6 Hips, Ridges, and Rakes

Trim tiles are designed to weatherproof hips and ridges; additional accessories similar to eave closures may be provided to close off hip and ridge tile pieces. Wood, plastic, or metal members should be installed to a roof deck in these areas to serve as a suitable attachment point for hip and ridge tile. *Figure 23* (see the black circle lettered *A*) depicts a ridge member that provides a substrate onto which ridge tile can be fastened. Hip and ridge tile may be secured with mechanical fasteners, adhesives, or a combination of those methods.

Most tile systems use accessory-piece rake tile along rake edges (*Figure 24*). Additional flashing may also be installed in these areas to protect underlying building components and promote water runoff.

2.2.0 Flashings

Clay and concrete tile roofs must have flashings installed to ensure water sheds off roofs without leaking through the systems. Leak barrier may be used at areas that require flashing, such as rakes, eaves, chimneys, pipes, and vents.

2.2.1 Perimeter Flashings

Perimeter weatherproofing may include flashing eaves and rakes in addition to using accessory eave- and rake-tile pieces. Drip edge metal is not always required for tile installation, but it may be appropriate to add depending on regional climate, manufacturer instructions, and local building codes. *Figure 25* depicts L-type drip edge metal installed on a clay tile roof.

2.2.2 Vertical-Surface Protrusions

Vertical-surface protrusions include walls, dormers, and chimneys. There are typically four types of vertical-surface flashings used in tile systems: apron flashing, channel or step flashing, cricket or backer flashing, and counterflashing. A layer of appropriate underlayment is often applied to these areas before flashing is installed.

Apron flashing is installed at headwalls, or the lowest of the (typically four) sides that compose a vertical-surface protrusion.

Channel flashing, base flashing, or step flashing should be installed where a roof area intersects vertical walls. Channel flashing, also called *pan flashing,* is used in most medium- and high-profile systems.

The backwall of a vertical-surface protrusion requires either backer flashing or cricket flashing. Backer flashing is used with vertical protrusions less than 24" (61 cm) wide, and they are often used with skylights. Cricket flashings are used at the upslope side of a chimney or with penetrations 24" (61 cm) or greater.

Counterflashing must be installed along the top edge of base, apron, step, cricket, and backer flashings to protect the base flashings and their fasteners. Any building material that properly water-laps the top of the flashing might act as the counterflashing.

Figure 26 details a vertical-surface protrusion with the appropriate vertical-surface flashings.

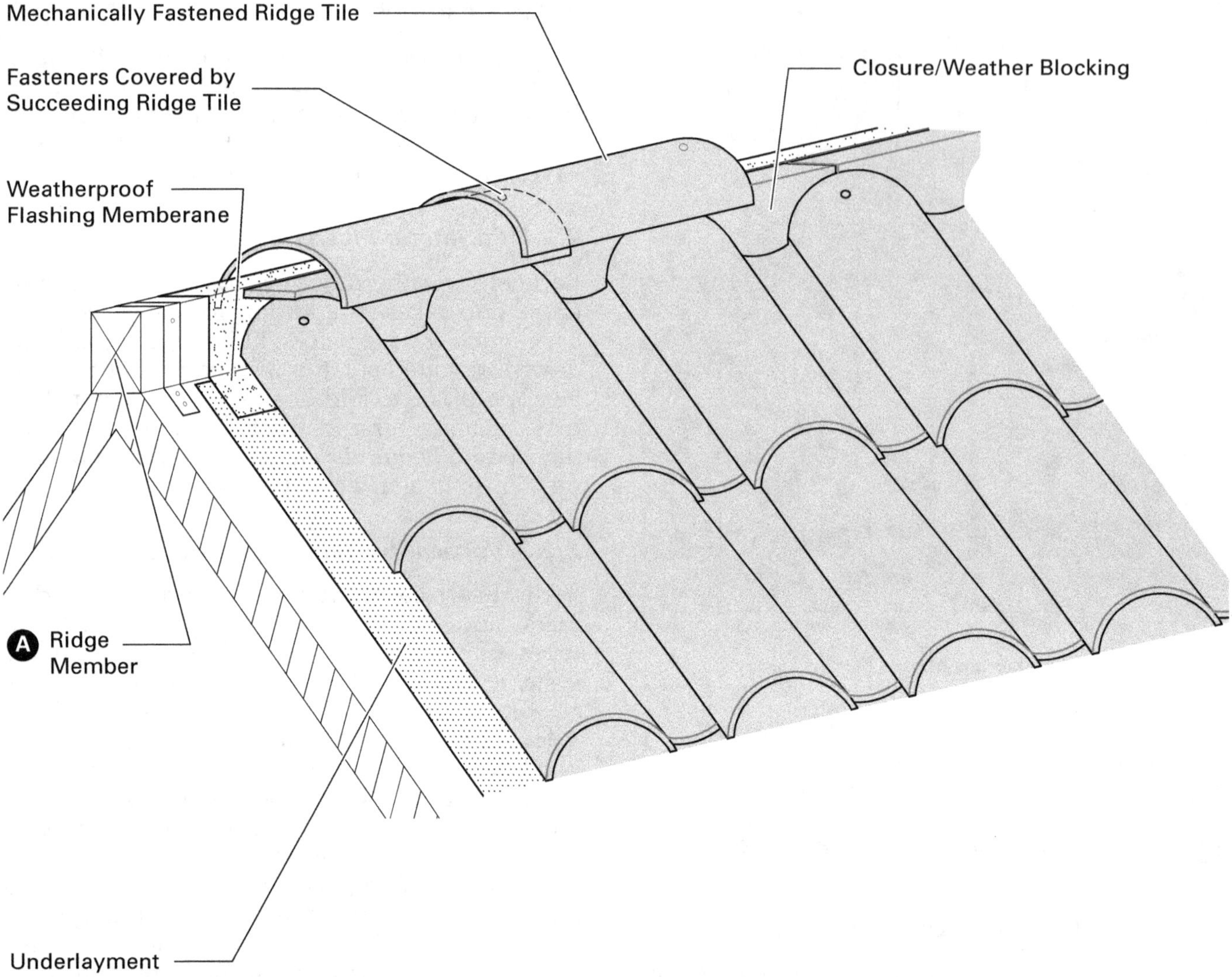

Figure 23 Ridge-tile installation.

2.2.3 Penetrations

Penetrations require flashing to weatherproof the area where they interrupt the roof's field. Penetrations include features such as vent pipes, exhaust vents, and exhaust fans.

Prefabricated boots and collars, made of soft metal that can be tightened around surface contours, are customized to fit around commonly found roof penetrations. These have flanges that attach to the roof membrane and cylindrical or rectangular sections that extend above the flange to fit around pipe- or box-shaped penetrations. These boots and collars are often installed under the tile and leak barrier on the upslope side of a penetration and extend over the tile at the downslope side (*Figure 27*). The flange may be set into asphalt roof cement or elastomeric flashing sealant to provide additional weatherproofing.

In some cases, primary metal flashing may be installed under the underlayment on the downslope side of a penetration. To ensure this primary flashing is weatherproofed, a secondary flashing should be installed at the tile where the upslope side of the flashing is under the tile and the downslope side of the flashing passes over the tile.

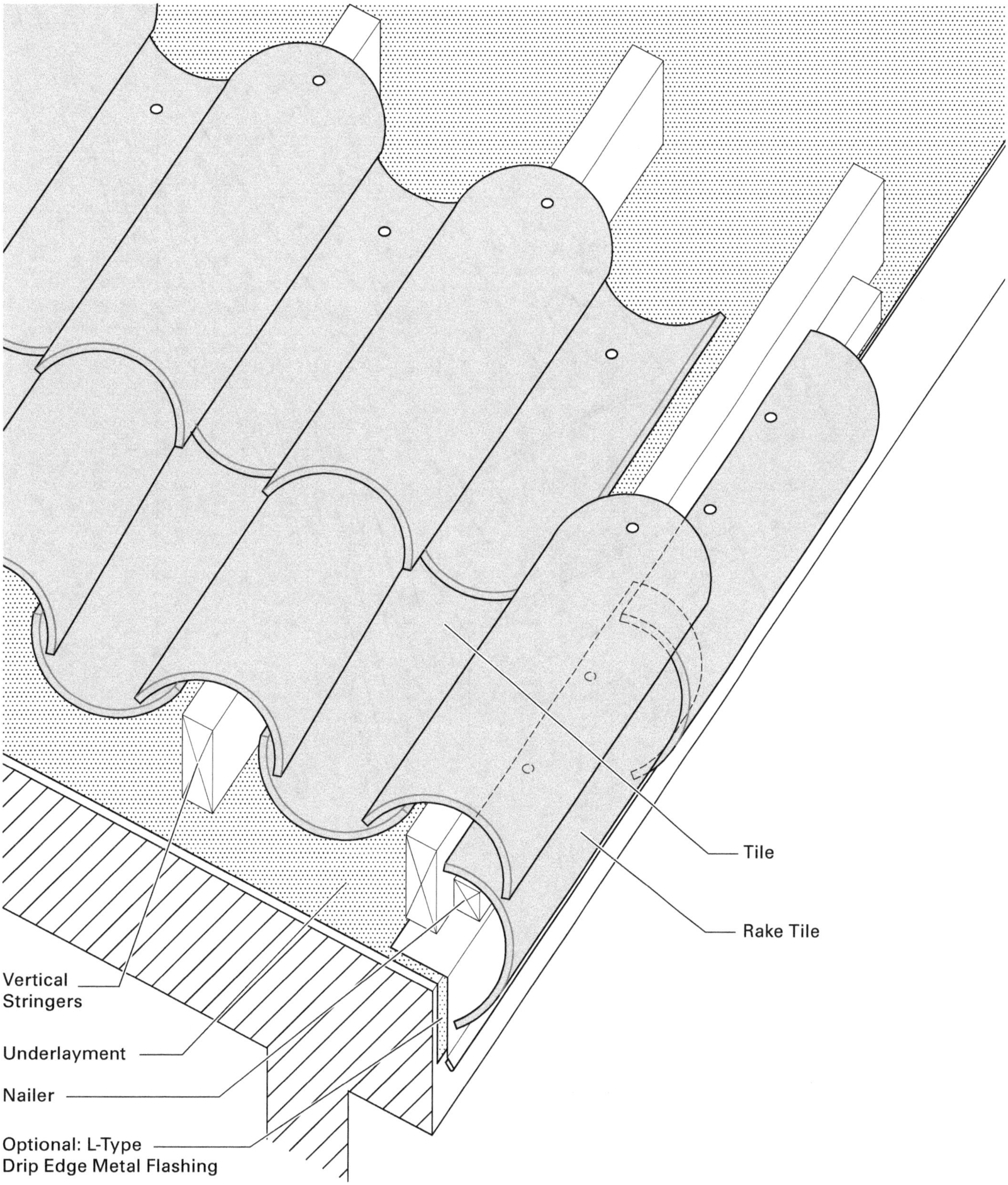

Figure 24 Rake tile.

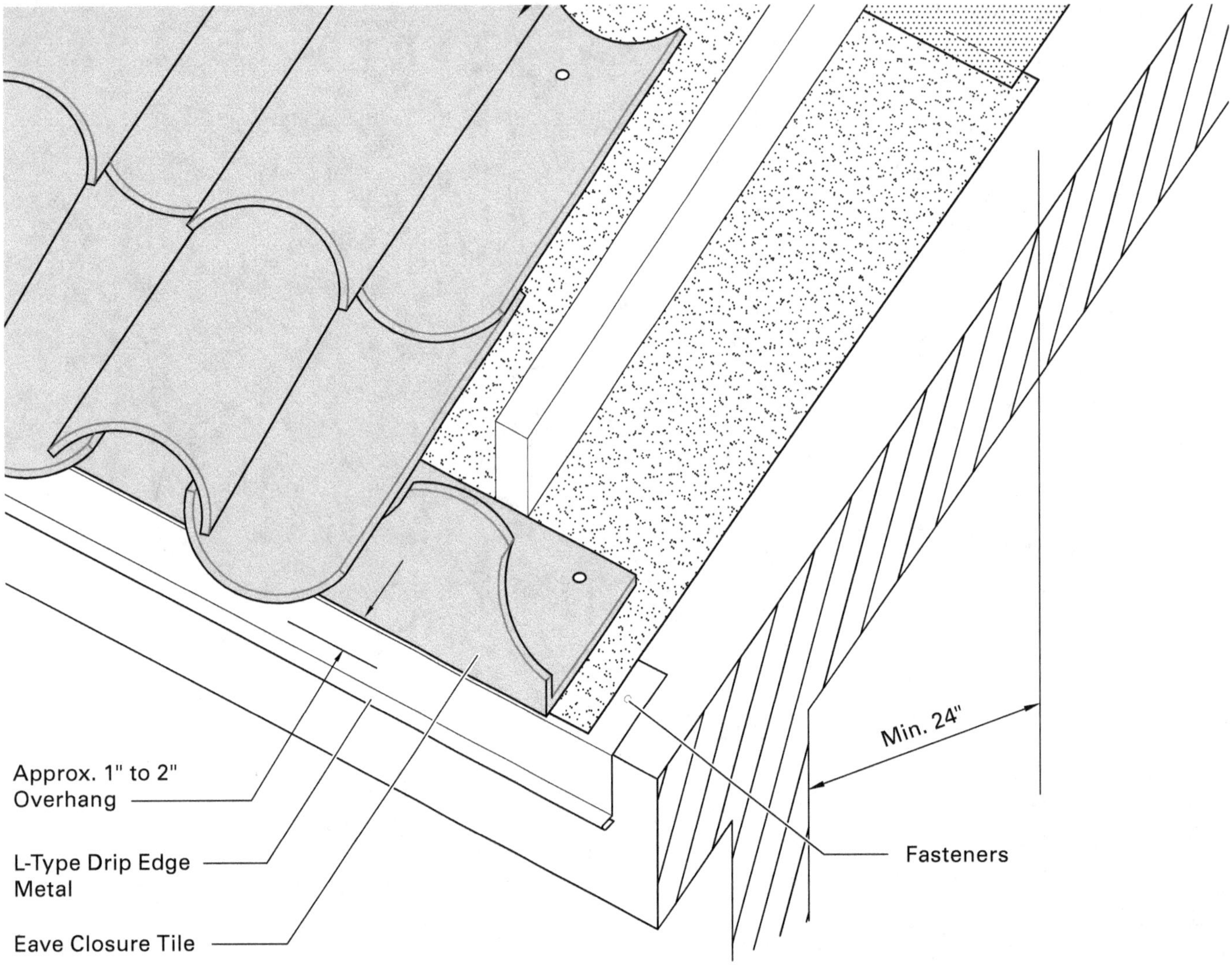

Figure 25 Drip edge metal.

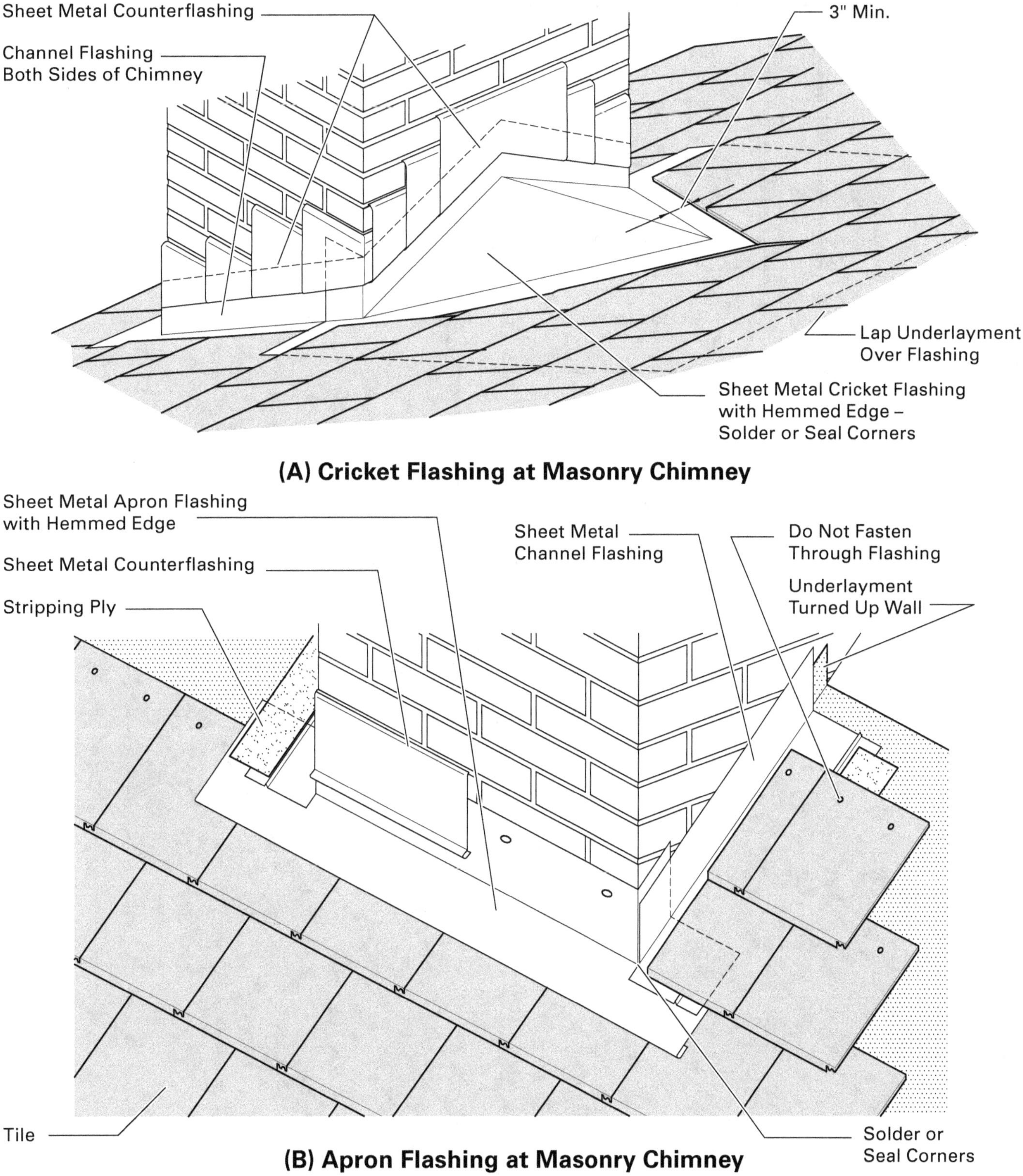

Figure 26 Vertical-surface flashings.

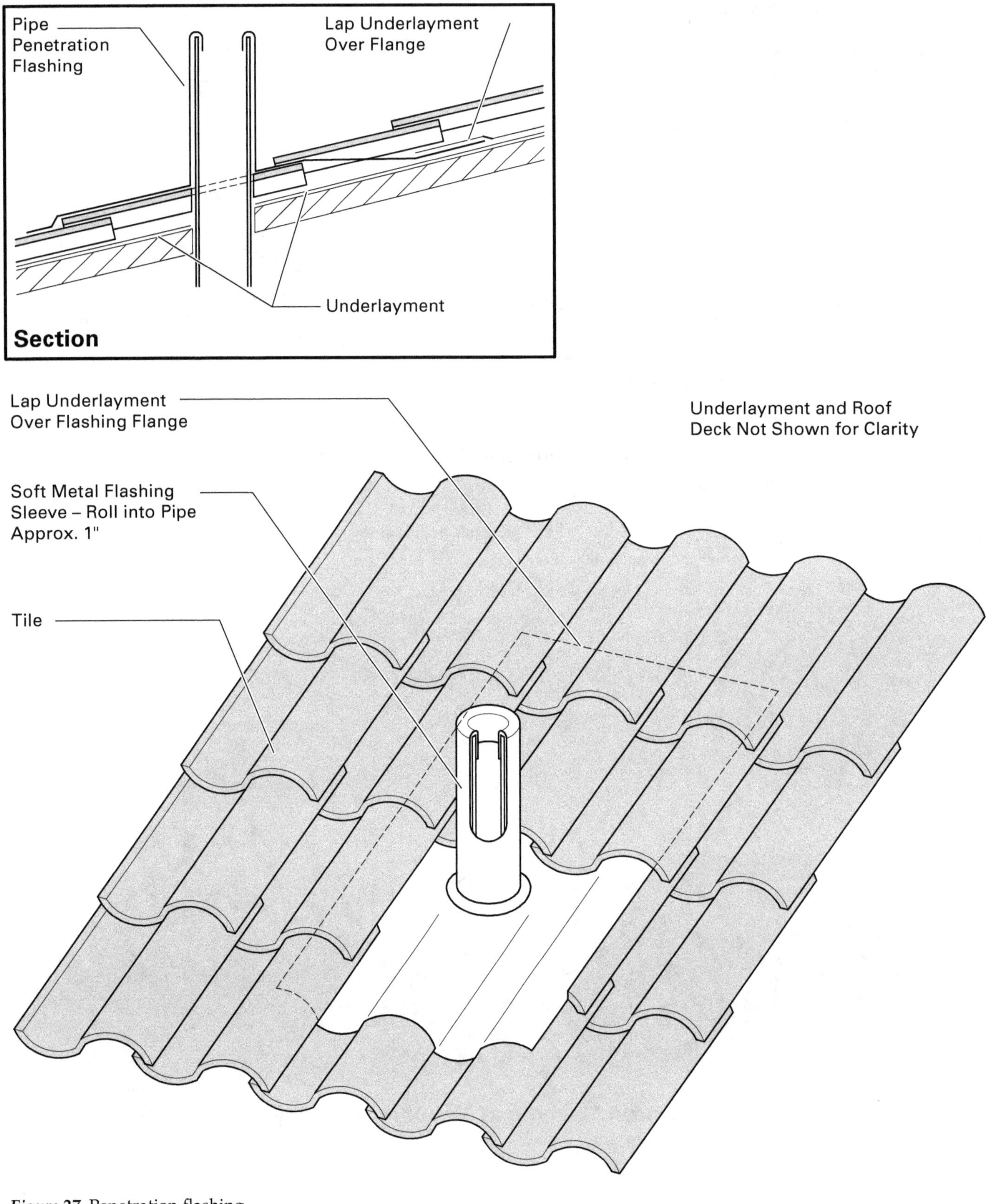

Figure 27 Penetration flashing.

2.3.0 Basic Repair

Along with understanding how to install a roof system, from underlayments to flashings, roofing professionals should notice and fix mistakes that may occur during tile installation.

A common repair is to tiles that are damaged by foot traffic during the installation process. Pay particular attention to the quality and alignment of the tile, accessory pieces, base flashings, and counterflashings. If a tile or flashing is damaged, it should be replaced with a new supply. *Figure 28* shows an example of a cracked tile in need of repair.

A common type of repair required at the end of tile installation is replacing a broken tile. If a tile is broken or incorrectly installed, it will likely need to be removed from the system completely. Be sure to look for less obvious damage in the adjacent tiles. Then, assume the underlayment is damaged until you visually determine otherwise.

You may need to use a hammer and flat bar to break up and remove pieces of a damaged tile. A slide bar is useful for removing tile that was adhered with foam or mortar. A small hand grinder may be used to cut off fasteners during tile replacement. In some cases, such as with underdriven screws, repairs can be made without reinstalling any tile.

Figure 28 Damaged tile.

Whether a tile is removed completely or repaired, be sure to seal any holes in the area with sealant. For tile installed on battens, reinstallation involves hanging a new tile on that batten. If a damaged tile is installed with foam, cut out the old piece of foam to slip in a new tile before adding new foam.

2.0.0 Section Review

1. Which is an example of a mechanical fastener?
   a. Nails
   b. Mortar
   c. Foam
   d. Sealant

2. In tile systems, there are only three types of flashings used at locations where tiles intersect a vertical surface.
   a. True
   b. False

3. To repair a broken tile in a batten system, roofing professionals will likely need to _____.
   a. flash the area with thin metal
   b. fix the tile with mortar or foam
   c. fix the tile with screws
   d. hang a new tile on the batten

Review Questions

1. Which roof system has been used the *longest*?
   a. Asphalt shingles
   b. Clay tile
   c. Concrete tile
   d. Metal shingles

2. Interlocking describes _____.
   a. a style of clay and concrete tile that has integrated connecting mechanisms on each tile piece that help link tile together during installation
   b. a style of clay and concrete tile that has a batten lug, typically on the backs of the tile pieces to help with hanging
   c. accessory pieces that help waterproof systems by covering the area where hips and/or ridges occur
   d. attachment methods that adhere, or stick, objects together, such as foam or mortar

3. Using manual tile chippers eliminates any exposure to silica dust.
   a. True
   b. False

4. Manual tile chippers are also referred to as _____.
   a. trowels
   b. portable mixers
   c. manual cutters
   d. masonry bits

5. Which of the following steps should be completed *first* in clay and concrete tile installation?
   a. Measure and install field tile.
   b. Apply underlayment.
   c. Ensure the roof deck is clean, dry, and in good condition.
   d. Install base flashings and eave closures.

6. Which of the following closes off the open space under tile at a roof's eave edge?
   a. Eave closures
   b. Field tile
   c. Starter courses
   d. Ridge tile

7. Which of the following valley types is open-valley installation?
   a. Tile is cut at an angle parallel to the valley and then mitered together.
   b. Tile is installed throughout the area with individual pieces of flashing under each tile course that runs along the centerline.
   c. Valley metal is installed, and tile is fixed over the metal flashing.
   d. Valley metal is installed, and tile is fixed on both sides of the metal flashing.

8. Which type of flashing is installed at head-walls, or the vertical face that faces the eave?
   a. Apron flashing
   b. Channel flashing
   c. Step flashing
   d. Cricket flashing

9. A vent pipe is an example of a roof penetration that may require flashing.
   a. True
   b. False

10. What type of repair is *most* commonly required at the end of the tile installation?
    a. Missing storm clips
    b. Battens too long
    c. Broken tile
    d. Underlayment laps too narrow

Trade Terms Introduced in This Module

Apex tile: An accessory piece that helps waterproof systems by covering the area where hips and/or ridges converge from different directions.

Batten-hung: A style of clay and concrete tile that has a batten lug, typically on the backs of the tile pieces to help hang the tile off batten systems.

Eave closures: Accessory pieces used to close off and flash the open space at a roof's perimeter edge. Also called *eave risers* for low-profile tile and *bird stops* for medium- to high-profile tile.

Exposure: The amount of tile surface exposed to the weather and not overlapped by the next course or surrounding tiles.

Interlocking: A style of clay and concrete tile that has integrated connecting mechanisms on each tile piece that help link tile together during installation.

Mechanical fasteners: Attachment methods that secure objects together through fastening. Examples include nails and screws.

Profiles: The ratio of a tile's height to its width; the lower the profile, the flatter the tile.

Tapered edge boards: Strips used to modify the angle where the roofing membrane meets an eave; uniformly props up tile courses starting from the eave.

Tile adhesives: Attachment methods that adhere, or stick, objects together. For example, foam or mortar.

Trim tile: An accessory piece provided by the manufacturer used to attractively weatherproof various areas of a roof, such as at ridges, hips, and rakes.

Additional Resources

This module presents thorough resources for task training. The following reference material is suggested for further study.

29 CFR 1926.1153(c)(1), Table 1, **www.ecfr.gov**.

National Roofing Contractors Association (NRCA), **www.nrca.net**.

NCCER Module 00103, *Introduction to Hand Tools*.

NCCER Module 00104, *Introduction to Power Tools*.

NCCER Module 16105, *Introduction to Steep-Slope Roofing*.

NCCER Module 16208, *Built-Up Roof Systems*.

NCCER Module 75901, *Fall Protection Orientation*.

NRCA Pocket Guide to Safety, NRCA. 2015. Rosemont, IL: National Roofing Contractors Association.

The NRCA Roofing Manual: Architectural Metal Flashing and Condensation and Air Leakage Control, NRCA. 2018. Rosemont, IL: National Roofing Contractors Association.

The NRCA Roofing Manual: Steep-slope Roof Systems, NRCA. 2017. Rosemont, IL: National Roofing Contractors Association.

The NRCA Safety Manual, NRCA. 2012. Rosemont, IL: National Roofing Contractors Association.

Occupational Outlook Handbook, Roofers, Bureau of Labor Statistics, US Department of Labor, **www.bls.gov/ooh/construction-and-extraction/roofers.htm**.

Occupational Safety and Health Administration (OSHA), US Department of Labor, **www.osha.gov**.

OSHA 3755-05, *Protecting Roofing Workers*. US Department of Labor. **www.osha.gov/Publications/OSHA3755.pdf**.

TRI Alliance Cold & Snow Concrete and Clay Tile Design Criteria for Cold and Snow Regions, Leland E. Gillan, P.E. and Terry Anderson. 2005. Chicago, IL: Tile Roofing Institute & Western States Roofing Contractors Association.

TRI Alliance Concrete and Clay Roof Tile Installation Manual. 2015. Chicago, IL: Tile Roofing Institute & Western States Roofing Contractors Association.

Figure Credits

iStock@nitimongkolchai, Module Opener

iStock@redtea, Figure 2

iStock@MileA, Figure 3

iStock@geargodz, Figure 4

roboriginal - stock.adobe.com, Figure 5

Eagle Roofing Products, Figures 7, 9

iStock@ewg3D, Figure 8

iStock@Emilija Randjelovic, Figure 10

tfoxfoto, Figure 13

Richard C. Littleton, Figure 14

iStock@DarthArt, Figure 15

iStock@ajt, Figure 16

Marshalltown, Figure 17

iStock@Bilanol, Figure 18

iStock@Michael Vi, Figure 28

Section Review Answer Key

Section 1.0.0

Answer	Section Reference	Objective
1. d	1.1.2	1a
2. a	1.2.4	1b

Section 2.0.0

Answer	Section Reference	Objective
1. a	2.1.2	2a
2. b	2.2.2	2b
3. d	2.3.0	2c

User Update

NCCER makes every effort to keep its textbooks up-to-date and free of technical errors. We appreciate your help in this process. If you find an error, a typographical mistake, or an inaccuracy in NCCER's curricula, please submit a User Update form by visiting **https://www.nccer.org/olf**. You can also scan the QR code using the camera on your phone or mobile device to access the form.

Wood Roof Systems

OVERVIEW

Wood is one of the oldest types of roof systems. Wood shingles were the most common roof covering in Colonial America due to the abundant availability of raw materials. Wood roof systems remain a viable and popular choice today, especially where a historic style is desired. Roofing professionals must be familiar with the types and features of wood roofs in order to perform quality and safe installations.

Module 16203

Trainees with successful module completions may be eligible for credentialing through the NCCER Registry. To learn more, go to **www.nccer.org** or contact us at 1.888.622.3720. Our website, **www.nccer.org**, has information on the latest product releases and training.

Your feedback is welcome. You may email your comments to **curriculum@nccer.org**, send general comments and inquiries to **info@nccer.org**, or fill in the User Update form at the back of this module.

This information is general in nature and intended for training purposes only. Actual performance of activities described in this manual requires compliance with all applicable operating, service, maintenance, and safety procedures under the direction of qualified personnel. References in this manual to patented or proprietary devices do not constitute a recommendation of their use.

16203 V1.0

16203
Wood Roof Systems

Objectives

Successful completion of this module prepares you to do the following:

1. Describe wood roof systems and components and identify the tools used to install them.
   a. List and describe types and styles of wood shingles and shakes.
   b. Identify the hand and power tools used to install wood roof systems.
2. Explain wood roof installation.
   a. Explain procedures for installing wood roof systems.
   b. Identify flashings used in wood roof systems.
   c. Describe basic repair methods used to correct mistakes during an installation.

Performance Tasks

Under supervision, you should be able to do the following:

1. Install the drip edge, underlayment, and rake edge.
2. Install the starter course shingle and first course of a shake roof system.
3. Install the first course of interlayment.
4. Install the second course of shake with the correct joint stagger and nail placement. Repeat interlayment and shake coursing to the ridge.
5. Install step flashing for each course of the shake at any side walls.
6. Install valley pan.
7. Cut and install valley shakes.
8. Install hip and ridge shakes.

Trade Terms

Bundle
Butt edge
Edge grain
Flat grain
Heartwood
Interlayments
Undercourse
Wood billet
Wood shakes
Wood shingles

Industry Recognized Credentials

If you are training through an NCCER-accredited sponsor, you may be eligible for credentials from NCCER's Registry. The ID number for this module is 16203. Note that this module may have been used in other NCCER curricula and may apply to other level completions. Contact NCCER's Registry at 1.888.622.3720 or go to **www.nccer.org** for more information.

You can also show off your industry-recognized credentials online with NCCER's digital badges. Transform your knowledge, skills, and achievements into badges that you can share across social media platforms, send to your network, and add to your resume. For more information, visit **www.nccer.org**.

Videos and Digital Resources

How to Access Resources

This craft has additional videos and resources to enhance your learning experience. To view these resources, scan the QR below. The videos and resources are separated by module.

You can scan this code using the camera on your phone or mobile device to view these videos and resources.

Contents

1.0.0 Wood Roof Systems 1
1.1.0 Types and Styles 1
1.2.0 Tools and Equipment 3
1.2.1 Roofing hammer 4
1.2.2 Nail gun 4
1.2.3 Nails 4
2.0.0 Installing Wood Roof Systems 6
2.1.0 Components and Installation Procedures 6
2.1.1 Underlayments and Interlayments 6
2.1.2 Wood Shake Installation Outline 7
2.1.3 Wood Shingle Installation Outline 10
2.1.4 Field Installation 11
2.1.5 Valleys 11
2.2.0 Flashings 13
2.2.1 Perimeter Flashings 13
2.2.2 Vertical-Surface Protrusions 13
2.2.3 Penetrations 14
2.3.0 Basic Repair 14

Figures

Figure 1 Wood shakes and shingles 2
Figure 2 Types of wood shakes 2
Figure 3 Types of wood shingles 3
Figure 4 Wood cuts 4
Figure 5 Roofing hammer 4
Figure 6 Nail gun 5
Figure 7 Nail types 5
Figure 8 Wood shingle roof system 6
Figure 9 Interlayment between shake courses over continuous deck 8
Figure 10 Interlayment between shake courses over a spaced deck 9
Figure 11 Wood shake installation with interlayment 10
Figure 12 Step flashing at wall 10
Figure 13 Wood shingle application with correctly offset joints 11
Figure 14 Wood shake coverage and exposure table 12
Figure 15 Wood shingle coverage and exposure table 13
Figure 16 Open valley 13
Figure 17 Closed valley 14
Figure 18 Drip edge flashing types 14
Figure 19 Sheet metal flashing components used at a chimney 15
Figure 20 Examples of kick-out roof flashing 15
Figure 21 Penetration flashings 16
Figure 22 Removing damaged shake with a shingle ripper 16

Section One

1.0.0 Wood Roof Systems

Objective

Describe wood roof systems and components and identify the tools used to install them.

a. List and describe types and styles of wood shingles and shakes.
b. Identify the hand and power tools used to install wood roof systems.

Trade Terms

Bundle: The measurement unit that mills use to sell shakes and shingles.

Butt edge: The bottom or exposed end of a shake or shingle.

Edge grain: A log cut perpendicular to the rings of the tree.

Flat grain: A log cut parallel to the tree's rings.

Heartwood: Wood located at the center of a log.

Wood shakes: Roof covering typically split from a wood billet and then resawed or shaped by a mill.

Wood shingles: Sawn products with uniform butt thickness relative to the specific length used as a roof covering.

Undercourse: A course of shakes or shingles installed beneath the covering course.

Wood billet: Wood blocks from which shakes or shingles are made.

Roof systems utilizing wood shakes or wood shingles are often chosen for their appearance. With their natural color tones and weathering characteristics, these roof systems enhance architectural design elements and add a rustic style to buildings.

With proper installation, care, and maintenance, wood roofs can provide attractive and long-term protection to a residence. Studying the history and best practices of this roof system helps roofing professionals understand how to work with wood shakes and shingles. Before installing wood roof systems, roofing professionals must also be familiar with the common types and styles of this system, as well as the tools used during installation.

WARNING! As with installing all roof systems, safety is a crucial consideration. For more information on roofing safety, review NCCER Module 16102, *Roofing Safety*.

1.1.0 Types and Styles

Wood roof systems come in two varieties: shakes and shingles (*Figure 1*). Wood shakes are typically split from a wood billet and then resawed or shaped by a mill. Shakes were commonly hand-split until the early 19th century, when the Industrial Revolution and introduction of sawmills made it possible to create sawn wooden shakes. Shakes and shingles are sold by the bundle.

The most common type of wood used to make wood shakes and shingles is Western Red Cedar. Cedar contains tannic acids that help prevent mold growth and termite infestation.

Wood shakes and shingles are often treated with chromated copper arsenate (CCA) or fire retardant. CCA is a preservative treatment that protects and extends the life of wood roof systems. Fire retardant *does not* completely prevent fires from starting; instead, it helps slow the burn if a wood roof catches fire. Whether wood used in a roof system is treated with CCA or fire retardant depends on the region and local building codes.

NOTE: Consult the Cedar Shake & Shingle Bureau for safety-related information about treatments used in wood roof systems.

There are four basic types of wood shakes (*Figure 2*):

- Hand-split and resawn shakes
- Taper-sawn shakes
- Taper-split shakes
- Straight-split shakes

A wood shingle is a sawn product, generally available in 16" (40.6 cm), 18" (45.7 cm), and 24" (61 cm) lengths with uniform butt edge thickness relative to the specific length. Wood shingles are referred to as *taper-sawn*, and there are two general categories: straight-sawn butt and fancy-sawn butt styles (*Figure 3*). Wood shakes and shingles are both still used as roofing materials, though shingles are more common.

Shakes   Shingles

Figure 1 Wood shakes and shingles.

Wood Shakes		
Name	Length and Thickness	Description
No. 1 Hand-Split and Resawn 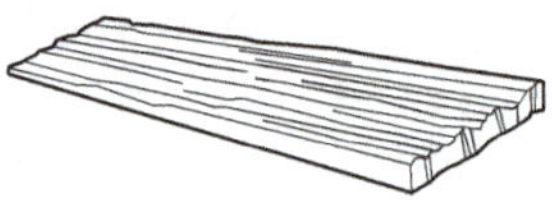	15" Starter 18" × ½" Mediums 18" × ¾" Heavies 24" × ⅜" Starter 24" × ½" Mediums 24" × ¾" Heavies	These shakes have split faces and sawn backs. The logs are first cut into desired length sections. Blanks of proper thickness are split from the section and then run diagonally through a bandsaw to produce two tapered shakes from each blank.
No. 1 Taper-Sawn 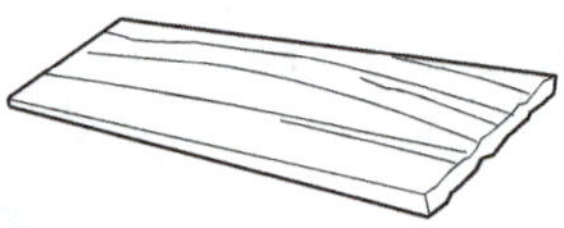	24" × ⅝" 18" × ⅝"	Taper-sawn shakes are sawn on both sides.
No. 1 Taper-Split 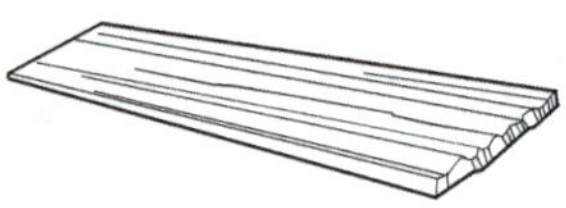	24" × ½"	These shakes are produced by hand using a sharp-bladed steel froe and a mallet. The natural taper is achieved by reversing the wood block, end-for-end, with each split.
No. 1 Straight-Split 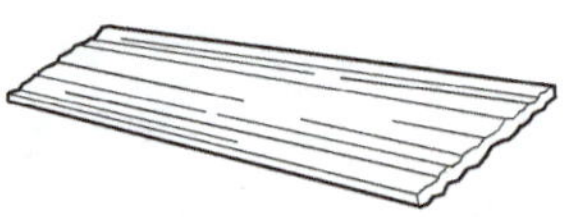	Length and Thickness Vary	Straight-split shakes are produced in the same manner as taper-split shakes, but because they are split from the same end of the block, the shakes acquire the same relative thickness throughout their lengths.
Notes: 1. Premium or grade no. 1 is recommended for roofing. 2. Lower grades (e.g., no. 2 and no. 3) are available and are used for starter or undercoursing.		

Figure 2 Types of wood shakes.

WARNING!

Gloves help prevent splinters from handling shakes and shingles. They also provide protection from skin irritations that can come from handling chemically treated products.

Wood shakes and shingles are natural products that contain naturally occurring features or defects that may not be desirable in a roof covering. Because of this, wood shakes and shingles are graded to categorize defects, which may compromise weathering resistance from the elements once installed. There are four grades:

- *No. 1* — This is top-grade, taken from the tree's **heartwood**. It is 100 percent **edge grain**, and contains no knots or other defects. *Edge grain* means the wood is cut perpendicular to the rings of the tree, resulting in greater durability. This grade is used for roofing and sidewalls. *Figure 4* shows the different types of wood cuts.
- *No. 2* — This grade is a **flat grain** cut with limited sapwood and has knots and defects above the part of the shingles that will show once installed. *Flat grain* means the wood is cut parallel to the tree's rings. This grade is often used for reroofing, walls, and as a starter course.
- *No. 3* — This grade can include sapwood and flat grain with limited knots and defects allowed above the part of the shingles that will show once installed. This grade is commonly used for garden sheds, walls of outbuildings, and gazebos.
- **Undercourse** — This is a utility-grade used for undercoursing double-coursed sidewalls only. It is not considered roofing material and is not to be used as a starter course for roofs.

1.2.0 Tools and Equipment

Roofing professionals must understand how to safely use the tools necessary to install wood roof systems. Along with understanding tool usage, roofing professionals are also responsible for ensuring their tools are clean and in good condition before using them to install wood roofs.

NOTE

This section highlights some of the tools you will encounter when installing wood roof systems. Other common tools used in wood roof installation, such as hammers and circular saws, are covered in greater detail in NCCER Module 00103, *Introduction to Hand Tools*, and NCCER Module 00104, *Introduction to Power Tools*.

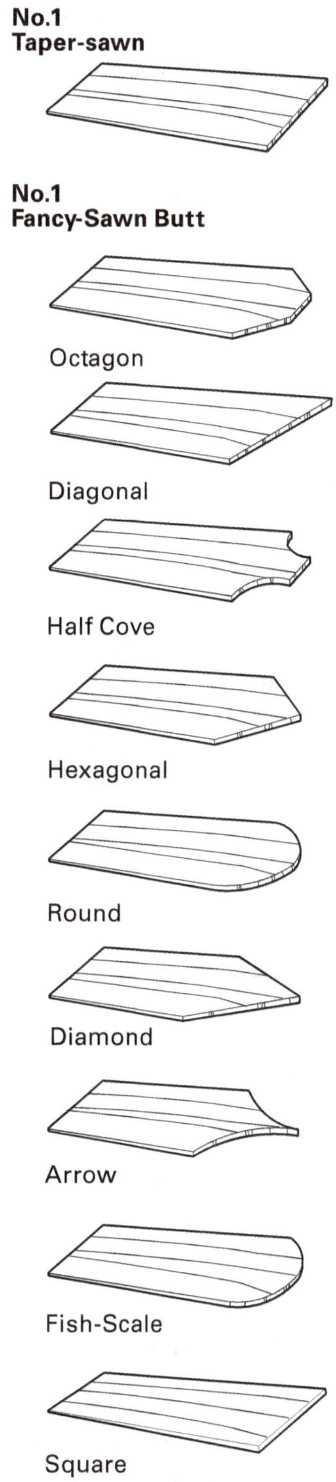

Figure 3 Types of wood shingles.

WARNING!

Dust released from cutting wood presents a safety hazard. Be sure to use the proper PPE, such as gloves and safety glasses.

DO NOT cut wood pieces unsafely; *DO NOT* use your legs to prop up wood pieces while you cut.

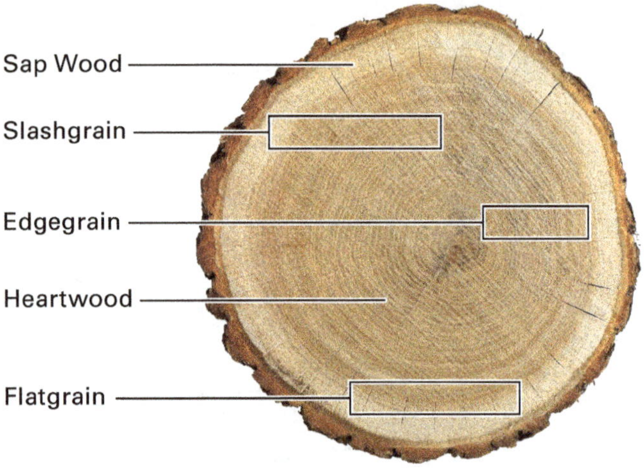

Figure 4 Wood cuts.

1.2.1 Roofing hammer

Roofing hammers (*Figure 5*), also called *roofing hatchets*, are specialized tools that help roofing professionals score and split shingles and shakes as well as drive nails into them. They have unique features to make this job more efficient, such as magnetized faces to help you pick up stray nails, a pounding head, an axe-like cutting edge on the other end, and even gauges that help you position shingles to the right exposure.

> **WARNING!** When you are carrying sharp, dangerous tools, be sure that you use a tool tether to avoid hazards falling off the roof.

1.2.2 Nail gun

A nail gun (*Figure 6*) can be used as an alternative to hand nailing wood roof systems, although special care must be taken. The pressure of the air compressor should be regulated to ensure proper installation.

The head of the nail should be driven flush with the surface of the shake or shingle being used. Underdriving the nail results in a shingle that is not sealed while overdriving can split the wood. Overdriven nails are a common problem in wood roof systems installed with a nail gun. Be careful not to use too much pressure during installation to avoid splitting the shake and creating a safety hazard.

> **WARNING!** To ensure proper air pressure, test nailing should be done before beginning installation. Pressure should also be monitored throughout the installation, especially when lengthening or shortening the air hose. It is also important to note that if the correct type of nail required for the specific shake or shingle cannot be found for the nail gun, hand nailing techniques must be used.

1.2.3 Nails

While nails are used in most roofing applications, care must be taken when choosing nails for wood roof systems. Fire-treated wood may require different nails (*Figure 7*) than pressure-treated installation. Some chemicals can cause a negative reaction that can compromise the integrity of the roof.

> **CAUTION** Keep in mind that some nails are not magnetic. Therefore, when you are using a magnet sweeper to clean a jobsite, be sure to keep an eye out for non-magnetic nails that need to be manually disposed.

Figure 5 Roofing hammer.

Figure 6 Nail gun.

Type of Shake or Shingle	Nail Type and Minimum Length
18" straight-split shakes	5d box, 1¾"
18" and 24" hand-split and re-sawn shakes	6d box, 2"
24" taper-split shakes	5d box, 1¾"
18" and 24" taper-sawn shakes	6d box, 2"
16" and 18" shingles	3d box, 1¼"
24" shingles	4d box, 1½"

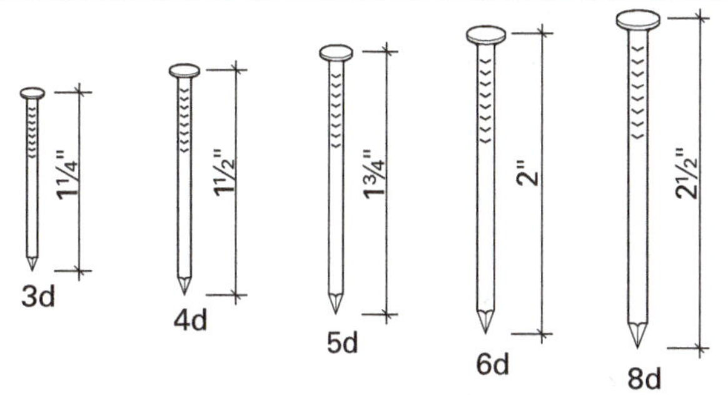

Notes:
1. When nails are specified, NRCA recommends box nails be specified in lieu of common nails.
2. Dimensions are approximate.

Figure 7 Nail types.

1.0.0 Section Review

1. In what way are wood shakes typically taken from a wood billet?
   a. Sawn
   b. Crosscut
   c. Split
   d. Ground

2. Systems that use fire-treated wood may require different nails than those that use pressure-treated wood.
   a. True
   b. False

Section Two

2.0.0 Installing Wood Roof Systems

Objective

Explain wood roof installation.

a. Explain procedures for installing wood roof systems.
b. Identify flashings used in wood roof systems.
c. Describe basic repair methods used to correct mistakes during an installation.

Performance Tasks

1. Install the drip edge, underlayment, and rake edge.
2. Install the starter course shingle and first course of a shake roof system.
3. Install the first course of interlayment.
4. Install the second course of shake with the correct joint stagger and nail placement. Repeat interlayment and shake coursing to the ridge.
5. Install step flashing for each course of the shake at any side walls.
6. Install valley pan.
7. Cut and install valley shakes.
8. Install hip and ridge shakes.

Trade Term

Interlayments: Underlayments installed between each course of shakes or shingles.

Wood roofs are known for their pleasant appearances and historical significance. Roofing professionals must know how to effectively install all the components of these roof systems, from underlayments to flashings.

2.1.0 Components and Installation Procedures

Installing a wood roof system (*Figure 8*) typically involves attaching underlayments, **interlayments**, drip edge, starter course shingles, and flashings. Once all required underlayments and starter courses have been added to a roof deck, roofing professionals must know how to safely install the rest of the shingles or shakes and apply the flashings.

Because there are so many types of wood roof systems, it is crucial to apply job-specific recommendations during installation. Each wood shingle or shake installation procedure may vary from installations you have encountered in the past.

2.1.1 Underlayments and Interlayments

Underlayments and interlayments provide additional water-shedding protection between the shakes or shingles and the roof deck. Whether underlayments and interlayments are used depends on system and deck type.

Underlayments

In wood roof systems, deck type helps determine how or if you should install underlayments. There are two main types of decks used in wood roof systems: a spaced deck and a continuous deck.

A spaced deck, also called *spaced sheathing*, does not use full-coverage underlayment. Installing full-coverage underlayment on a spaced deck can cause ventilation and performance issues. However, a continuous deck, also called *continuous sheathing*, requires underlayment installation to help weatherproof the system.

> **NOTE**
> Leak barrier should still be installed at valleys and ridges, even when the roof field has a spaced deck.

Figure 8 Wood shingle roof system.

When applicable, the following underlayment options are used during wood roof installation, depending on region, slope, and deck type:

- *Single-layer underlayment* — Underlayment materials, such as #30 ASTM D226 Type II or #30 ASTM D4869 Type IV installed in a single layer.
- *Double-layer underlayment* — Underlayment materials installed in two layers. This method may combine a variety of compatible underlayment types and installation methods.
- *Leak Barrier* — Also called *self-adhering sheets* and, in colder climates, *ice dam protection*. These underlayment sheets have adhesive bitumen applied at the factory and provide labor savings with installation on the job.

CAUTION

Leak barrier can cause serious issues in a system's ventilation if it is not installed properly. Always abide by local building-code requirements when installing roof systems.

Interlayments

Shakes are longer, thicker, and less uniform in their dimensions. They do not overlap as neatly as shingles due to their uneven surfaces. This may result in gaps both under and on the sides of the shakes that can allow rain, snow, and debris to penetrate through the courses.

To help with weatherproofing, roofing professionals install layers of sheets, called *interlayment*, between courses of shakes to seal out water and debris. Interlayments may also be referred to as *shake liner* or *shake felt*. Interlayments are typically made of asphalt-saturated felts.

When installing shakes, interlayment should be used between shake courses both over a continuous deck (*Figure 9*) and over a spaced deck (*Figure 10*). Interlayment should be installed between the starter course and the first course, and then between every subsequent course.

WARNING!

Spaced decks greatly increase the risk of falls. Be sure you are donning all proper safety equipment and fall protection when working on these roofs.

When wood shakes are installed over spaced decks, it is not recommended to use a separate underlayment in addition to the interlayment. This is to help ventilate the bottom side of the roof system and aid in the drying of the shakes.

Interlayment should be installed with its bottom edge positioned at a distance equal to twice the shake exposure above the butt-edge line of every course of shakes. Interlayment always should be included in a wood shake roof system, whether it is installed over continuous or spaced roof decks.

For interlayment, the Cedar Shake and Shingle Bureau recommends using an 18" (45.7 cm) wide strip of #30 *ASTM D226 Type II* or #30 *ASTM D4869 Type IV* roofing felt laid over the top portion of shakes and extending on to the deck. Check with your local building code official for exact specifications in your area.

2.1.2 Wood Shake Installation Outline

The following provides a general outline for wood shake installations:

CAUTION

This outline is intended only to overview the general installation sequence. Always check applicable project documentation and building-code requirements for specific guidelines.

Step 1 Don all proper PPE for the job. Ensure that the appropriate fall protection equipment is used or in place. Depending on the project and environment, this may be a personal fall arrest system (PFAS) with approved anchorage, or appropriate guardrails or scaffolding.

Step 2 Ensure that the roof deck is clean, dry, and in good condition.

Step 3 Install the roof's underlayment, drip edge metal, and valley metal. Underlayment is only required when the system is installed over a continuous deck.

Step 4 Starting at the roof's eave, install the starter course and the first course of shakes using the appropriate exposure. Each shake should be fastened with two fasteners.

CAUTION

Be sure the fasteners you use are compatible with the system being installed. All metal fasteners added to the roof must be compatible with the chemicals used in the system.

Step 5 Interlayment should be installed between the first course and the second course, and then between every subsequent course of shakes (*Figure 11*).

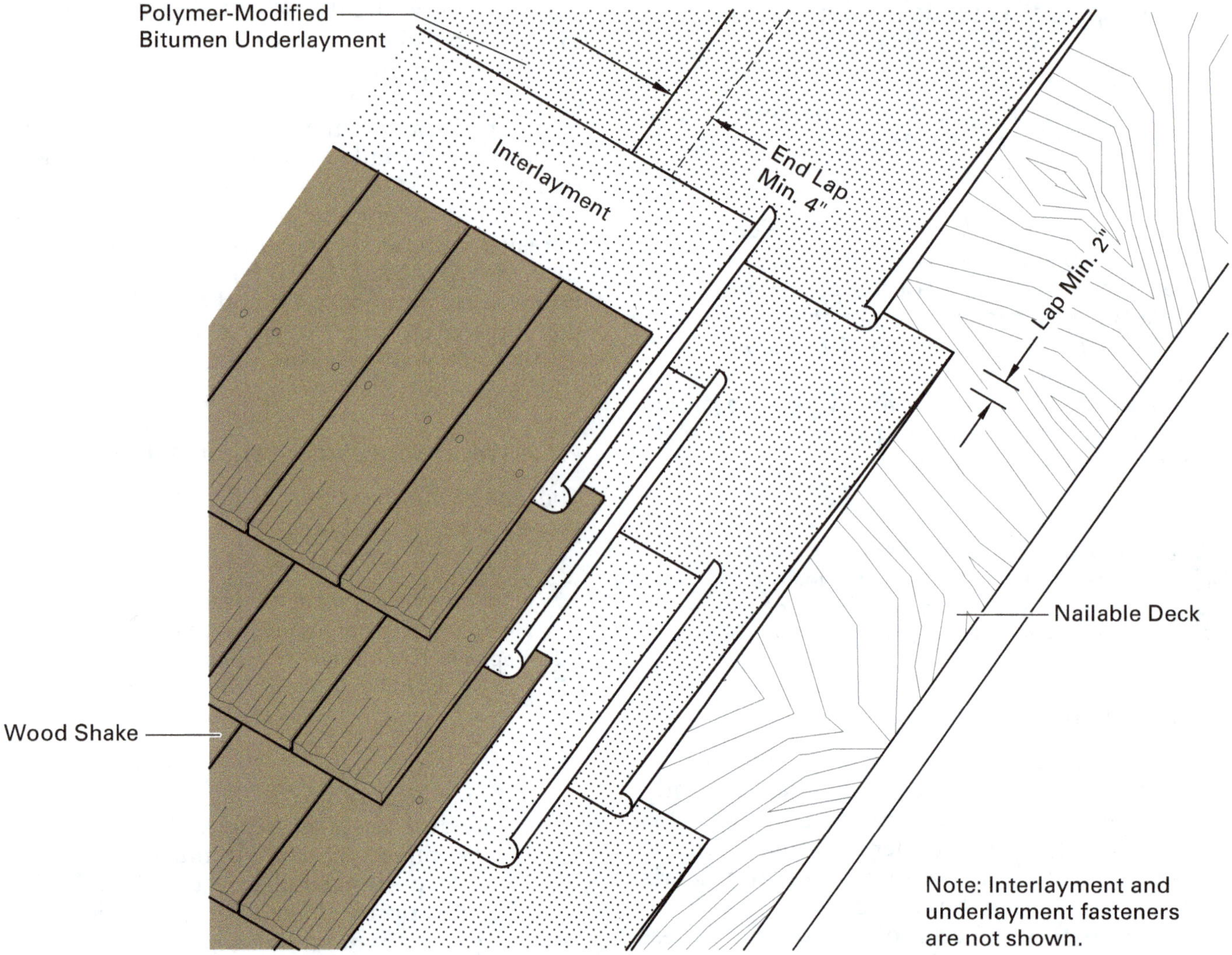

Wood Roof System with Single Layer of Mechanically Attached Underlayment Over Continuous Sheathing (Wood Shake and Polymer-Modified Bitumen Underlayment Shown)

Figure 9 Interlayment between shake courses over continuous deck.

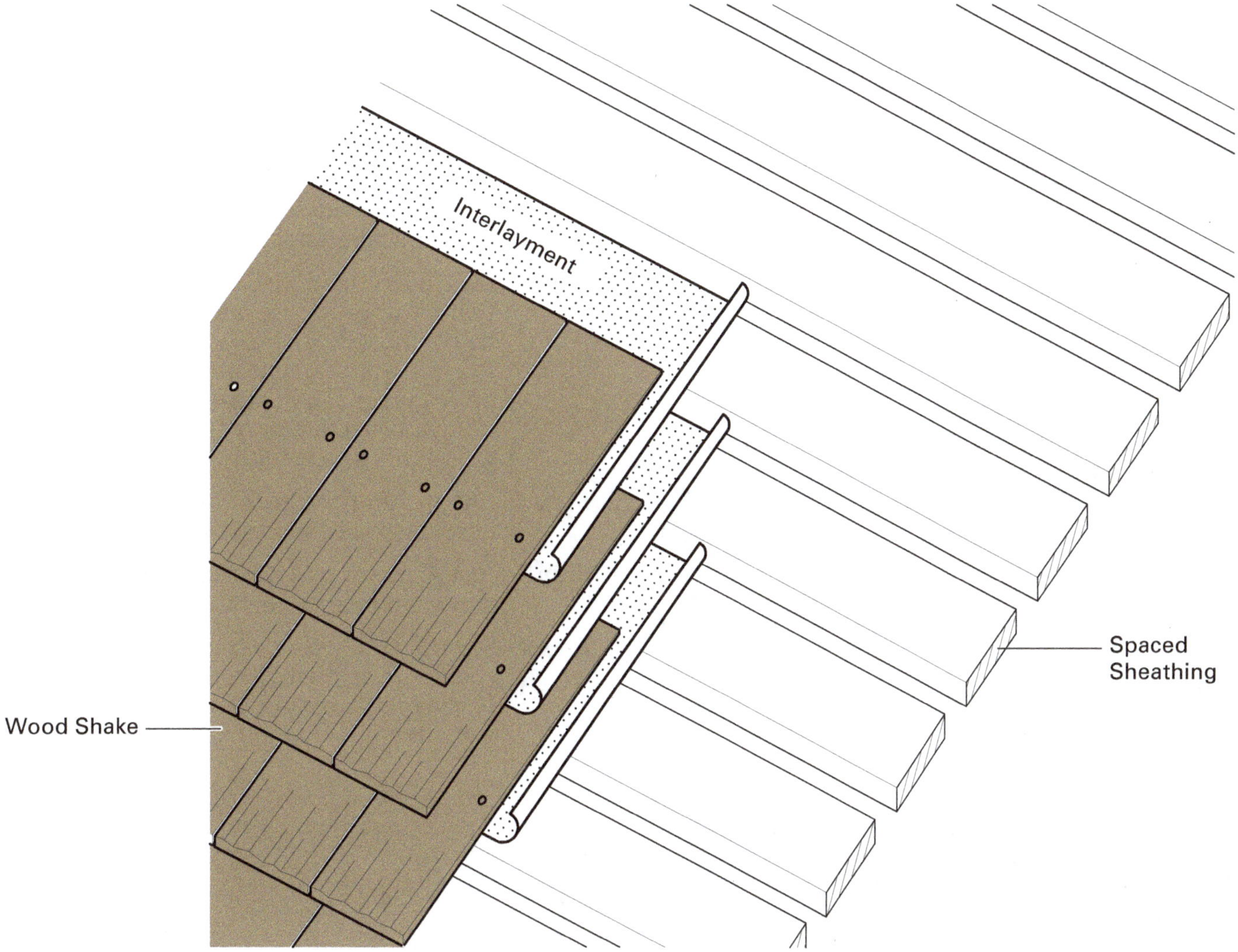

Wood Roof System Over Spaced Sheathing (Wood Shake Shown)

Figure 10 Interlayment between shake courses over a spaced deck.

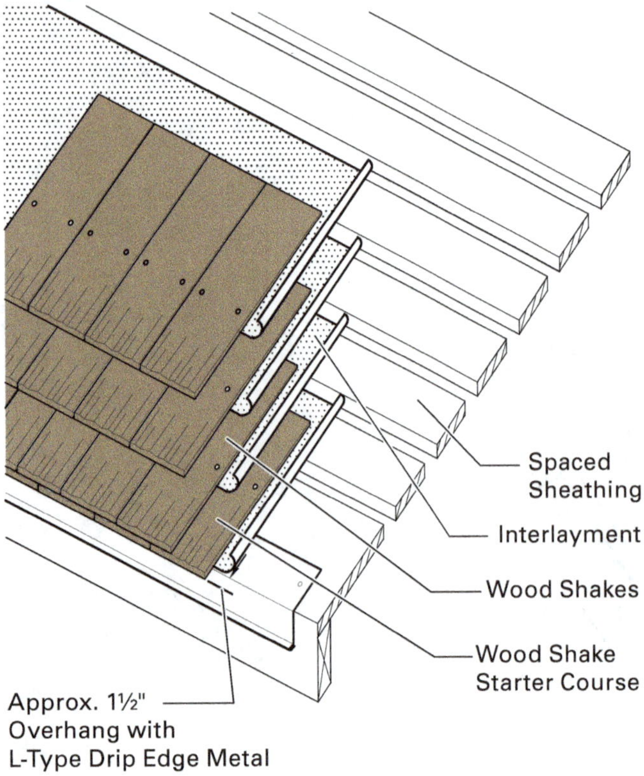

Figure 11 Wood shake installation with interlayment.

> **NOTE**
> To be more efficient, rows of interlayment can be laid out ahead of time in the right sequence and pattern. During installation, the shakes are slid under each row of felt. Alternatively, layers of interlayment and courses of shingles may be installed in an alternating sequence, working as you go, rather than laying out all interlayment ahead of time.

Step 6 On top of the starter course, measure and install the first course of shakes from one direction (either left-to-right or right-to-left). Local building codes must be consulted for the specific exposure and offset requirements of each specific system.

Step 7 Continue to install the remaining courses of shakes following the appropriate offsets and exposures.

Step 8 As you install the field, also apply flashings at penetrations and walls (*Figure 12*).

Step 9 Install hip and ridge shakes. These may be premanufactured or made on the jobsite.

Step 10 Install any remaining flashings.

Step 11 Inspect the completed roof and clean the jobsite. Keep an eye out for damaged shakes or flashings that may need replacing.

2.1.3 Wood Shingle Installation Outline

The following provides a general outline for wood shingle installations:

Step 1 Don all proper PPE for the job. Ensure that the appropriate fall protection equipment is used or in place. Depending on the project and environment, this may be a personal fall arrest system (PFAS) with approved anchorage, or appropriate guardrails or scaffolding.

Step 2 Ensure that the roof deck is clean, dry, and in good condition.

Step 3 Install any roof underlayment, drip edge metal, and valley metal.

Step 4 Starting at the roof's eave, install a starter course using the recommended exposure. Each shingle should be fastened with two fasteners.

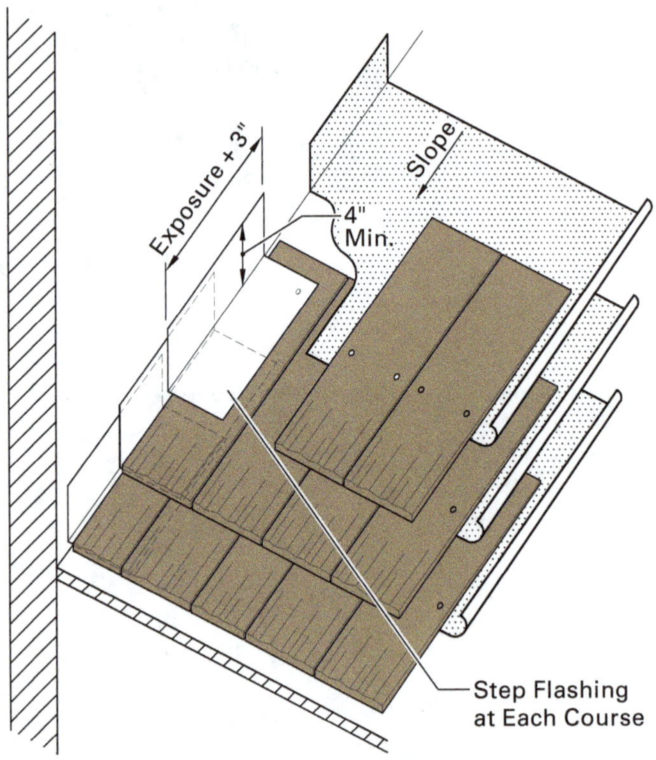

Note: Wall siding/cladding and building wrap not shown for clarity.

Figure 12 Step flashing at wall.

Step 5 On top of the starter course, measure and install the first course of shingles from one direction (either left-to-right or right-to-left). Local building codes must be consulted for the specific exposure and offset requirements of each specific system (*Figure 13*).

Step 6 Continue to install the remaining courses of shingles following the appropriate offsets and exposures.

Step 7 As you install the field, also apply flashings at penetrations and walls.

Step 8 Install hip and ridge shingles. These may be premanufactured or made on the jobsite.

Step 9 Install any remaining flashings.

Step 10 Inspect the completed roof and clean the jobsite. Keep an eye out for damaged shingles or flashings that may need replacing.

2.1.4 Field Installation

Wood roofs must be installed with the correct exposure to ensure the system effectively sheds water. The exposure of wood shakes and shingles is the portion of the wood that is not covered by the course above—that is, the part of the wood shake or shingle that is exposed to weather.

Wood shakes and shingles also have many possible exposures and configurations based on the length and grade of the shake or shingle being used. There are specific guidelines for maximum exposure for individual products based on deck slope and required headlap.

Common wood shake types, their lengths and thicknesses, and approximate exposures are shown in *Figure 14*. Common wood shingle types, their approximate lengths and thicknesses, and approximate area coverages for the weather exposure are shown in *Figure 15*.

2.1.5 Valleys

Valleys are the V-shaped angle that forms where the downslopes of two roof fields intersect. This area is particularly vulnerable to leakage since water will shed in its direction. The two methods of valley installation used in wood roof systems are open valley and closed (or *closed-cut*) valley.

It is important to avoid placing fasteners near the center of a valley to prevent obstruction. Fasteners should be kept a minimum of 8" (20 cm) away from the center of a valley. In some cases, this can be difficult because some of the wood roofing units may be narrower than necessary to facilitate keeping fasteners far enough away from the valley's centerline. Because of this, it is best to save the widest wood shakes and shingles specifically for valley construction.

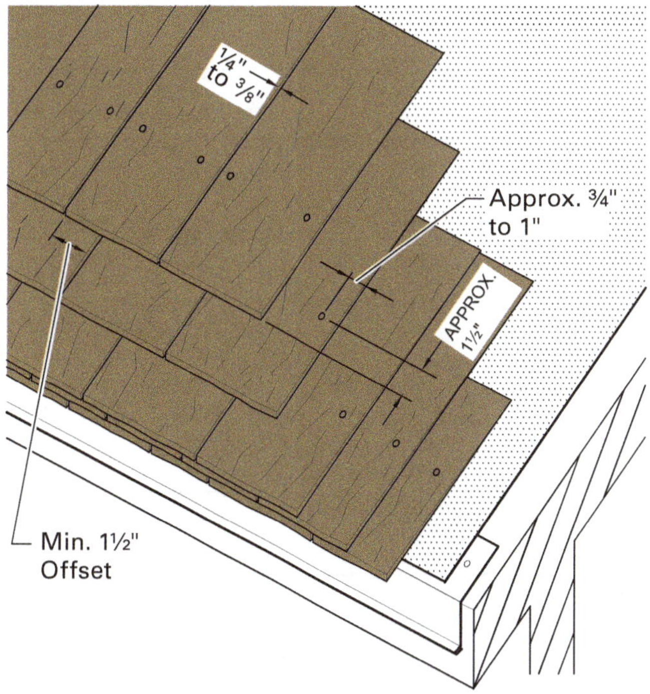

Wood Shingle Roof System

Figure 13 Wood shingle application with correctly offset joints.

Open-Valley Installation

Open valleys are typically constructed by installing 8' to 10' (2.4 m to 3 m) lengths of 24" (61 cm) wide valley metal from the low point to the high point in the valley. Wood shakes and wood shingles are then lapped onto both sides of the valley metal, leaving a clear space to help water shed down the valley (*Figure 16*).

Closed-Valley Installation

Closed valleys, also called *closed-cut valleys*, are lined with sheet metal valley flashing under the wood shakes or shingles. Wood roofing shakes or shingles are brought together over the metal liner at the valley centerline and mitered along the length of the valley to form a closed valley (*Figure 17*).

Shake Coverage					
Shake Grade, Type, Length and Thickness in Inches		Weather Exposures in Inches			
		5½	7½	8½	10
		Approximate Coverage in Sq. Ft. for Five Bundles When Shakes are Applied with an Average ½" Spacing Per Weather Exposures Above			
No. 1 Hand-Split & Re-Sawn 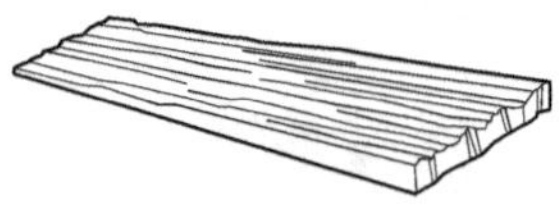	Heavies 24 × ¾	—	75(a)	85	100(b)
	Mediums 24 × ½	—	75(a)	85	100(b)
	Heavies 18 × ¾	75(a)	100(b)	—	—
	Mediums 18 × ½	75(a)	100(b)	—	—
No. 1 Taper-Sawn 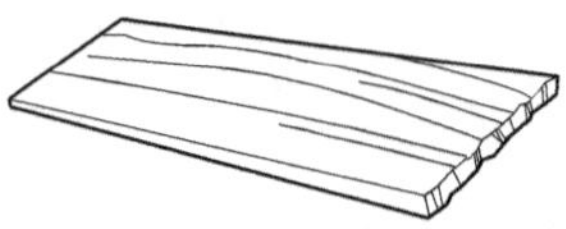	24 × ⅝	—	75(b)	85	100(b)
	18 × ⅝	75(a)	100(b)	—	—
No. 1 Taper-Split 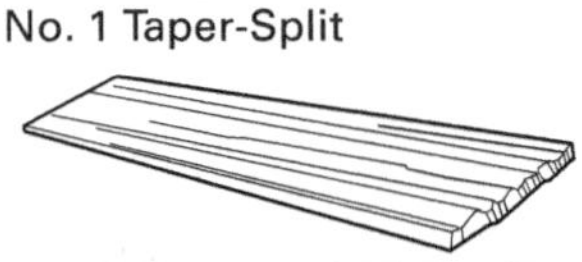	24 × ½	—	75(a)	85	100(b)
No. 1 Straight-Split 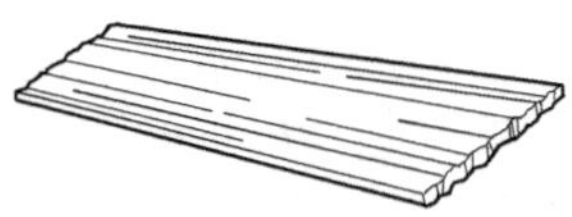	24 × ⅜	—	75(a)	85	100(b)
	18 × ⅜	65(a)	90(b)	—	—
No. 1 Hand-Split 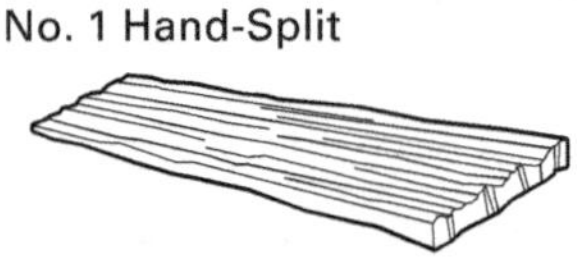	24 × ⅜	—	75(c)	—	—
No. 2 Starter Course 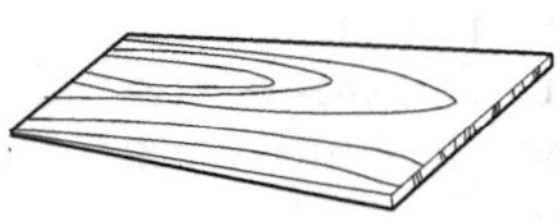	15 × ⅜	Use With Shakes Applied Not Over 7½" Weather Exposure			
	24 × ⅜	Use With Shakes Applied Not Over 10" Weather Exposure			

(a) Maximum Recommended Weather Exposure for Triple Coverage Roof Construction
(b) Maximum Recommended Weather Exposure for Double Coverage Roof Construction
(c) Maximum Recommended Weather Exposure

Figure 14 Wood shake coverage and exposure table.

Shingle Coverage										
Shingles	Length of No. 1 Shingles, in Inches	Weather Exposures in Inches								
		3½	4	4½	5	5½	6	6½	7	7½
		Approximate Coverage in Sq. Ft. of 1 Square (4 Bundles) of Shingles Per Weather Exposures Above								
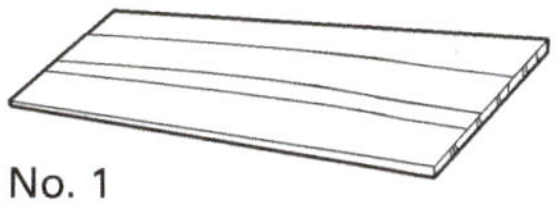	16	70	80	90	100*	—	—	—	—	—
No. 1	18	—	72.5	81.5	90.5	100*	—	—	—	—
	24	—	—	—	—	73.5	80	86.5	93	100*
	* Maximum Exposure Recommended for Roofs									

Figure 15 Wood shingle coverage and exposure table.

Closed valleys are not recommended in areas of high humidity or where debris accumulates from surrounding trees, preventing water from shedding down and off the roof. In wood roof systems, closed valleys are installed far less often than open valleys.

2.2.0 Flashings

Wood roofs must have flashings installed to ensure water sheds off roofs without leaking through the systems. Most flashings are made of metal. Before flashings are applied, any appropriate sequence of leak barrier or other underlayment should be applied to a roof deck around roof penetrations.

CAUTION

Be sure to use similar metals throughout the roof system. For example, if the roof uses copper gutters, it is best to also use copper for flashings and fasteners.

2.2.1 Perimeter Flashings

Eaves and rakes comprise the perimeter of a roof where the field terminates. These areas require drip edge flashings that come in both T- and L-type profiles (*Figure 18*). They can come premanufactured or can be made on the jobsite. Always follow regional codes for guidance on how to fasten drip edge in each particular system.

NOTE

Consult local building codes and the Cedar Shake & Shingle Bureau for instructions on whether underlayment should be installed under or over the drip edge. Depending on what region you are in, you should install perimeter metal with a leak barrier.

2.2.2 Vertical-Surface Protrusions

Flashings at vertical protrusions (such as chimneys, walls, or curbs) should have a relatively smooth substrate on the roof plane and vertical plane, up to a sufficient height according to industry standards. It may also be possible to use soft metal flashings formed to follow the substrate's contours.

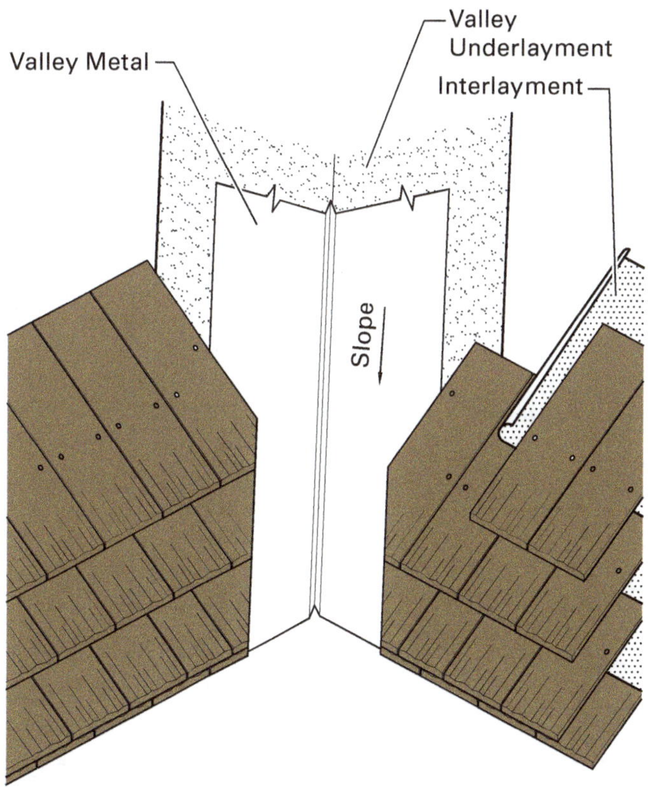

Note: Field underlayment and roof deck not shown for clarity.

Figure 16 Open valley.

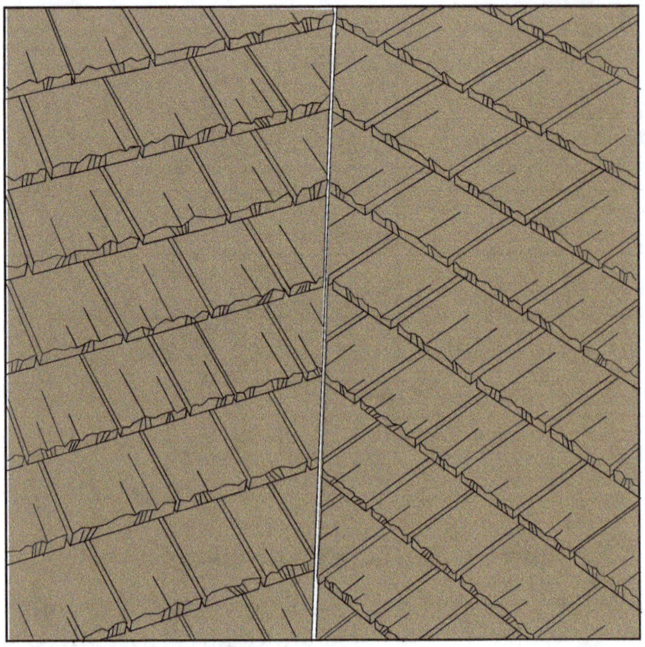

Figure 17 Closed valley.

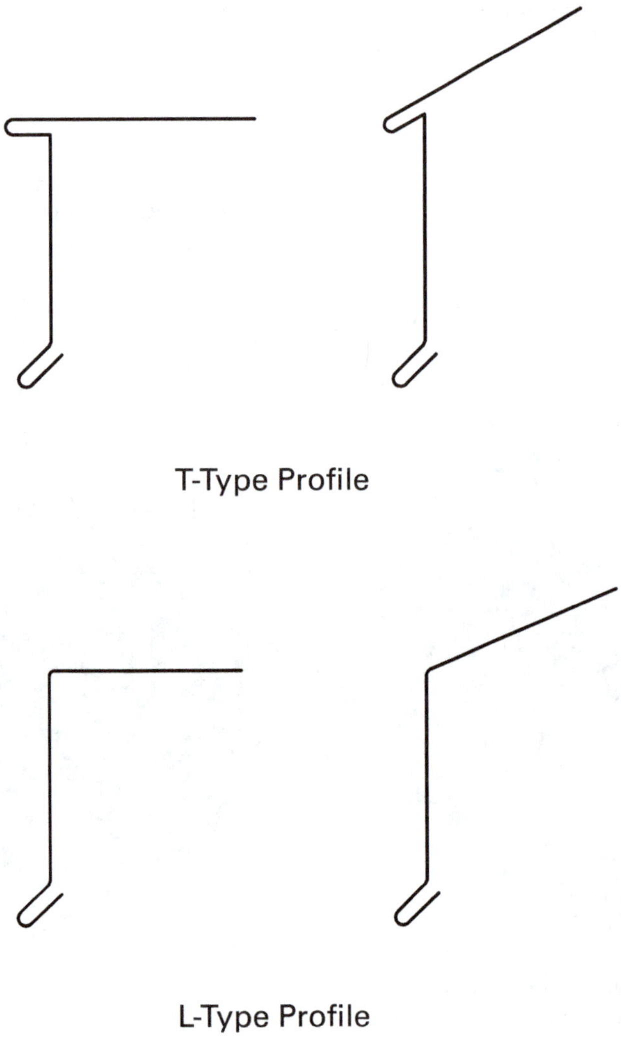

Figure 18 Drip edge flashing types.

The following are the four types of flashings used at locations where wood shakes or shingles intersect a vertical surface:

- *Apron flashings* — Weatherproof the area where the roof meets a headwall.
- *Step flashings* — Should be applied on wood systems that intersect a vertical sidewall.
- *Backer/cricket flashings* — Should be installed where a wood roof intersects the upslope side of a vertical roof penetration. Backer flashings are used on penetrations that are 24" (61 cm) wide or less, and cricket flashings are used on penetrations greater than 24" (61 cm).
- *Counterflashing* — Should be applied to flashings, such as apron, step, cricket, and backer flashings to protect the flashings' top edges from water intrusion.

Figure 19 illustrates the location of apron flashing, step flashing, cricket flashing, and counterflashing on a chimney protrusion.

Attention must be paid to the bottommost step flashing, where an eave intersects a continuous vertical surface, to make sure water sheds away from the wall covering. The use of a *kick-out flashing* at this intersection is recommended (*Figure 20*).

2.2.3 Penetrations

Small penetrations in wood shake and wood shingle roof systems (such as vent pipe penetrations, exhaust vents, exhaust fans, furnace or water heater flue pipes, electrical standpipes, and others) need to be flashed to prevent water leakage. This is accomplished by using some type of a flat flange that extends around a penetration (*Figure 21*). This flashing is installed under the wood shingles or shakes on the upslope side of the penetration and extends down on top of the shingles or shakes at the downslope side.

2.3.0 Basic Repair

A roof's lifespan depends on many things, such as climate, exposure to the elements, and slope of the roof. The most common repairs that are required on wood roof systems are to transitional areas, such as around chimneys, vent pipes, and other penetrations.

Strong storms, weathering, and aging can cause shakes and shingles to become warped, cracked, detached, or broken. These shingles must be replaced. Wind and rain can further erode the damaged shingles and cause them to be dislodged from the roof. Shingles that have detached from the roof can simply be re-nailed. Take care to seal

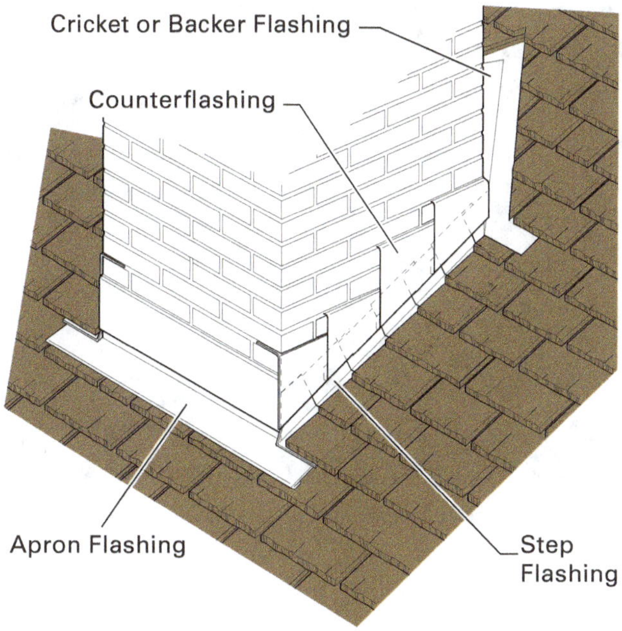

Figure 19 Sheet metal flashing components used at a chimney.

the newly applied nail heads. Missing shingles must also be replaced.

The following provides a general outline for replacing wood shakes:

Step 1 Don all proper PPE for the job. Ensure that the appropriate fall protection equipment is used or in place. Depending on the project and environment, this may be a personal fall arrest system (PFAS) with approved anchorage, or appropriate guardrails or scaffolding.

Step 2 Use a hammer and a wood chisel to split the damaged shake along the grain of the wood and pull out as many pieces as possible.

Step 3 Use a shingle ripper slid up under the shake and a hammer to cut the nails off as shown in *Figure 22*. Alternately, a hacksaw can be used.

Step 4 Measure and cut the replacement shake, being sure to leave $^1/_2$" (12.5 mm) for expansion.

Step 5 Using a wood block to protect the bottom of the shake, tap the replacement into place with a hammer. Leave $^1/_4$" (6 mm) on both sides and tap the shake in, stopping $^3/_4$" (19 mm) before it is flush with the shakes on either side.

CAUTION

Each shingle should have four nails going through it. When reinstalling shakes and shingles, you may not be able to remove the top two nails buried under the higher course. If you can only reach two nails, remove those. Then, trim the top of the replacement shake or shingle back before attaching it with two more nails.

Step 6 Drive two roofing nails in at a slightly upward angle, just below the butt edge of the shake above it. Then, again using a wood block, tap the replacement the final $^3/_4$" (19 mm) into place and conceal the nails.

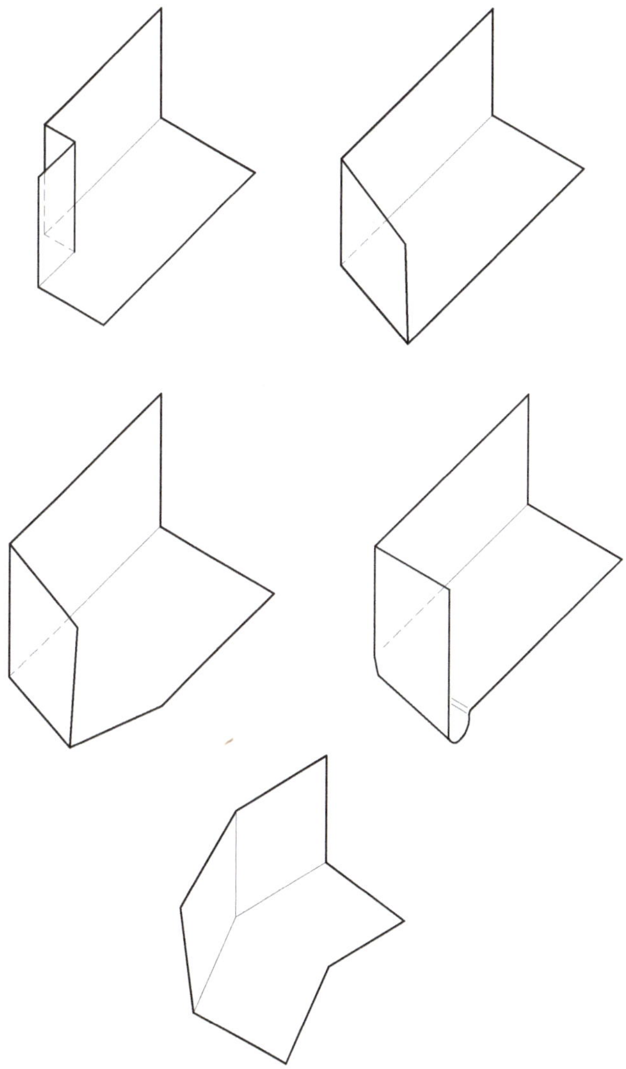

Figure 20 Examples of kick-out roof flashing.

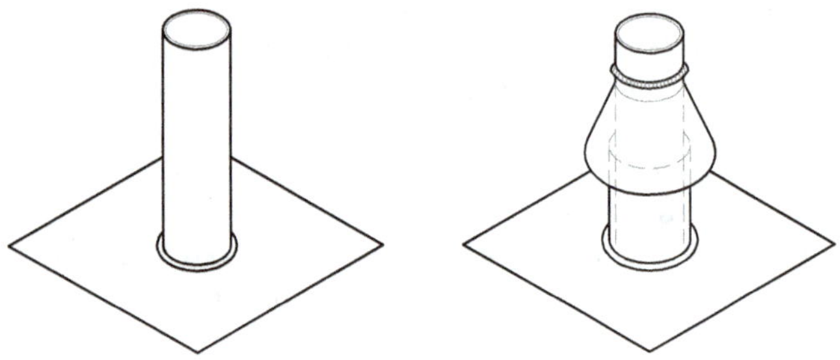

Figure 21 Penetration flashings.

Figure 22 Removing damaged shake with a shingle ripper.

2.0.0 Section Review

1. To seal out water and debris, which of the following is installed between courses of shakes?
   a. Underlayment
   b. Flashing
   c. Drip edge
   d. Interlayment

2. Backer flashings are used on penetrations that are 24" (61 cm) wide or less.
   a. True
   b. False

3. Warped, broken, and missing shingles *must* be _____.
   a. replaced
   b. covered
   c. split
   d. sawed

Review Questions

1. Wood roof systems utilizing shakes or shingles are often chosen for their _____.
   a. cost
   b. appearance
   c. ease of installation
   d. color options

2. Until the early 19th century, shakes were commonly _____.
   a. sawed
   b. milled
   c. hand-split
   d. whittled

3. Roofing hammers are sometimes called *roofing hatchets*.
   a. True
   b. False

4. Underdriven nails are considered the *most* common problem in wood roof systems installed with a nail gun.
   a. True
   b. False

5. Shakes and shingles can be installed on roofs with decks that are spaced or _____.
   a. continuous
   b. close
   c. vertical
   d. offset

6. Another phrase for interlayment is _____.
   a. slope felt
   b. shake liner
   c. leak barrier
   d. underlayment

7. When installing wood shakes, which of the following should be installed between the starter course and the first course, and then between every subsequent course?
   a. Underlayment
   b. Flashing
   c. Interlayment
   d. Ridge caps

8. Interlayment is typically made of _____.
   a. asphalt-saturated felts
   b. synthetic sheets
   c. self-adhering polymer-modified bitumen sheets
   d. polymer-modified bitumen sheets

9. Which type of flashing should be used to weatherproof where a wood roof system meets a headwall?
   a. Backer flashing
   b. Step flashing
   c. Apron flashing
   d. Cricket flashing

10. When removing a damaged shake or shingle, what should be used to split the shake or shingle along the grain?
    a. Circular saw
    b. Nail gun
    c. Shingle ripper
    d. Wood chisel

Trade Terms Introduced in This Module

Bundle: The measurement unit that mills use to sell shakes and shingles.

Butt edge: The bottom or exposed end of a shake or shingle.

Edge grain: A log cut perpendicular to the rings of the tree.

Flat grain: A log cut parallel to the tree's rings.

Heartwood: Wood located at the center of a log.

Interlayments: Underlayments installed between each course of shakes or shingles.

Wood shakes: Roof covering typically split from a wood billet and then resawed or shaped by a mill.

Wood shingles: Sawn products with uniform butt thickness relative to the specific length used as a roof covering.

Undercourse: A course of shakes or shingles installed beneath the covering course.

Wood billet: Wood blocks from which shakes or shingles are made.

Additional Resources

This module presents thorough resources for task training. The following reference material is suggested for further study.

Cedar Shake & Shingle Bureau, **www.cedarbureau.org**.

National Roofing Contractors Association (NRCA), **www.nrca.net**.

Occupational Safety and Health Administration (OSHA), US Department of Labor, **www.osha.gov**.

NCCER Module 16102, *Roofing Safety*.

NCCER Module 16105, *Introduction to Steep-Slope Roofing*.

NCCER Module 75901, *Fall Protection Orientation*.

NRCA Pocket Guide to Safety, NRCA. 2015. Rosemont, IL: National Roofing Contractors Association.

The NRCA Roofing Manual: Architectural Metal Flashing and Condensation and Air Leakage Control, NRCA. 2018. Rosemont, IL: National Roofing Contractors Association.

The NRCA Roofing Manual: Steep-slope Roof Systems, NRCA. 2017. Rosemont, IL: National Roofing Contractors Association.

The NRCA Safety Manual, NRCA. 2012. Rosemont, IL: National Roofing Contractors Association.

Figure Credits

Photo courtesy of Deer Park Roofing, Inc., Figure 1 (Shakes)

iStock@pancaketom, Figure 1 (Shingles)

Courtesy of the National Roofing Contractors Assoc., Figures 2–3, 7, 9–17, 19–21

iStock@artisteer, Figure 4

Image property of Stanley Black & Decker. Used with permission., Figures 5, 6

iStock@Rena-Marie, Figure 8

Ray Tschoepe, Courtesy Old-House Journal, Figure 22

Section Review Answer Key

Section 1.0.0

Answer	Section Reference	Objective
1. c	1.1.0	1a
2. a	1.2.3	1b

Section 2.0.0

Answer	Section Reference	Objective
1. d	2.1.1	2a
2. a	2.2.2	2b
3. a	2.3.0	2c

User Update

NCCER makes every effort to keep its textbooks up-to-date and free of technical errors. We appreciate your help in this process. If you find an error, a typographical mistake, or an inaccuracy in NCCER's curricula, please submit a User Update form by visiting **https://www.nccer.org/olf**. You can also scan the QR code using the camera on your phone or mobile device to access the form.

Slate Roof Systems

Overview

Slate roof systems are known for their longevity, durability, and aesthetics. Slate roofs have a rich history with roots in Europe. They are often found on castles and other militaristic structures. Today, a slate roof is a unique, long-term option for steep-slope buildings able to withstand the heavy load of slate systems. To safely install these systems, roofing professionals must be familiar with the properties of slate, the tools used during installation, and some common installation methods.

Module 16204

Trainees with successful module completions may be eligible for credentialing through the NCCER Registry. To learn more, go to **www.nccer.org** or contact us at 1.888.622.3720. Our website, **www.nccer.org**, has information on the latest product releases and training.

Your feedback is welcome. You may email your comments to **curriculum@nccer.org**, send general comments and inquiries to **info@nccer.org**, or fill in the User Update form at the back of this module.

This information is general in nature and intended for training purposes only. Actual performance of activities described in this manual requires compliance with all applicable operating, service, maintenance, and safety procedures under the direction of qualified personnel. References in this manual to patented or proprietary devices do not constitute a recommendation of their use.

16204 V1.0

16204
Slate Roof Systems

Objectives

Successful completion of this module prepares you to do the following:

1. Explain slate roof systems and components and identify the tools used to install them.
   a. Describe methods and styles of slate installation.
   b. Identify tools used to install slate roof systems.
2. Explain slate roof installation.
   a. Explain procedures for installing slate roof systems.
   b. Identify flashings used in slate roof systems.
   c. Describe basic repair methods used to correct mistakes during an installation.

Performance Tasks

Under supervision, you should be able to do the following:

1. Install an underlayment, drip edge metal, the cant strip, the starter course, and the first course.
2. Install slate with the proper head lap and side lap.
3. Flash the sidewalls on each row.
4. Establish the correct joint stagger and nail placement.
5. Install valley flashing.
6. Install valley slate.
7. Install hip and ridge slate.

Trade Terms

Bib
Cant strip
Grade
Point loading
Square

Industry Recognized Credentials

If you are training through an NCCER-accredited sponsor, you may be eligible for credentials from NCCER's Registry. The ID number for this module is 16204. Note that this module may have been used in other NCCER curricula and may apply to other level completions. Contact NCCER's Registry at 1.888.622.3720 or go to **www.nccer.org** for more information.

You can also show off your industry-recognized credentials online with NCCER's digital badges. Transform your knowledge, skills, and achievements into badges that you can share across social media platforms, send to your network, and add to your resume. For more information, visit **www.nccer.org**.

Videos and Digital Resources

How to Access Resources

This craft has additional videos and resources to enhance your learning experience. To view these resources, scan the QR below. The videos and resources are separated by module.

You can scan this code using the camera on your phone or mobile device to view these videos and resources.

Contents

1.0.0 Background and Tools 1
- 1.1.0 Types and Styles 1
  - 1.1.1 Methods and Styles 1
- 1.2.0 Tools 2
  - 1.2.1 Hammers 3
  - 1.2.2 Stakes 3
  - 1.2.3 Cutters 3
  - 1.2.4 Rippers 3

2.0.0 Installing Slate Roof Systems 5
- 2.1.0 Materials and Procedures 5
  - 2.1.1 Underlayments 5
  - 2.1.2 Attachment Methods 6
  - 2.1.3 Installation Outline 6
  - 2.1.4 Field Installation 7
  - 2.1.5 Valley Installation 7
- 2.2.0 Flashings 9
  - 2.2.1 Vertical-Surface Flashings 9
  - 2.2.2 Penetrations 9
- 2.3.0 Basic Repair 10

Figures and Tables

Figure 1 Slate roof on Brittany Parliament 2
Figure 2 Gray slate 2
Figure 3 Slate hammer 3
Figure 4 Slate stake 3
Figure 5 Slate cutter 3
Figure 6 Slate ripper 3
Figure 7 Proper nailing of slate 6
Figure 8 Drip edge with integrated cant 7
Figure 9 Slate installed in offset pattern 7
Figure 10 Example of a starter course of slate 8
Figure 11 Open valley 8
Figure 12 Closed valley 9
Figure 13 Flashing components used at a chimney 9
Figure 14 Various types of penetration flashings 10

Table 1 Slate Exposures 8

SECTION ONE

1.0.0 BACKGROUND AND TOOLS

Objective

Explain slate roof systems and components and identify the tools used to install them.

a. Describe methods and styles of slate installation.
b. Identify tools used to install slate roof systems.

Trade Terms

Grade: Classification of materials by quality or specific type.

Square: The amount of slate it takes to cover 100 square feet of roofing surface when installed with the appropriate head lap.

With deep roots in modern history, slate roofs are renowned for their aesthetics, longevity, and durability (*Figure 1*). Slate is naturally occurring and comes from sedimentary particles of clay and silt that accumulate and build up into beds of rock over millions of years.

Slate is carefully extracted from a quarry and moved to a mill that cuts, splits, and trims the raw material into pieces of roof covering. Slate is assigned a grade, or classification of the material by quality or specific type.

CAUTION

Slate roofs can only be installed on roof structures that are designed to support the heavy weight of these systems.

NOTE

Throughout NCCER's Roofing program, the term *steep-slope* refers to a roof slope of above 3:12. This differs from OSHA's definition of steep-slope roofs (slope above 4:12). When installing or repairing roof systems, always check company, local, state, and federal codes. If any of the applicable codes contradict each other, follow the strictest one.

1.1.0 Types and Styles

Slate roof systems are available in a variety of colors, thicknesses, weights, sizes, and other physical properties.

In the United States, slate is sold by the square. A square is measured by the amount of slate it takes to cover 100 ft^2 (9 m^2) of roofing surface when laid with the appropriate head laps. A square of slate may vary in weight from 700 lb to 8,000 lb (318 kg to 3,629 kg) depending on the thickness of each slate.

1.1.1 Methods and Styles

Slate roof systems can be divided into three general methods of installation:

- *Standard* — Slate is typically $^3/_{16}$" to $^3/_8$" (4.8 mm to 9.5 mm) thick. This system is used to form a variety of steep-slope roof designs. The most common slate has uniform lengths and widths, and the butt ends of the slate are usually square.
- *Graduated* — Use a variety of textures, colors, and thicknesses to create a unique, variable appearance. The thickest, widest, and longest slates are attached along the roof's eaves. The slate decreases in size, thickness, and exposure as the courses move upslope toward ridges to create a graduated effect. This type makes slate roofs appear larger.
- *Textured* — Sometimes referred to as *textural*, these systems use a mixture of thicknesses throughout a single roof. This system type generally has a rougher texture than standard slate.

Purchasing Considerations

To purchase the correct amount of slate to cover an entire roof, it is important to consider the head laps required in addition to the steepness of the roof. If an order is placed that does not calculate these considerations, not enough slate will be delivered to complete the project.

Slate may also break during installation, so it is important to ensure you have extra pieces to make any necessary replacements. Consult with the slate quarry to ask how much breakage you can expect during installation to plan replacement pieces accordingly.

Figure 1 Slate roof on Brittany Parliament.

Slate comes in many colors, allowing for various aesthetic options. Common slate colors include black, green, gray (*Figure 2*), gray-green, mottled purple, mottled green, purple, and red.

Manufacturers offer these colors in both weathering and unfading options. Weathering slate may change colors after being exposed to the elements over time. Unfading slate better holds its original color, better maintaining the look of the system over time.

The National Slate Association established standards for commercial slate in the 1800s. These standards involved grading slate by color, texture, straightness, thickness, mineral composition, and weight.

Since the 1950s, ASTM International has taken over slate standards. *ASTM C406* includes three grades of slate:

- *Grade S1* — Expected service life of more than 75 years.
- *Grade S2* — Expected service life of 40–75 years.
- *Grade S3* — Expected service life of 20–40 years.

Figure 2 Gray slate.

1.2.0 Tools

Slate systems use specialized tools for safer and more effective installation.

> **NOTE**
> Additional tools used in slate roof installation are covered in greater detail in NCCER Module 00103, *Introduction to Hand Tools*, and NCCER Module 00104, *Introduction to Power Tools*.

1.2.1 Hammers

Slate hammers (*Figure 3*) are designed specifically for installing slate roof systems. Similar to standard hammers, the head is used to drive nails. This hammer also has a point for punching slate and a knife edge on its shaft for making cuts in slate.

1.2.2 Stakes

Slate stakes (*Figure 4*) are used with slate hammers to cut slate on the field. These T-shaped tools are stuck into a roof deck to provide support to the underside of slate where a hole or punch is to be made.

1.2.3 Cutters

Slate cutters (*Figure 5*) are designed like a small lever with an arm that serves as a cutting blade. Slate cutters may be used to punch holes in slate or to cut concave curves.

Figure 3 Slate hammer.

Figure 4 Slate stake.

1.2.4 Rippers

Slate rippers (*Figure 6*) are handy for removing broken pieces of slate. Hook this tool around nails to loosen or cut them by force. Slate rippers have handles that are designed to be struck with a hammer to assist with nail removal.

> **WARNING!**
>
> Always follow safety precautions, such as wearing gloves, safety glasses, and a mask, when working with these tools.

Figure 5 Slate cutter.

Figure 6 Slate ripper.

1.0.0 Section Review

1. In the United States, slate is sold by the foot.
   a. True
   b. False
2. Which of the following tools is designed like a small lever with an arm that serves as a cutting blade?
   a. Slate hammer
   b. Slate stake
   c. Slate cutter
   d. Slate ripper

SECTION TWO

2.0.0 INSTALLING SLATE ROOF SYSTEMS

Objective

Explain slate roof installation.

a. Explain procedures for installing slate roof systems.
b. Identify flashings used in slate roof systems.
c. Describe basic repair methods used to correct mistakes during an installation.

Performance Tasks

1. Install an underlayment, drip edge metal, the cant strip, the starter course, and the first course.
2. Install slate with the proper head lap and side lap.
3. Flash the sidewalls on each row.
4. Establish the correct joint stagger and nail placement.
5. Install valley flashing.
6. Install valley slate.
7. Install hip and ridge slate.

Trade Terms

Bib: Flashing that covers and weatherproofs the area where a new nail is used to replace individual slates with a new supply.

Cant strip: An angled strip installed at the roof's eaves that raises the butt end of the starter course to match same angle as the above slate courses. Also referred to as *cant*.

Point loading: The act of spacing out roofing materials in different areas on the roof so that all the weight of the roofing materials is not in one place.

Slate roofs are among the longest lasting and most attractive systems. Roofing professionals must know how to install all the components of these systems safely and effectively.

CAUTION

Buildings that can support a slate roof must also be able to hold the weight of the scaffolding and staging of materials during the installation process.

2.1.0 Materials and Procedures

Application methods for slate roof systems vary between different projects and regions. Roofing professionals should be familiar with general installation procedures and accessories so they can safely and knowledgeably begin installing slate roof systems.

However, it is critical to consult any applicable literature, including quarry instructions, project specifications, and building codes, before and during installation. It is also important to familiarize yourself with the types of flashings installed in these systems, as well as some basics of slate repair that may come in handy on the jobsite.

WARNING!

When staging materials for installation, do not stack all supplies in one section of the roof. Putting too much material in one area can lead to roof collapse. Spacing out roofing materials in different areas on the roof so that all the weight of the roofing materials is not in one place is called **point loading**.

2.1.1 Underlayments

Slate roof systems are compatible with several underlayment types. Underlayment for slate roof systems is typically heavier than underlayment used for other systems. Along with determining the type of system selected for installation, regional weather patterns and slope also influence the type of underlayment that should be used in each project. Roofing professionals must make sure a deck is clean, dry, and in good condition before applying underlayment.

Underlayment used in slate systems performs better when it is compatible with the long life of slate roof systems. The following underlayment options are commonly used during slate roof installation, depending on region and slope:

- *Single-layer underlayment* — Underlayment materials, such as #30 felt, #40 felt, #43 felt, or synthetic underlayment installed in a single layer.
- *Double-layer underlayment* — Underlayment materials installed in two layers. This method may combine a variety of underlayment types and installation methods, such as mechanically attaching a first layer of #30 felt and adhering a cap sheet as the second layer.
- *Leak barrier* — Also called *self-adhering sheets* and, in colder climates, *ice dam protection*. These underlayment sheets have adhesive bitumen applied at the factory and provide labor savings with installation on the job.

CAUTION

Leak barrier can cause serious issues in a system's ventilation if it is not installed properly. Always abide by manufacturer instructions and local building code requirements when installing roof systems.

2.1.2 Attachment Methods

Roofing professionals may use a variety of attachment methods to install slate roof systems. The best attachment method for a given project depends on factors such as regional weather, deck type, slope, building codes, and quarry recommendations.

Slate is usually mechanically attached with fasteners, such as nails. In this type of installation, each piece of slate typically has two holes at the top. This allows fasteners to be nailed through and into the deck below (*Figure 7*). Nails should be long enough to go through the slate and penetrate the substrate underneath.

Be sure to drive nails straight and not too tight that the slate breaks. The slate should hang on the nail, and the nail head must sit flush with the surface of the slate without being too loose or too tight.

WARNING!

Before installation, the bottom of the deck should be inspected to make sure no electrical currents will disrupt installation on top of the roof.

Because of the long service life of slate, roof system designers must choose fasteners carefully. The fastener material must have a similar service life to the roof system so that the slate does not outlast the fasteners. Examples include copper and stainless steel.

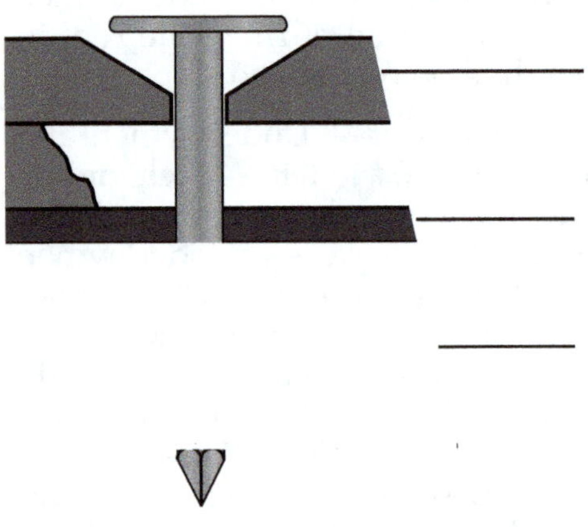

Figure 7 Proper nailing of slate.

2.1.3 Installation Outline

The following steps provide a general procedure for slate installation:

CAUTION

This outline is intended only to overview the general installation sequence. It cannot be used in place of quarry instructions, project specifications, or building code requirements.

Step 1 Don all proper PPE for the job. Ensure that the appropriate fall protection equipment is used or in place. Depending on the project and environment, this may be a personal fall arrest system (PFAS) with approved anchorage, or appropriate guardrails or scaffolding.

Step 2 Ensure the roof deck is clean, dry, and in good condition.

Step 3 Install the underlayment.

Step 4 Install a walking platform using roof jacks and boards as you proceed up the roof.

WARNING!

It is not recommended to walk on a slate roof. Slate may break or be slippery when wet.

Step 5 Install drip edge and **cant strip**. The cant strip, also referred to as *cant*, is sometimes integrated in the drip edge metal, as shown in *Figure 8*.

Step 6 Snap chalk lines to help you achieve the necessary minimum head lap and consistent exposures from the eave to the rake.

Step 7 Ensure slate is in good condition and install starter course.

NOTE

Occasionally, roofing professionals may need to punch holes in slate prior to nailing.

Step 8 Install next courses of slate, ensuring each course is installed above the previous one in an offset, water-shedding pattern (*Figure 9*). Ensure installation meets the minimum offset pattern required by the project specifications.

Figure 8 Drip edge with integrated cant.

Step 9 Install valley metal as you install your courses of slate.

Step 10 Install ridges and hips.

Step 11 As you install slate, install any other flashings needed.

Step 12 Inspect and clean the jobsite so it is left safe and tidy. Keep an eye out for any broken slate that needs replacing.

2.1.4 Field Installation

Slate must be installed with the proper head lap to ensure the roof system correctly sheds water. The exposure of slate is the portion of the slate that is not covered by the course above or the pieces on either of its sides—that is, the part of the slate that is exposed to weather. To calculate exposure, subtract the system's required head lap from the length of the slate and divide that resulting number by 2. *Table 1* shows proper exposures for various lengths of slate if all are to be set with a 3" (7.5 cm) head lap.

> **NOTE**
> Always consult applicable documentation and project specifications for installation details and exposures.

Before the first course of slate is installed, a cant strip and row of starter slates is applied directly over the underlayment or water and ice dam protection membrane along the eaves of the roof system. The cant strip creates the necessary angle for the subsequent starter course. The starter course helps shed water that may leak between the slates in the first course.

The lower edge of the starter course should extend past the roof's eaves and rakes a minimum of 1" (2.5 cm) to help direct runoff away from the fascia board and other underlying building components, as shown in *Figure 10*. In some cases when gutters are used, the overhang may be reduced to less than 1" (2.5 cm).

2.1.5 Valley Installation

Valleys are the V-shaped angle that forms where the downslopes of two roof fields intersect. This area is particularly vulnerable to leakage since water will shed in its direction. The two typical methods of valley installation used in slate roof systems are open valley and closed-cut valley.

> **CAUTION**
> All valley construction must avoid placing fasteners and penetrations in the valley. This is to ensure there is a clear pathway for water to shed off the roof after it collects in valleys with no obstructions.

Open Valley

An open valley (*Figure 11*) is constructed by installing 8' to 10' (2.4 m to 3 m) lengths of corrosion-resistant metal from the low point to the high point in the valley. The slates (and the underlayment in some regions) are lapped onto the flange on each side of the valley metal. This leaves a clear space between the roofing material to channel runoff water down the valley.

Figure 9 Slate installed in offset pattern.

Table 1 Slate Exposures

Length of Slate	Slate Exposure
24" (61 cm)	$10\frac{1}{2}$" (26.5 cm)
22" (56 cm)	$9\frac{1}{2}$" (24 cm)
20" (51 cm)	$8\frac{1}{2}$" (21.5 cm)
18" (45.5 cm)	$7\frac{1}{2}$" (19 cm)
16" (40.5 cm)	$6\frac{1}{2}$" (16.5 cm)
14" (35.5 cm)	$5\frac{1}{2}$" (14 cm)
12" (30.5 cm)	$4\frac{1}{2}$" (11.5 cm)
10" (25.5 cm)	$3\frac{1}{2}$" (9 cm)

Closed-Cut Valley

In a closed-cut valley, also called a *closed valley* (*Figure 12*), slates on both sides are cut at an angle parallel to the centerline of the valley and are butted together.

Valley metal used in closed-valley installation is typically shorter in length than the 8' to 10' (2.4 m to 3 m) valley metal used in open-valley installation. The size of closed-cut valley metal is determined by the length of the slate being used. This is because the valley metal must overlap the top part of the slate and then extend up the roof 1" to 2" (2.5 cm to 5 cm) above the slate so that it can be nailed directly into the deck.

In a closed valley, each course of slate should be weatherproofed with an appropriately sized piece of valley metal laid over and extending past the top of the slate. The next course of slate should be installed over that piece of valley metal so that the metal is no longer visible. Then, the top of that slate course should be overlapped with another piece of valley metal that extends past the top of the slate so it can be nailed directly to the deck. This should be followed by another slate course.

This pattern continues to the top of the valley so that the closed valley is stepped up in pieces and weatherproofed through each course. Be sure that slate is lined up and even across the valley and that no metal flashing pieces show through the courses.

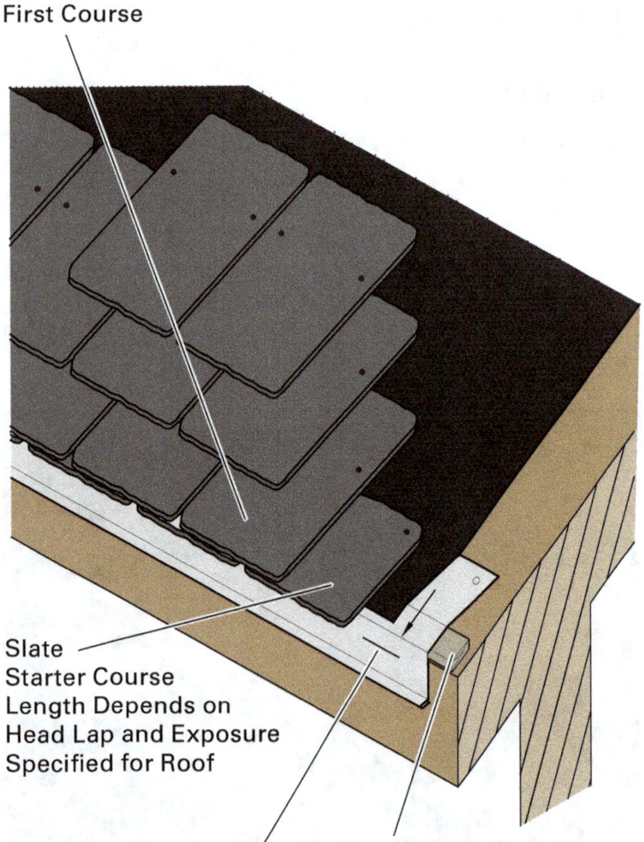

Figure 10 Example of a starter course of slate.

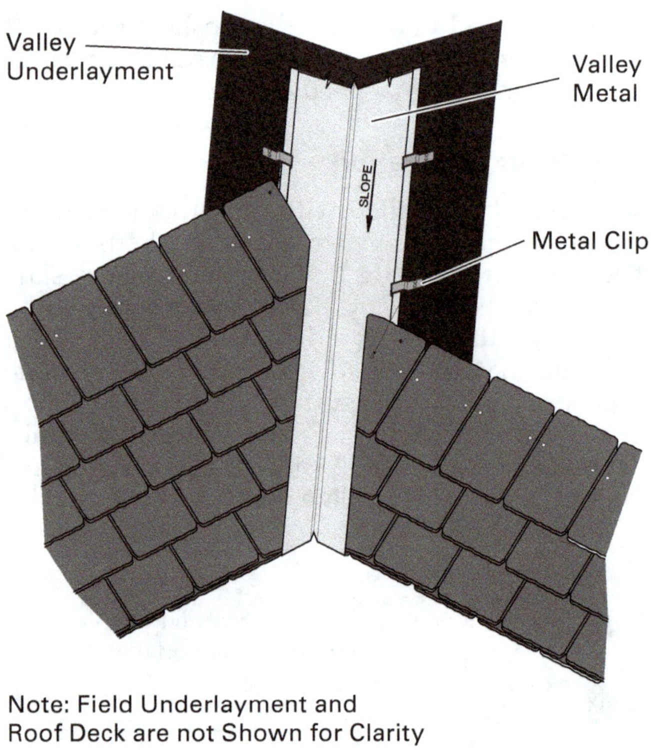

Figure 11 Open valley.

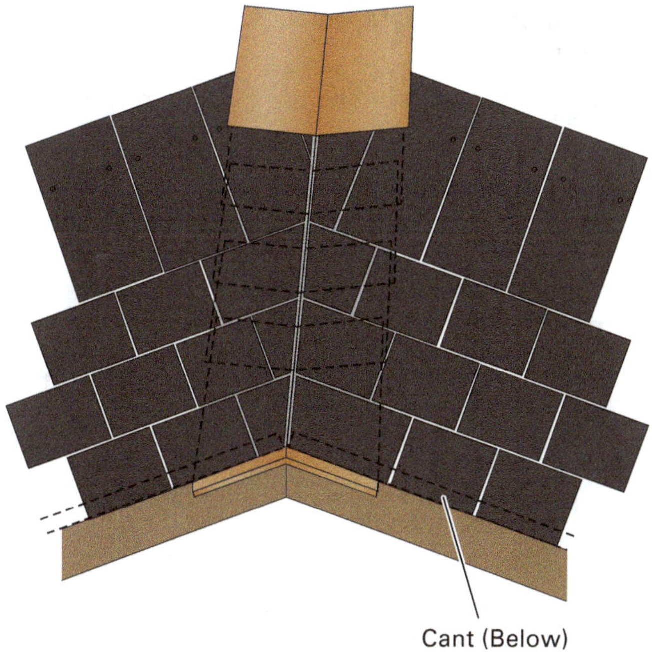

Figure 12 Closed valley.

2.2.0 Flashings

Slate roofs must have flashings installed to ensure water sheds off roofs without leaking through the systems.

When installing slate roof systems, it is important to use flashings with a similar service life to slate. Some of these include copper, stainless steel, lead-coated copper, lead, and zinc. In general, avoid using galvanized steel with slate roof systems. To prevent corrosion, galvanized nails should not be used with copper flashings.

Generally, before flashings are applied, a layer of leak barrier or other appropriate underlayment should be applied to a roof deck around roof penetrations.

> **WARNING!**
> When interacting with lead-containing materials, be sure to wear the proper PPE and follow all applicable safety guidelines.

2.2.1 Vertical-Surface Flashings

Flashings at vertical protrusions (such as chimneys, walls, or curbs) should have a relatively smooth substrate on the roof plane and vertical plane, up to a sufficient height, to receive the metal flashing. Rough or contoured vertical surfaces, such as cut stone and rough timber, should be provided with a flush substrate above the roof line made to accept the vertical flashing and/or counterflashing. It may also be possible to use soft metal flashings formed to follow the substrate's contours.

The following are the four types of flashings used at locations where metal shingles intersect a vertical surface:

- *Apron flashings* — Weatherproof the area where the roof meets a headwall.
- *Channel/step flashings* — Should be applied on slate systems that intersect a vertical sidewall.
- *Backer/cricket flashings* — Should be installed where slate roof intersects the upslope side of a vertical roof penetration. Backer flashings are used on penetrations that are 24" (61 cm) wide or less, and cricket flashings are used on penetrations greater than 24" (61 cm).
- *Counterflashing* — Should be applied over base flashings, such as apron, step, channel, cricket, and backer flashings to protect the base flashings' top edges from water intrusion.

Figure 13 illustrates the location of apron flashing, step flashing, cricket flashing, and counterflashing on a chimney protrusion.

2.2.2 Penetrations

There are many small penetrations that need to be flashed into slate roof systems, such as vent pipe penetrations, exhaust vents, exhaust fans, furnace or water heater flue pipes, and electrical

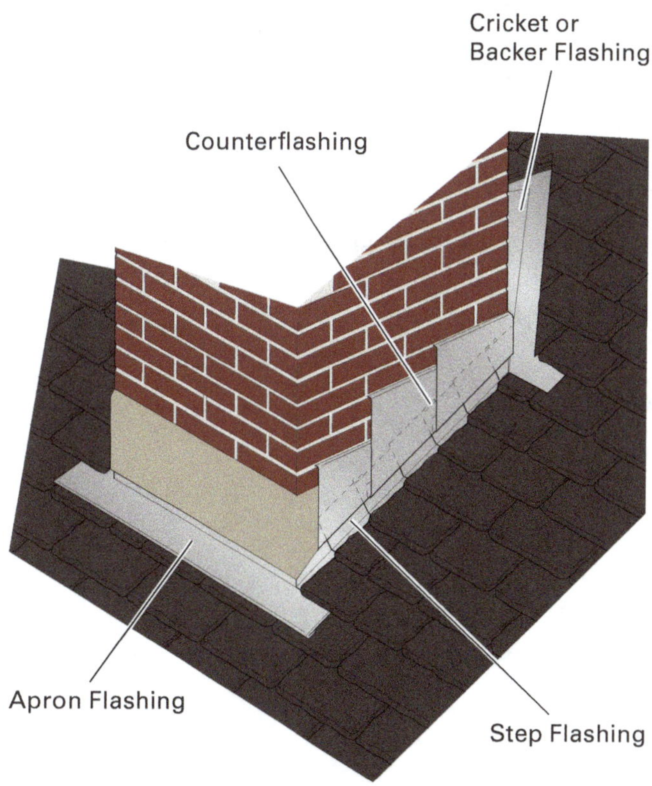

Figure 13 Flashing components used at a chimney.

standpipes (*Figure 14*). This is typically accomplished with the use of some type of flat flange that extends around the penetration. This piece is installed under the slate and underlayment on the upslope side of the flange and extends down on top of the slate at the downslope side. Attached and sealed to the flange is a cylinder or rectangular box used to seal around a penetration. The flange can be set into asphalt roof cement or elastomeric flashing sealant for additional weatherproofing protection. These flashing components often are supplied by other trades but may be installed by a roofing contractor.

2.3.0 Basic Repair

Along with understanding how to install a roof system, from underlayment to flashings, roofing professionals should be able to notice and fix mistakes that may occur during installation. Pay particular attention to the quality and alignment of the slate, accessory pieces, base flashings, and counterflashings. If the slate or flashing is damaged, it should be replaced.

The following is a general procedure for replacing slate:

Step 1 Remove old nails with a slate ripper.

Step 2 Remove damaged slate.

Step 3 Put in new slate.

Step 4 Punch or drill a new hole between two slate pieces above.

Step 5 Install nail.

Step 6 Install bib.

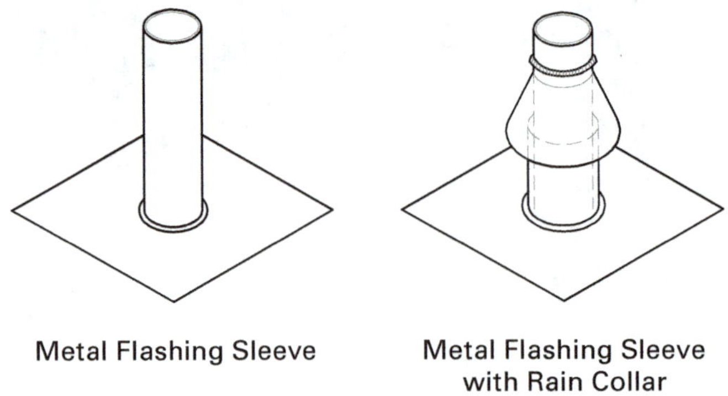

Figure 14 Various types of penetration flashings.

2.0.0 Section Review

1. To calculate correct exposure, subtract 1" (2.5 cm) for head lap and multiply that resulting number by two.
   a. True
   b. False
2. It is okay to use galvanized nails with copper flashings.
   a. True
   b. False

Review Questions

1. In the United States, slate is sold by the _____.
   a. square
   b. yard
   c. foot
   d. pound

2. Which type of slate installation uses a variety of textures, colors, and thicknesses of slate, typically decreasing the exposure as the courses move upslope?
   a. Standard
   b. Tapered
   c. Graduated
   d. Textured

3. Which grade of slate lasts the *longest*?
   a. S0
   b. S1
   c. S2
   d. S3

4. Which of the following is a T-shaped tool that is stuck into the roof deck to provide support to the underside of slate where a hole is to be made?
   a. Slate ripper
   b. Slate cutter
   c. Slate hammer
   d. Slate stake

Figure Credit: North Country Slate

Figure RQ01

5. Which tool is shown in *Figure RQ01*?
   a. Slate ripper
   b. Slate cutter
   c. Slate hammer
   d. Stake

6. When staging materials for installation, you should _____.
   a. stack all the roofing materials in one section so they are out of the way
   b. categorize roofing materials by weight
   c. space roofing materials out in different areas to distribute the weight
   d. leave all roofing materials on the ground

7. When installing a slate roof system, ensure a *minimum* head lap of _____.
   a. 1" (2.5 cm)
   b. 2" (5 cm)
   c. 3" (7.5 cm)
   d. 1' (30 cm)

8. When installing slate roof systems, it is important to use flashings that _____.
   a. are coated
   b. are made of galvanized steel
   c. match the color scheme of the roof
   d. have the same service life as the slate

9. Which type of flashing is used on the downslope side of a chimney?
   a. Backer flashing
   b. Cricket flashing
   c. Step flashing
   d. Apron flashing

10. If a slate is damaged, you should _____.
    a. ignore it
    b. fill in the damaged area with roof cement
    c. replace it
    d. cover it with underlayment

Trade Terms Introduced in This Module

Bib: Flashing that covers and weatherproofs the area where a new nail is used to replace individual slates with a new supply.

Cant strip: An angled strip installed at the roof's eaves that raises the butt end of the starter course to match same angle as the above slate courses. Also referred to as *cant*.

Grade: Classification of materials by quality or specific type.

Point loading: The act of spacing out roofing materials in different areas on the roof so that all the weight of the roofing materials is not in one place.

Square: The amount of slate it takes to cover 100 square feet of roofing surface when installed with the appropriate head lap.

Additional Resources

This module presents thorough resources for task training. The following reference material is suggested for further study.

National Roofing Contractors Association (NRCA), **www.nrca.net**.

National Slate Association, **https://www.slateassociation.org**.

NCCER Module 00103, *Introduction to Hand Tools*.

NCCER Module 00104, *Introduction to Power Tools*.

NCCER Module 16105, *Introduction to Steep-Slope Roofing*.

NCCER Module 75901, *Fall Protection Orientation*.

NRCA Pocket Guide to Safety, NRCA. 2015. Rosemont, IL: National Roofing Contractors Association.

The NRCA Roofing Manual: Architectural Metal Flashing and Condensation and Air Leakage Control, NRCA. 2018. Rosemont, IL: National Roofing Contractors Association.

The NRCA Roofing Manual: Steep-slope Roof Systems, NRCA. 2017. Rosemont, IL: National Roofing Contractors Association.

The NRCA Safety Manual, NRCA. 2012. Rosemont, IL: National Roofing Contractors Association.

Occupational Safety and Health Administration (OSHA), US Department of Labor, **www.osha.gov**.

OSHA 3755-05, *Protecting Roofing Workers*. US Department of Labor. **www.osha.gov/Publications/OSHA3755.pdf**.

Figure Credits

iStock@Gregory_DUBUS, Module Opener

iStock@excentric_01, Figure 1

iStock@PaulMaguire, Figure 2

North Country Slate, Figures 3, 6

Gary Howes, Figures 4–5

Courtesy of the National Roofing Contractors Assoc., Figures 7, 10–11, 13–14

Patrick D Murphy Co., Inc., Architects, Figure 9

Section Review Answer Key

Section 1.0.0

Answer	Section Reference	Objective
1. b	1.1.0	1a
2. c	1.2.0	1b

Section 2.0.0

Answer	Section Reference	Objective
1. b	2.1.4	2a
2. b	2.2.0	2b

User Update

NCCER makes every effort to keep its textbooks up-to-date and free of technical errors. We appreciate your help in this process. If you find an error, a typographical mistake, or an inaccuracy in NCCER's curricula, please submit a User Update form by visiting **https://www.nccer.org/olf**. You can also scan the QR code using the camera on your phone or mobile device to access the form.

Metal Roof Systems

Overview

Metal roofing was in use prior to the rise of the Roman Empire. Metal roof systems are known for their aesthetic characteristics and long-term performance. When installing metal roofs, roofing professionals must remember key features, such as metal's natural expansion and contraction when exposed to changing temperatures. To install these systems safely and optimally, it is important to understand their unique characteristics.

Module 16205

Trainees with successful module completions may be eligible for credentialing through the NCCER Registry. To learn more, go to **www.nccer.org** or contact us at 1.888.622.3720. Our website, **www.nccer.org**, has information on the latest product releases and training.

Your feedback is welcome. You may email your comments to **curriculum@nccer.org**, send general comments and inquiries to **info@nccer.org**, or fill in the User Update form at the back of this module.

This information is general in nature and intended for training purposes only. Actual performance of activities described in this manual requires compliance with all applicable operating, service, maintenance, and safety procedures under the direction of qualified personnel. References in this manual to patented or proprietary devices do not constitute a recommendation of their use.

16205 V1.0

16205 METAL ROOF SYSTEMS

Objectives

Successful completion of this module prepares you to do the following:

1. Summarize features of metal roof systems and the tools used to install them.
   a. Describe the seam types, methods, and tools used to secure metal roof systems.
2. Explain how to install metal roof systems.
   a. Describe the installation process and flashing of architectural metal roof systems.
   b. Describe the installation process and flashing of structural metal roof systems.
   c. Describe the installation process and flashing of metal shingle systems.

Performance Tasks

Under supervision, you should be able to do the following:

1. Don all appropriate PPE for architectural metal panel installation.
2. Install manufacturer-recommended underlayment and any necessary base flashings, valley metal, and perimeter metal to a wood deck.
3. Properly measure, cut, clip, fit, and install architectural metal panels according to manufacturer instructions.
4. Complete any remaining flashings, such as penetration, wall, and head flashings.

Trade Terms

Architectural metal roofs
Fixed points
Flux
Galvanic corrosion
Metal shingle roofs
Solder
Soldering
Standing-seam roofs
Structural metal roofs

Industry Recognized Credentials

If you are training through an NCCER-accredited sponsor, you may be eligible for credentials from NCCER's Registry. The ID number for this module is 16205. Note that this module may have been used in other NCCER curricula and may apply to other level completions. Contact NCCER's Registry at 1.888.622.3720 or go to **www.nccer.org** for more information.

You can also show off your industry-recognized credentials online with NCCER's digital badges. Transform your knowledge, skills, and achievements into badges that you can share across social media platforms, send to your network, and add to your resume. For more information, visit **www.nccer.org**.

Videos and Digital Resources

How to Access Resources

This craft has additional videos and resources to enhance your learning experience. To view these resources, scan the QR below. The videos and resources are separated by module.

You can scan this code using the camera on your phone or mobile device to view these videos and resources.

Contents

1.0.0 Metal Roof Components and Tools 1
- 1.1.0 Metal Roof Components 2
  - 1.1.1 Seams 3
  - 1.1.2 Attachment 5
  - 1.1.3 Hand and Power Tools 7

2.0.0 Metal Roof Installation 13
- 2.1.0 Architectural Metal Roofs 13
  - 2.1.1 Flashings 15
- 2.2.0 Structural Metal Roofs 16
- 2.3.0 Metal Shingle Roofs 19
  - 2.3.1 Flashings 20

Figures

Figure 1 The Pantheon 2
Figure 2 Notre Dame 3
Figure 3 Cut-resistant gloves with Kevlar® 3
Figure 4 Common examples of standing seams 4
Figure 5 Flat-seam roof with soldered joints 4
Figure 6 Flat-seam roof with nonsoldered, interlocking panels 5
Figure 7 Examples of metal shingle designs 5
Figure 8 Bermuda horizontal panels with hooked seams 6
Figure 9 Panel layout before attachment 6
Figure 10 Possible fixed-point locations 7
Figure 11 Expansion clip 7
Figure 12 Roof attached with concealed fasteners 8
Figure 13 Roof attached with exposed fasteners 9
Figure 14 Continuous cleat 9
Figure 15 Tinner's hammer 9
Figure 16 Rivet gun and pop rivets 10
Figure 17 Installing a pop rivet 11
Figure 18 Basic soldering iron 12
Figure 19 Self-contained soldering iron kit 12
Figure 20 Steep-slope architectural metal roof 14
Figure 21 Example of a rake edge 15
Figure 22 Typical valley profile for architectural metal roofing 15
Figure 23 Premanufactured pipe flashing centered on panel 16
Figure 24 HVAC unit curb on a sloped roof 16
Figure 25 Exhaust fan curb custom-fabricated for a standing-seam roof 17
Figure 26 Ridge cap installed over panel ends 18
Figure 27 Structural metal roof 18
Figure 28 Blanket insulation installed between purlins 18
Figure 29 Structural metal panel profiles 19
Figure 30 Metal shingle roof 19
Figure 31 Underlayment centered in valley 20
Figure 32 Penetration flashing 20
Figure 33 Metal-shingle channel flashing 21
Figure 34 Backer flashing 21
Figure 35 Cricket flashing 21
Figure 36 Counterflashing 21

SECTION ONE

1.0.0 METAL ROOF COMPONENTS AND TOOLS

Objective

Summarize features of metal roof systems and the tools used to install them.

a. Describe the seam types, methods, and tools used to secure metal roof systems.

Trade Terms

Architectural metal roofs: A metal panel system constructed of 26-gauge or thinner metals used in steep-slope applications that require a continuous or closely spaced substrate underneath the system for support.

Fixed points: Points of attachment on steep-slope metal roof systems that secure the system and represent an obstacle to the natural expansion and contraction of metal. A fixed point for a panel is usually located at a roof's ridges or hips, or at the eaves or valleys.

Flux: A chemical substance that helps remove and prevents the formation of oxides on metal surfaces as they are heated for soldering, and acts as a vehicle to help solder flow more readily.

Metal shingle roofs: Steep-slope metal roof systems installed in overlapping courses like traditional shingles.

Solder: A metal alloy often containing tin and lead that is melted using a hot iron to seal seams and join metals.

Soldering: A non-fusion process used to bond metal objects together with solder. See *solder*.

Standing-seam roofs: Metal roofs with a panel seam made by turning up the edges of adjacent metal panels and interlocking them in a variety of ways. Common to both architectural and structural metal roofs.

Structural metal roofs: A metal panel system constructed of 24-gauge or thicker metals used primarily in low-slope applications but also for steep-slope applications.

Metal roof systems offer a great range of aesthetic and performance characteristics. Metal roofs have been used throughout the course of history—from Ancient Rome's Pantheon (*Figure 1*) to Paris's Notre Dame (*Figure 2*). Although the shape and structural integrity of the Pantheon roof relies on several ancient concrete mixtures, the concrete was originally covered with copper and bronze. Despite the many varieties of metal roofs, they can all be characterized by their durability.

In the 1960s, modern **standing-seam roofs** were developed as an option for steep- and low-slope systems. Standing-seam roofs include **architectural metal roofs** and **structural metal roofs**. The words *standing seam* could be accurately inserted into both of those terms. However, they have not been inserted in this module to avoid long, repetitive names. Metal structures are made stronger by joining panels at angles to each other, thus stiffening the roof surface and helping it resist buckling under load.

While architectural metal roofs are used only for steep-slope applications, structural metal roofs are used for low-slope applications. Structural panels can also be used in steep-slope systems if they are installed over a continuous or closely spaced substrate. Architectural panels cannot be used in low-slope applications since they are made from thinner material.

Metal shingle roofs are steep-slope systems that can mimic the look of many other materials. Like traditional asphalt shingles, they are installed in overlapping courses to shed water from the roof.

> **WARNING!**
>
> Safety is a crucial consideration in metal roof installation. For more information on roofing safety, review NCCER Module 16102, *Roofing Safety*.

> **NOTE**
>
> Throughout NCCER's Roofing program, the term *steep-slope* refers to a roof slope greater than 3:12. This differs from OSHA's definition of steep-slope roofs (slope greater than 4:12).
>
> When installing or repairing roof systems, always check company, local, state, and federal codes. If any of the applicable codes contradict each other, follow the strictest one.

Figure 1 The Pantheon.

1.1.0 Metal Roof Components

Metal roof panels are available in a variety of widths, profiles, and seam types. Metal panels are made from metals such as steel, aluminum, and copper. *Surfacing* refers to the finish applied for protection and to add a pleasing look to the roof. Surfacing options include the following:

- *Naturally weathering* — Includes metals that do not require a paint finish but develop their own unique appearance, such as copper, zinc, and lead.
- *Paint and coatings* — Paint and coatings can be applied after installation, but many finishes and textures can be added during the manufacturing process to mimic other materials.

> **NOTE**
> Properties of metal that affect its use in roof systems are presented in NCCER Module 16107, *Sheet Metal in Roofing.*

Since metal transfers heat more readily than other materials, the underlayments must be able to withstand higher temperatures. The underlayment should be what is considered a high-temperature product. Some common underlayment products have a temperature rating of 140°F (60°C). This is too low for metal roofing. High-temperature products are typically rated for temperatures in the range of 250°F (~120°C). Although metal roofs do increase in temperature quickly, they also cool more quickly for the same reason, helping to transfer heat out of the substrate once the sun goes down.

The possibility for injuries is always present when working with sheet metal. Understanding the hazard can help decrease the risks. Some of the most common injuries are cuts, lacerations and punctures, strains and sprains, and burns. Personnel should wear appropriate personal protective equipment (PPE) when operating equipment, including safety goggles or glasses, cut-resistant gloves, long-sleeved shirts, and long pants. Shoes should be appropriate for the work, such as leather uppers, and have soles that provide good grip and traction. Gloves containing Kevlar® (*Figure 3*) are ideal for working with sheet metal. Avoid wearing loose-fitting clothes or jewelry when working with metal roofing materials or around equipment.

Figure 2 Notre Dame.

1.1.1 Seams

As mentioned previously, standing-seam roofs may be structural or architectural in nature. They have seams that run vertically along the slope of the roof. These seams are designed to interlock or attach panels together to form a continuous, reliable, water-shedding system.

There are many standing-seam designs used in metal roofing. Seams may be formed mechanically or by hand, depending on the project. Some common examples of standing seams can be seen in *Figure 4*.

Flat seams can be soldered (*Figure 5*) or joined using interlocking edges (*Figure 6*). They can also be welded, but it is rarely done. **Soldering** is the process of bonding metal objects together by melting an alloy called **solder**. Solder often contains lead and tin, although other metals may also be part of the alloy. The metals have a significantly lower melting point than the metals they bond, such as steel and copper. Roof seams can be soldered to make sure they are weatherproof.

Figure 3 Cut-resistant gloves with Kevlar®.

On steep-slope roofs, metal shingles often interlock and may also be attached with clips. Metal shingles are available in various sizes and profiles (*Figure 7*). Metal shingle panels are simply large versions of smaller individual shingles, perhaps appearing as a dozen or so individual shingles on a single piece.

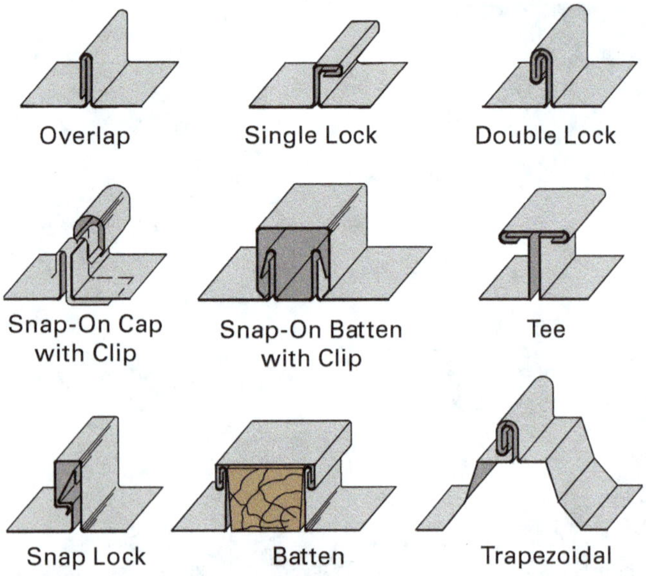

Figure 4 Common examples of standing seams.

The following are three basic panel-to-panel seaming techniques:

- *Mechanical seam* — Created by hand brakes, tongs, or powered seaming devices. These tools are used to turn a seam completely over itself to create a weatherproof joint.
- *Snap-together seam* — Includes snap-lock seams and snap-on seams. Snap-lock seams directly connect to one another, whereas snap-on seams use a separate cap or batten that snaps onto a panel's rib or clip to complete the seam. Snap-together seams may have poor water resistance and should only be used for water-shedding systems since they are not weatherproof.

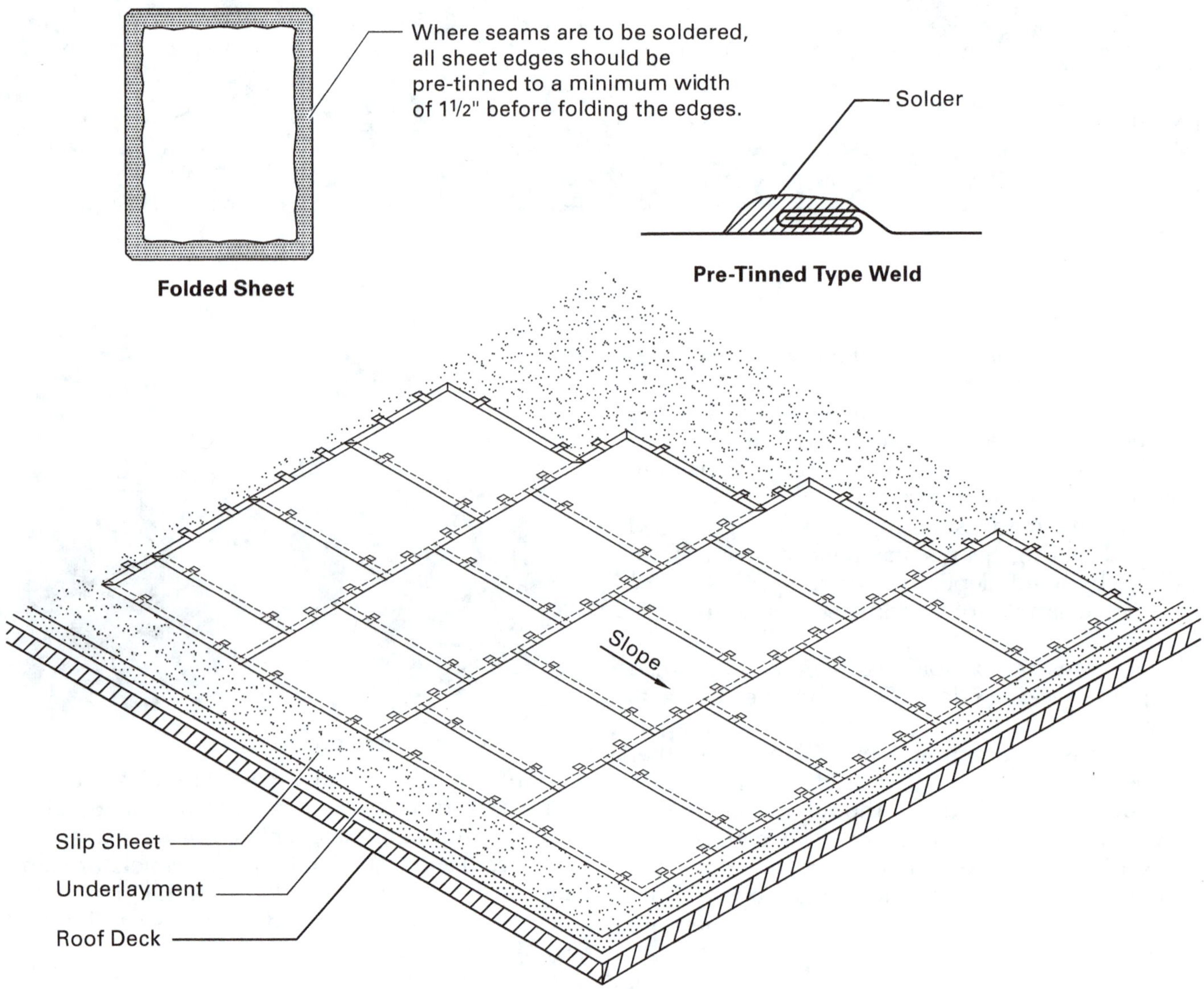

Figure 5 Flat-seam roof with soldered joints.

- *Hooked seam* — May be soldered or sealed depending on slope and manufacturer instructions. Hooked seams are used in some architectural metal systems and installed in shingle fashion to shed water. Systems that use hooked seams include Bermuda horizontal panels (*Figure 8*).

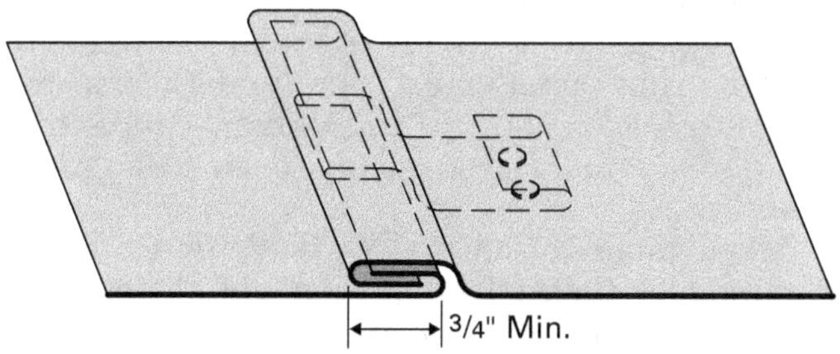

Figure 6 Flat-seam roof with nonsoldered, interlocking panels.

1.1.2 Attachment

Mechanical fasteners, such as roofing nails and screws, are usually used to secure metal roof systems. Adhesives, such as mortar and foam, do not allow for the expansion and contraction of metal and are not reliable as a means of attaching metal roofing materials.

Various loads must be considered in the weight of a roof system. *Dead loads* are weight from objects on the roof that are permanent, such as air conditioning equipment and roofing materials. *Live loads* come from temporary weight added to the structure, such as craftworkers working on the roof or snowfall. Mechanical fasteners must allow for the movement of metal panels as live loads come and go, as well as for thermal growth due to temperature changes.

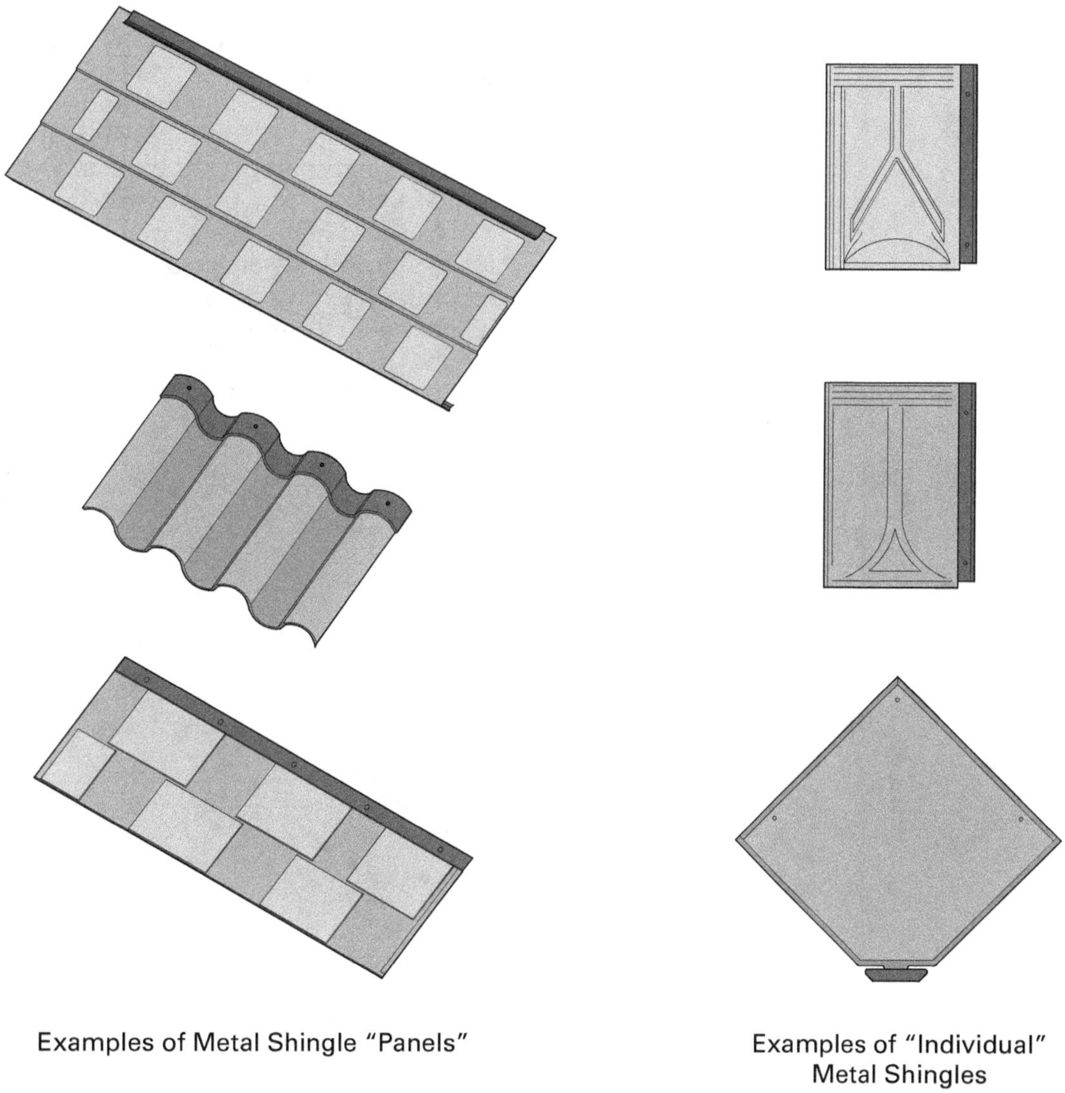

Figure 7 Examples of metal shingle designs.

Figure 8 Bermuda horizontal panels with hooked seams.

Metal roof systems should be laid out before they are installed to verify correct panel or shingle size and number. Using chalk lines can help ensure that panels are in their correct positions before they are attached to a roof substrate (*Figure 9*). Note the chalk line shown in *Figure 9*. Applying and following accurate chalk lines when laying out the roof is essential. They ensure that the roof panels are properly aligned to the structure and square to each other. Panels that are out of square are obvious even to a casual observer.

Metal panel systems use **fixed points** to accommodate thermal expansion and loads. You should not have more than one fixed point per panel. Fixed points of attachment are usually located at the hip and ridge or the eave and valleys of a roof (*Figure 10*). A fixed point represents an obstacle to thermal growth in its direction. Too many fixed points and the metal would have no choice but to buckle as it expands.

Metal shingles have various attachment methods, including screws, nails, expansion clips (*Figure 11*), interlocking tabs, and installation over batten systems. Always consult and follow the product manufacturer's instructions to ensure you are using the proper means to attach the shingles and have an acceptable substrate.

Mechanical fasteners and connections can be concealed, exposed, or interlocking. Concealed fasteners, such as screws, clips, and nail strips, can be hidden by special caps or coping over the seam, as shown in *Figure 12*. Exposed fasteners are through-fastened attachments that can be seen on the surface (*Figure 13*). Interlocking folds and edges include the use of continuous hook strips and cleats (*Figure 14*).

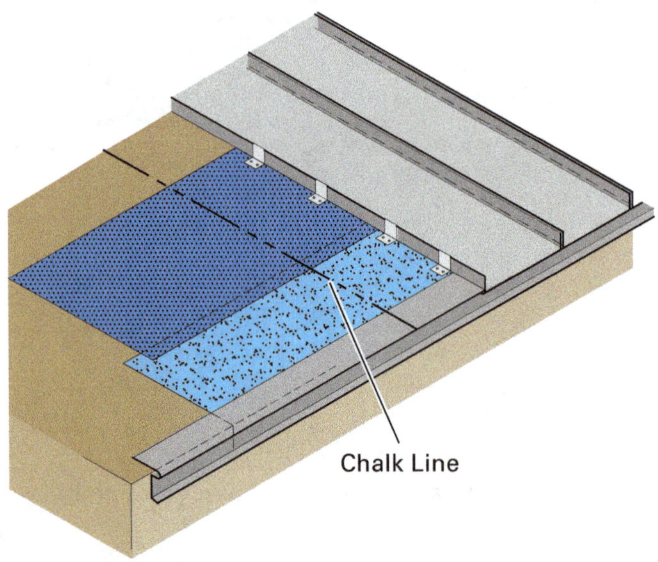

Figure 9 Panel layout before attachment.

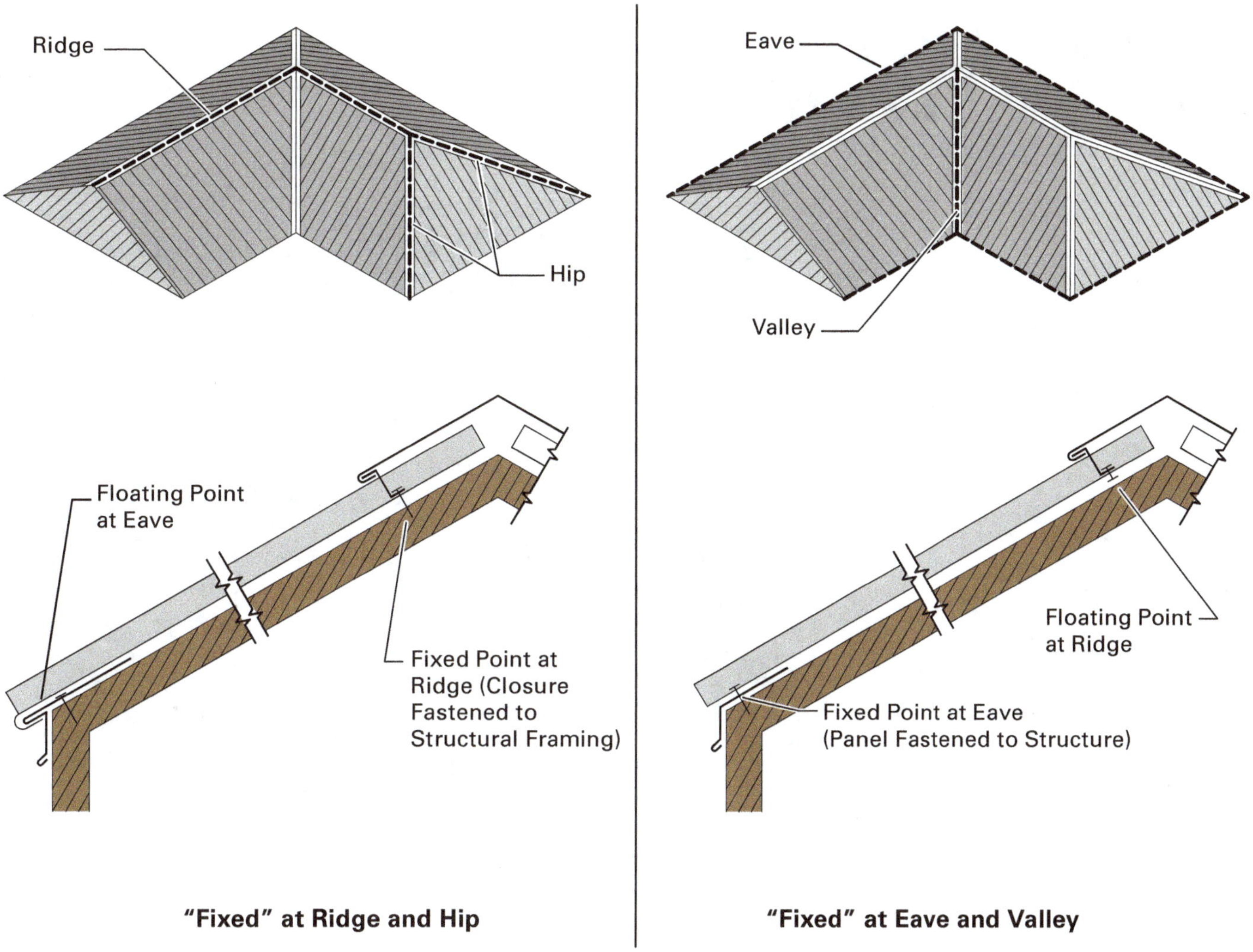

Figure 10 Possible fixed-point locations.

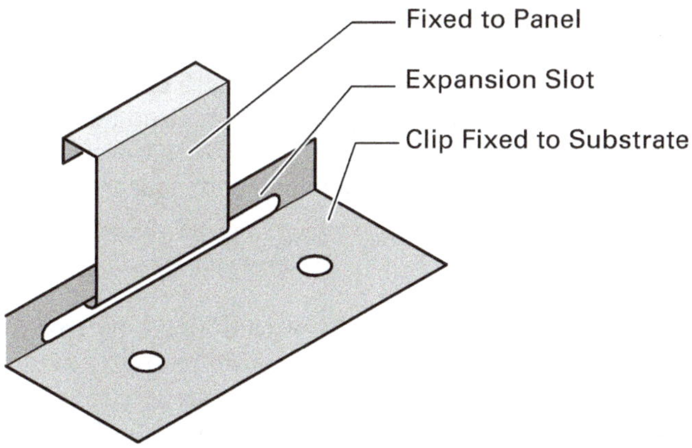

Figure 11 Expansion clip.

1.1.3 Hand and Power Tools

Roofing professionals must know how to safely utilize several common tools used with metal roofing materials.

> **NOTE**
>
> Tools used in metal roof installation are covered in greater detail in NCCER Module 00103, *Introduction to Hand Tools*; NCCER Module 00104, *Introduction to Power Tools*; NCCER Module 16107, *Sheet Metal in Roofing*; and NCCER Module 04102, *Sheet Metal Tools and Equipment*.

Figure 12 Roof attached with concealed fasteners.

Continuing Your Education as a Craft Professional

This curriculum cannot possibly teach you all the things you need to know about the many roof materials and applications you will encounter. At the end of a long and successful career, you will still be learning.

Throughout this module, trainees are directed to read and follow the manufacturer's instructions for a roofing product before installation begins. Some of the best training you can hope to receive is found in these documents. As you encounter new products, study all the available information, including marketing information. Understand the materials' strengths and weaknesses and ask questions when you need deeper understanding. This is one way to get ahead, become a resource to those around you, and move into a position of leadership in your chosen craft.

Tinner's Hammer

Tinner's hammers (*Figure 15*), also called *riveting hammers*, have been widely used in sheet metal work for generations. These tools have a square head with a slightly rounded face and beveled edges. They are ideal for closing seams and setting metal edges.

Rivet Gun

A rivet gun (*Figure 16*) is a tool used to install fasteners called *rivets*. Rivets come in many sizes, including those used with heavy structural steel and hammered to create a mushroom head. Small rivets used for roofing materials are often referred to as *pop rivets*, as the tool makes a single popping sound as a rivet is installed. Powered rivet guns may be pneumatic, cordless electric, or gas-powered using a fuel cell and ignition source.

Figure 13 Roof attached with exposed fasteners.

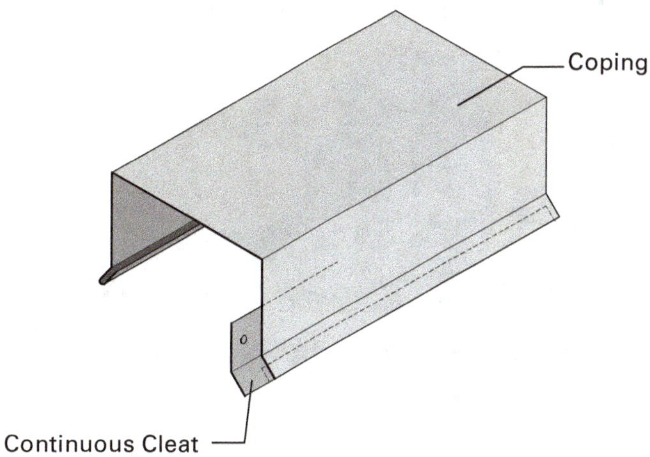

Figure 14 Continuous cleat.

Figure 15 Tinner's hammer.

(A) Rivet Gun

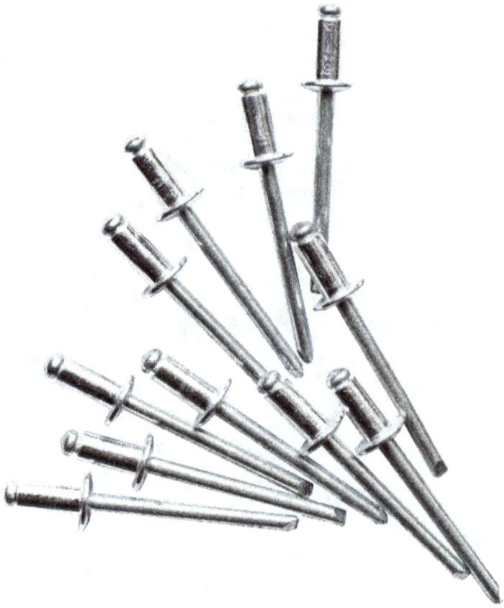

(B) Pop Rivets

Figure 16 Rivet gun and pop rivets.

Did You Know?

Rosie the Riveter

As scores of men joined the war effort in World War II, they left essential positions open in factories across the country. Industries that supported the war effort with their production were particularly needy. Factory workers, particularly in the aircraft industry, were desperately needed. Women were not yet permitted in combat roles, and although most were not yet accustomed to industrial work, they represented the only opportunity to rebuild a dwindling workforce.

Rosie the Riveter was the centerpiece of a marketing campaign to recruit women for defense-related industries. Aircraft skins were riveted in place, and thousands of rivets had to be installed to assemble a single aircraft, leading to the choice of Rosie's craft. The recruiting effort worked well, as the percentage of women in the workforce increased by nearly 10 percent in the span of five years. Without their entry into the workforce, the war, as well as life on the home front, may have gone in another direction altogether.

Although Rosie was essentially fictitious, different women have been named as the likely inspiration for this iconic representative of feminine strength and spirit.

Figure Credit: iStock@KingWu

Hand-operated rivet guns are fine for small tasks, but a power tool is much better for repetitive work.

Pop rivets are available in several diameters, such as $^{3}/_{32}$", $^{1}/_{8}$", $^{5}/_{32}$", and $^{3}/_{16}$". Like the rivet gun shown in *Figure 16*, most rivet guns come with an assortment of small chucks that fit different rivet diameters. To change chucks, you simply unscrew the size you need from the body of the tool and screw it into the tip.

To install a pop rivet, refer to *Figure 17* and follow these steps:

Step 1 Select a rivet diameter and length that is appropriate for the material thickness. A rivet that is too long will not properly secure the pieces. If it is too short, the expanded body of the installed rivet will not pass all the way through the components.

Step 2 Properly align the parts to be joined. To install the first few rivets, use locking pliers or a similar tool to hold them in position.

Step 3 Consult a manufacturer's chart or the rivet packaging to find the drill size needed. A tight fit is needed, but the body of the rivet must be able to pass through the hole. A numbered or lettered drill size is often listed. For example, a $^{1}/_{8}$" rivet requires a drill size of #30. Using the proper drill size, drill a hole through the parts to be joined.

Step 4 Insert the rivet mandrel into the tool, then push the body of the rivet through the hole.

Step 5 Actuate the tool by squeezing the handles together. The tool grips the mandrel and pulls the rounded end toward it, expanding the body to secure the two pieces.

Step 6 As the rivet tightens against the material, the load increases and the mandrel breaks free of the rivet. Collect the mandrels and dispose of them properly.

Guillotine Shears

Guillotine shears are often used in metal roofing to create precise cuts in metal panels. They may be foot powered, hand powered, or mechanically powered. These tools can also cut profiled metal shingles, including those with stone coatings that allow them to resemble similar materials.

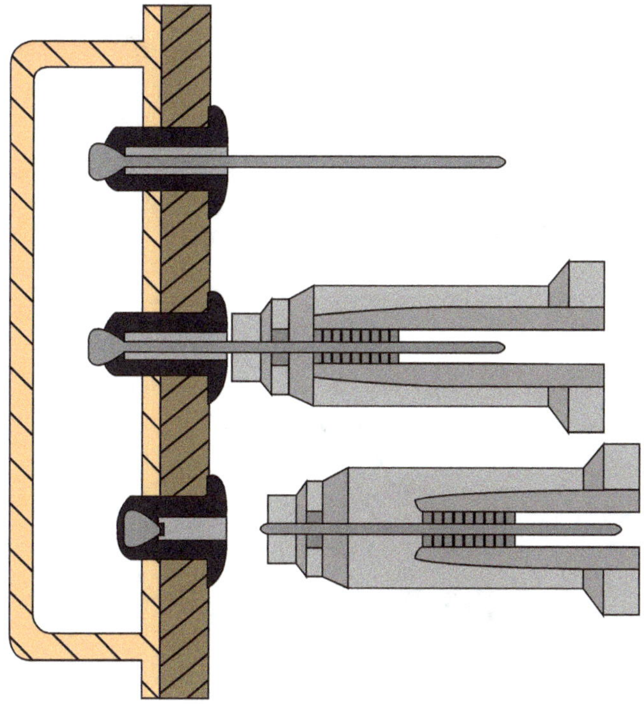

Figure 17 Installing a pop rivet.

They are sharp and very powerful—always be sure you are wearing the appropriate PPE before use. However, it is important to point out that gloves can't protect you from the power of a shear if your fingers are placed under the blade. Do not use a guillotine shear unless all the guards are in place and think carefully before actuating the tool to make a cut.

Soldering Irons

Although you may not be called upon to solder seams, it is helpful to be familiar with soldering tools. A tool commonly used in soldering sheet metal is the soldering iron (*Figure 18*). Because a traditional soldering iron has a copper head, it may be referred to as a *soldering copper*.

A traditional soldering iron is a steel rod with a handle on one end and a copper head on the other. The soldering iron performs two functions. It heats the metal to the melting point of the solder (but well below the melting point of the metal), and the heat carries molten solder along the seam to be joined.

Basic soldering irons are sold by head weight. More mass provides the ability to absorb and transfer more heat. The common range is from $^{1}/_{4}$ lb up to 3 lb. They are heated either in a portable furnace designed for soldering irons, or with a torch.

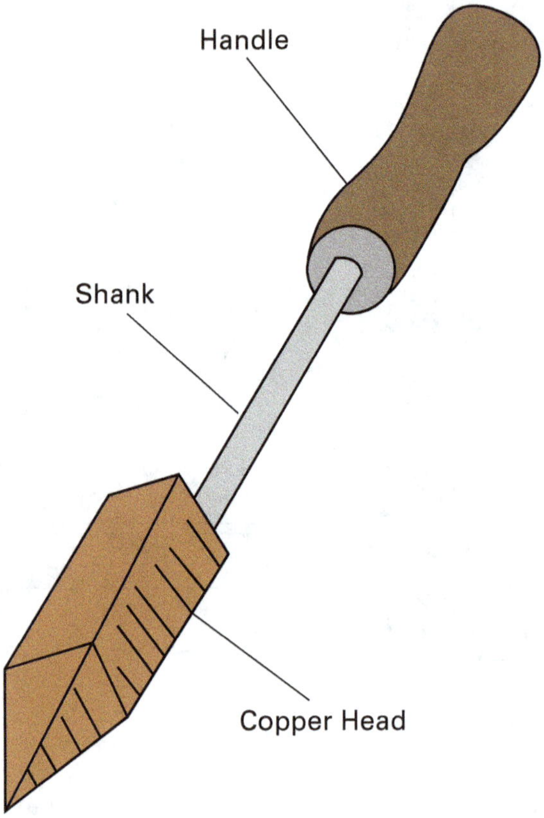

Figure 18 Basic soldering iron.

Self-contained soldering irons (*Figure 19*) have an integral torch. The flame from the torch heats the iron. This allows the soldering process to be continuous, instead of taking constant breaks to heat the iron. Soldering torch kits consist of a fuel gas cylinder; a hose to deliver the gas; a regulator to control the flow of gas; the burner head; and the soldering head. Gases used to fuel a soldering torch include oxyfuel, natural gas, and propane. The size of the flame is controlled by the gas flow regulator. These tools significantly enhance productivity.

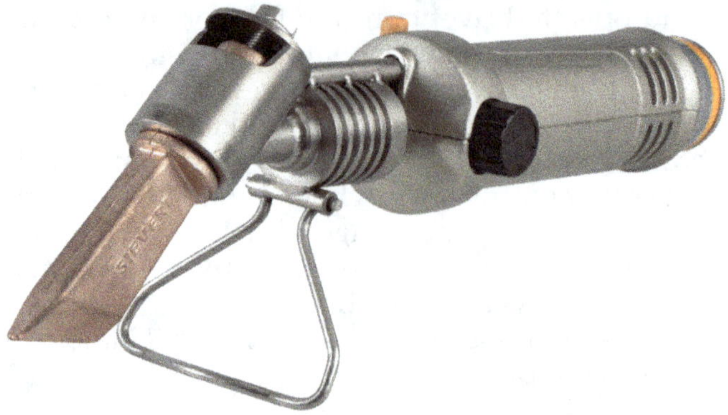

Figure 19 Self-contained soldering iron kit.

The torch flame is directed toward the soldering head, but it can heat large areas of metal quickly if the flame is too large. The high temperature of the flame may also burn the flux. Carefully control the flame to produce only the amount of heat needed to maintain a proper tip temperature. Avoid applying the flame to the workpiece itself.

Soldering does require some practice to control the iron and get the solder to move into and fill spaces between the components, forming an effective bond. Remember, though, that soldering isn't welding. Soldering does not "fuse" metal together as welding does, which requires the base metals to be melted. Welding is a fusion process, while soldering is a non-fusion bonding process.

WARNING!

Working with heated irons and open flames on a roof can be extremely hazardous. Wear the appropriate PPE, take all necessary precautions, and ensure that there are no flammable materials near the work area. Ensure that an appropriate fire extinguisher is nearby. Some jobsites may require prior permission to use an open flame, and a fire watch may be required.

1.0.0 Section Review

1. Live roof loads include things like snowfall and working roofing professionals.

   a. True
   b. False

SECTION TWO

2.0.0 METAL ROOF INSTALLATION

Objective

Explain how to install metal roofing systems.

a. Describe the installation process and flashing of architectural metal roofing systems.
b. Describe the installation process and flashing of structural metal roofing systems.
c. Describe the installation process and flashing of metal shingle systems.

Performance Tasks

1. Don all appropriate PPE for architectural metal panel installation.
2. Install manufacturer-recommended underlayment and any necessary base flashings, valley metal, and perimeter metal to a wood deck.
3. Properly measure, cut, clip, fit, and install architectural metal panels according to manufacturer instructions.
4. Complete any remaining flashings, such as penetration, wall, and head flashings.

Trade Term

Galvanic corrosion: A form of corrosion (rusting) resulting from the creation of an electric current flowing between dissimilar metals in contact with each other.

The most common metal roofing materials include architectural metal panels, structural metal panels, and metal shingles. This section overviews some features of these systems and provides general information on how they are installed.

There are two things to consider during the installation that affect all types of metal roofing. All metals naturally expand and contract as the temperature changes. Thermal movement must be considered throughout a metal roof installation. Metal roofs are also subject to corrosion. It is crucial that the various metals used in a single roof system are compatible with one another to reduce the potential for galvanic corrosion and structural damage.

NOTE

Properties of metal that affect roofing are presented in NCCER Module 16107, *Sheet Metal in Roofing.*

2.1.0 Architectural Metal Roofs

Architectural metal roofs (*Figure 20*) are water-shedding systems that are most effective on steep slopes. For lower slopes, climactic conditions such as snow, ice, and debris accumulation must be carefully considered, and it may be necessary to use a sealant on all seams. Live loads like snow are often too much for architectural roofing in a low-slope application.

Architectural metal systems must be installed over a continuous or closely spaced roof deck to ensure adequate support. The panel may be made from these metals or alloys that combine several metals:

- Copper
- Terne
- Aluminum
- Stainless Steel
- Zinc
- Tin
- Lead

Architectural roof panels are constructed of metal, 26 gauge or thinner. Metal panels classified as structural panels (24 gauge or thicker) may also be used in architectural applications as long as they are installed over a continuous or closely spaced roof deck.

CAUTION

Roofing professionals must carefully install metal roof systems and know when to fasten system components either directly to the deck or to other parts of the system. Always consult the manufacturer's instructions as well as local building codes to be sure you are correctly installing each system.

The following steps provide a basic overview of the architectural metal roof installation process:

Step 1 Don all proper PPE for the job. Also ensure that the appropriate fall protection equipment is used or in place. Depending on the project and environment, this may be a personal fall arrest system (PFAS) with approved anchorage, or appropriate guardrails or scaffolding.

Figure 20 Steep-slope architectural metal roof.

Step 2 Be sure that the roof deck is clean, dry, and in good condition.

Step 3 Apply the manufacturer-recommended underlayment to the roof deck in compliance with underlayment instructions and local building codes.

Step 4 Lay out the panels using a tape measure and chalk line to help with alignment. Decide the best placement for panels that considers the manufacturer's instructions, local building codes, and the flashings and penetrations you must accommodate. Know where your starting point is and where your ending seam will be.

Step 5 Apply any necessary base flashings, valley metal, and perimeter metal. Note that any flashing that must go beneath the panels must be installed first, while some flashing is completed following panel installation. Manufacturers may use different terms for various flashing applications, so it is important to follow their drawings and instructions.

Step 6 Measure, cut, fit, and clip each panel according to the manufacturer's specifications and layout positioning. You will need to cut some panels to fit using a guillotine shear or other tools. Compound or aviation snips are used for intricate work and small pieces of flashing.

WARNING!

Always watch your step while installing metal roofing. Most panels have paint or a similar surface finish. Careless footsteps can buckle or mar the appearance of the panels. Many manufacturers apply a plastic coating or a release agent to panels for temporary surface protection. If present, leave these protective coatings or liners in place until there is no longer foot traffic on the roof. Also ensure that you are wearing appropriate footwear that protects both you and the roof surface.

Step 7 After you have installed the panels, complete any remaining flashings, such as penetration, wall, and head flashings. Apply hip and ridge capping.

Step 8 Apply finishes to your system by adding any necessary caulk or sealant and using manufacturer-provided touch-up paint on minor scuffs and scratches.

Step 9 Once foot traffic on the roof is no longer a factor, remove the protective plastic coating or release liners from each panel.

Step 10 Clean up the jobsite to ensure it is safe and tidy.

CAUTION

Throughout the roof installation process, take care of the materials you handle. Be sure any chemicals used during system installation are handled safely and do not cause discoloration or other appearance issues in the system. The final appearance of a metal roof system is very important to the building owner, and your work should reflect your commitment to delivering high-quality results.

2.1.1 Flashings

Metal panel systems use flashings to prevent water from leaking through penetrations and interruptions in the assembly. The gauge of flashing metal should be equal to or heavier than the metal used in the roof panels to ensure system integrity.

Eave and rake flashings (*Figure 21*), such as a T-type drip edge, may serve as a securement point for a roof panel. These perimeter flashings are typically installed under metal panels and extend down the face of a wall.

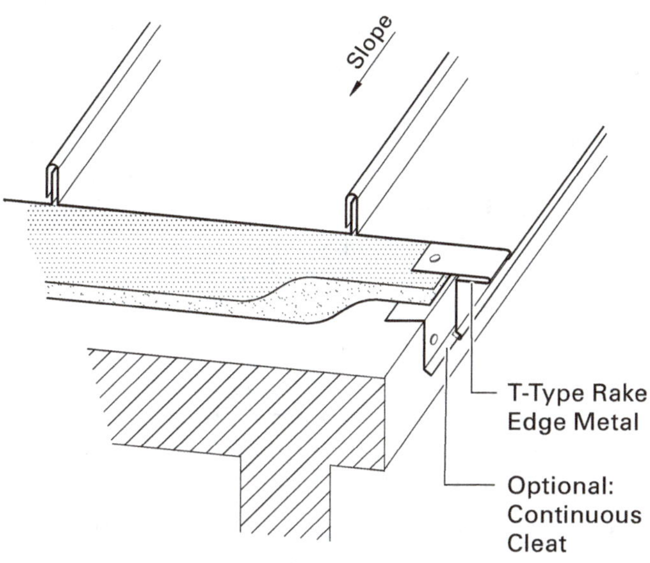

Figure 21 Example of a rake edge.

Metal valley-flashing installation must consider how laps are joined to resist loads and effectively shed water from the system. A raised center rib in the valley can help control water runoff as shown in *Figure 22*. If clips are used to attach a valley, each clip must be able to accept two fasteners, as shown at the bottom of *Figure 22*. Requirements for valley installation, however, vary from one system to another. Always consult and follow the product manufacturer's instructions.

Small penetration flashings, such as pipes and vents, are best centered within a panel to make them easier to flash (*Figure 23*). However, the position of the panel cannot always coincide with the location of the penetration. While it is not ideal, penetrations through seams will occur and must be accommodated with the proper flashing materials and techniques.

Penetration flashings are often prefabricated and sent to a jobsite. Be sure to use flashing fasteners and materials that allow for slight movement of the roof and penetrating component due to thermal expansion.

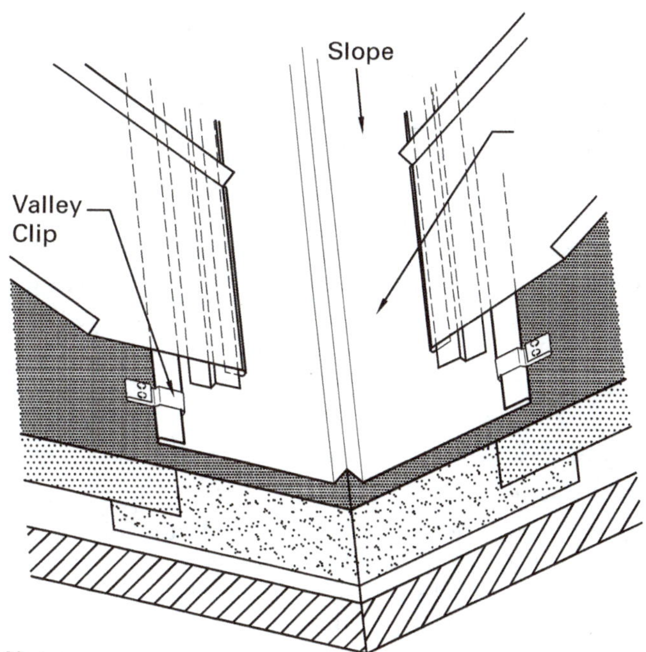

Note:

1. This detail is for solderable metals only.
2. Minimum roof slope for this valley construction is 3:12.
3. Dimensions for valley metal width may vary according to the panel lengths and geographic considerations.

Figure 22 Typical valley profile for architectural metal roofing.

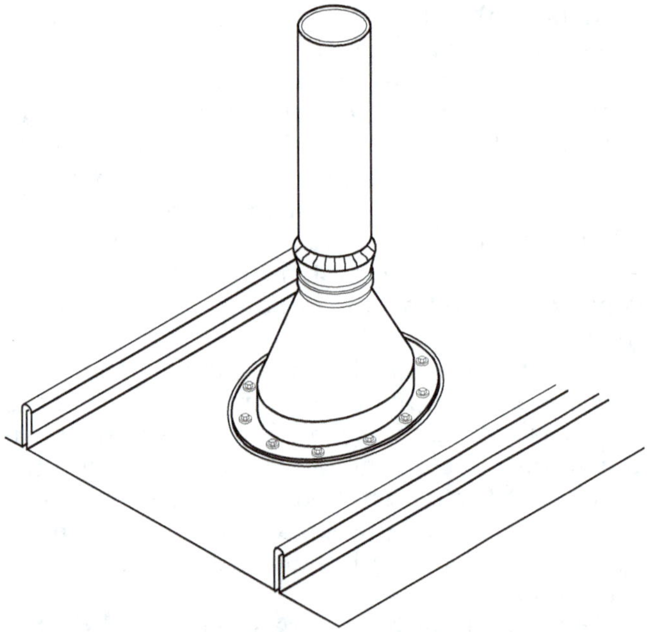

Figure 23 Premanufactured pipe flashing centered on panel.

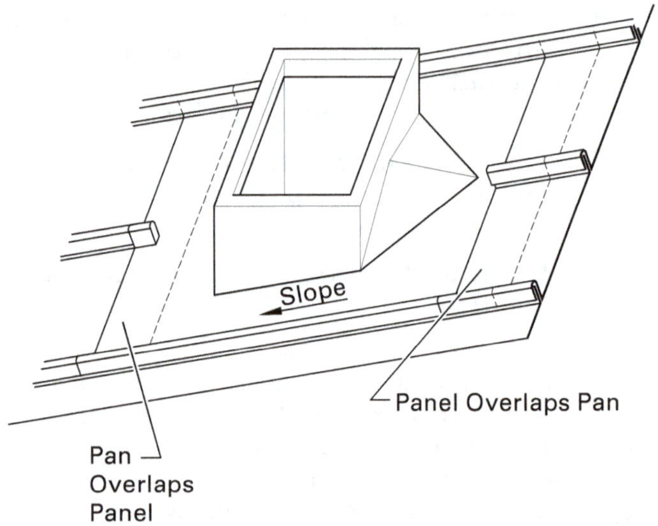

Figure 24 HVAC unit curb on a sloped roof.

Larger penetrations, such as those required for HVAC units, require the use of a curb, sleeve, or flat sheet without seams located in the flashing area (*Figure 24*). When installing larger roof curbs, be sure any approaching panel ribs stop short of the curb to allow room for water to run around and down the roof. Equipment curbs can be manufactured to fit the spacing and standing seams of a specific roofing product for a neat, easily flashed installation (*Figure 25*).

Flash hips and ridges to provide a closure for the top edges of metal panels. Metal caps and panels should be properly installed together to allow for water runoff. To help with water-shedding, roofing professionals may turn panel pans up at hips and ridges to provide a secondary closure beneath the caps (*Figure 26*).

2.2.0 Structural Metal Roofs

Structural metal roofs (*Figure 27*) have the strength to run across joists of purlins without being supported by a continuous or closely spaced roof deck, even in low-slope applications. They are generally weatherproof systems that do not require an underlayment. Although structural metal panels may be used in steep-slope applications, architectural metal panels cannot be used in low-slope applications due to their reduced strength.

Structural metal panels are made from metals such as steel, stainless steel, and aluminum. They get their strength from the heavier metal gauge as well as standing seams designed for greater rigidity.

Insulation beneath the panels is often required, depending on the structure. Building codes and engineers determine what kind and amount of insulation is necessary. Structural metal roofs frequently have fiberglass blanket insulation with a plastic facer installed between the purlins (*Figure 28*).

Figure 29 shows some common profiles found in structural metal panels. Stiffening ribs, each constructed with multiple bends, make the metal more rigid, reduce distortion in the surface, and help prevent *oil canning*.

> **NOTE**
> Oil canning is discussed in greater detail in NCCER Module 16107, *Sheet Metal in Roofing*.

The following steps provide an overview of a low-slope, structural metal roof installation process:

Step 1 Don all proper PPE for the job. Also ensure that the appropriate fall protection equipment is used or in place. Depending on the project and environment, this may be a personal fall arrest system (PFAS) with approved anchorage, or appropriate guardrails or scaffolding.

Step 2 Lay out and install insulation on top of battens or between purlins in compliance with manufacturer instructions and local building codes.

Step 3 Map your layout before you begin the installation to be sure you have correct panel spacing. Use chalk lines to ensure panels are square to each other and the structure.

5V Crimped Panels

5V crimped panels have been used in steep-slope applications for many years, although they are not typically used in areas with extreme weather or heavy snow loads. Unlike other architectural panels, 5V panels have no standing seam. This makes them weaker under load, but they can be installed very quickly.

The name comes from the profile of the panel as viewed from the edge—five inverted Vs, with an effective width of 24" (~60 cm). Each V is ½" (12.7 mm) tall. The fasteners remain exposed and are driven through the panel at the peak of each V. Fasteners may also be driven in the center of the flats for added security. Overtightening the screws does tend to distort the panel, so care must be used when driving the fasteners.

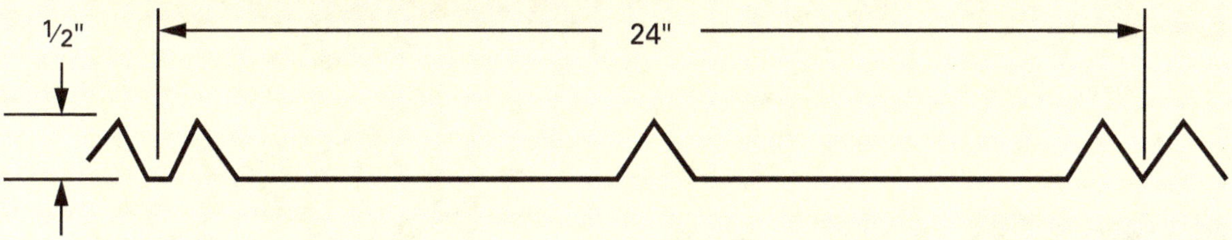

Step 4 Begin installing panels from one direction (either left-to-right or right-to-left). Work your way across the rows as you measure, cut, fit, and attach each panel. Be sure you are correctly anchoring each panel to allow for thermal movement as the manufacturer directs.

Step 5 Install any necessary flashings.

Step 6 Neatly apply sealants to your system as necessary. Apply manufacturer-provided touch-up paint to scratches or other minor defects.

Step 7 Once foot traffic is no longer necessary, remove any protective plastic coating or release liners from the panels.

Step 8 Clean up the jobsite to ensure it is safe and tidy.

Figure 25 Exhaust fan curb custom-fabricated for a standing-seam roof.

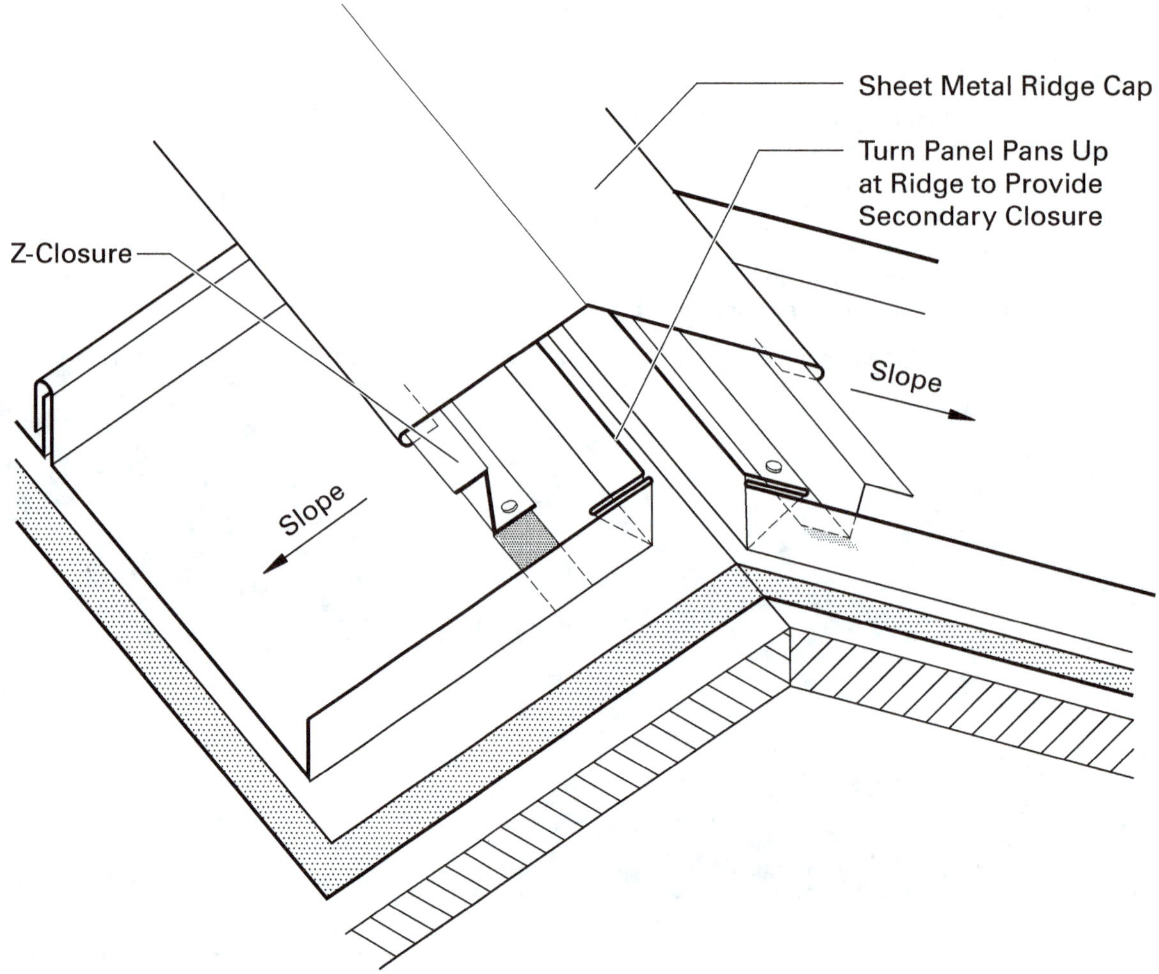

Figure 26 Ridge cap installed over panel ends.

Figure 27 Structural metal roof.

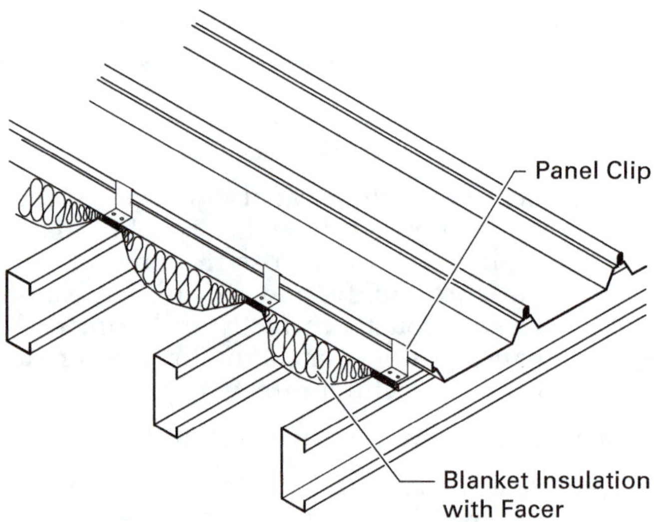

Figure 28 Blanket insulation installed between purlins.

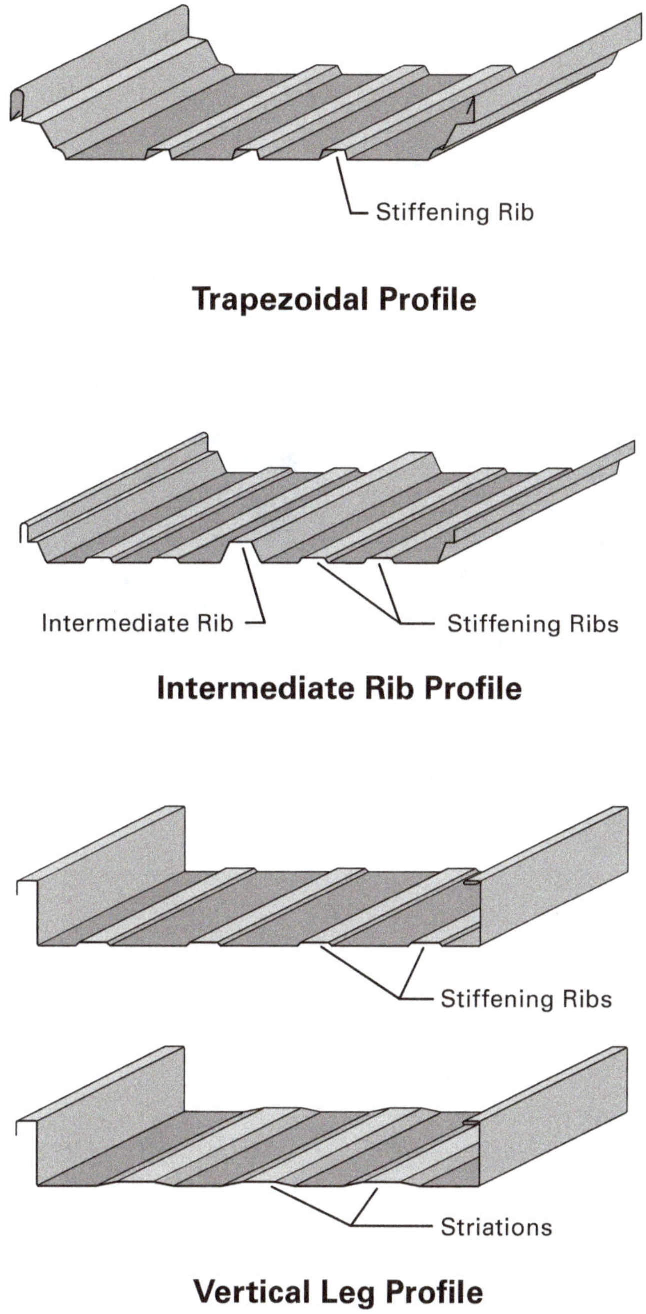

Figure 29 Structural metal panel profiles.

2.3.0 Metal Shingle Roofs

Metal shingles are overlapping, water-shedding components used on steep-slope roofs (*Figure 30*). Metal shingles are a lightweight alternative to asphalt shingles, slate, clay and concrete tile, and wood. They can also mimic the color and texture of these materials. As shown in *Figure 30*, various clay tile colors are popular. The coatings are also protective as well as appealing. Metal shingles may be made from aluminum, copper, stainless steel, galvanized steel, or zinc.

Figure 30 Metal shingle roof.

Metal shingles are primarily installed on a continuous substrate. In some cases, they may be secured to battens placed beneath them. They may be secured using screws, nails, clips, or tabs. Always consult the manufacturer's instructions to ensure you are using the correct attachment method for the product.

Metal shingles also come in a wide variety of sizes. Some are designed for interlocking with each other, across the top and possibly the ends.

A metal shingle system is usually placed using an offset pattern recommended by its manufacturer. The following steps provide an overview of the installation process:

Step 1 Don all proper PPE for the job. Also ensure that the appropriate fall protection equipment is used or in place. Depending on the project and environment, this may be a personal fall arrest system (PFAS) with approved anchorage, or appropriate guardrails or scaffolding.

Step 2 Be sure that the roof deck is clean, dry, and in good condition.

Step 3 Apply the manufacturer-recommended underlayment to the roof deck in compliance with underlayment instructions and local building codes.

Step 4 Map your layout before you begin the installation to be sure it is horizontally accurate. Use chalk lines and a measuring tape to ensure proper alignment.

Step 5 Apply any necessary base flashings, such as drip-edge and valley metal. Most manufacturers provide eave metal and cleat to anchor the starter courses.

Step 6 Measure, cut, fit, and fasten your metal shingles. Interlock the shingles together as needed before you fasten them. Be sure you are correctly fastening each panel to allow for thermal movement, although thermal growth is not as pronounced as it may be when working with larger metal panels.

Step 7 After you have installed the metal shingles, complete any remaining flashings, such as penetration, wall, head, and base flashings. Apply the hip and ridge caps.

Step 8 Neatly apply sealants to your system as necessary. Apply manufacturer-provided touch-up paint to scratches or other minor defects.

Step 9 Once foot traffic is no longer necessary, remove any protective plastic coating or release liners from the panels.

Step 10 Clean up the jobsite to ensure it is safe and tidy.

2.3.1 Flashings

Like most roofing systems, metal shingle systems use flashings to prevent water leakage. Drip-edge metal should be applied to the roof's perimeter and secured directly to the deck at the eaves. Depending on the system, metal shingles may be designed to clip or lock onto drip-edge flashings.

Center underlayment in the valley (*Figure 31*) before installing valley metal. Valley metal should be installed from the low point to the high point of the valley, so that the upper pieces overlap the lower ones.

With the valley metal in place, metal shingles are then lapped onto both sides of it, typically leaving appropriate space in the center for unobstructed water runoff. Manufacturers supply preformed valley metal for metal shingle systems.

Penetration flashings, such as vent pipes, must also be installed. An example of a metal flashing sleeve commonly used in metal shingle roofing is shown in *Figure 32*.

The following are four types of flashings used at locations where metal shingles intersect a vertical surface:

- *Apron flashings* — Weatherproof the area where the roof meets a headwall.
- *Channel flashings or step flashings* — Applied for metal shingle systems that intersect a vertical sidewall (*Figure 33*).

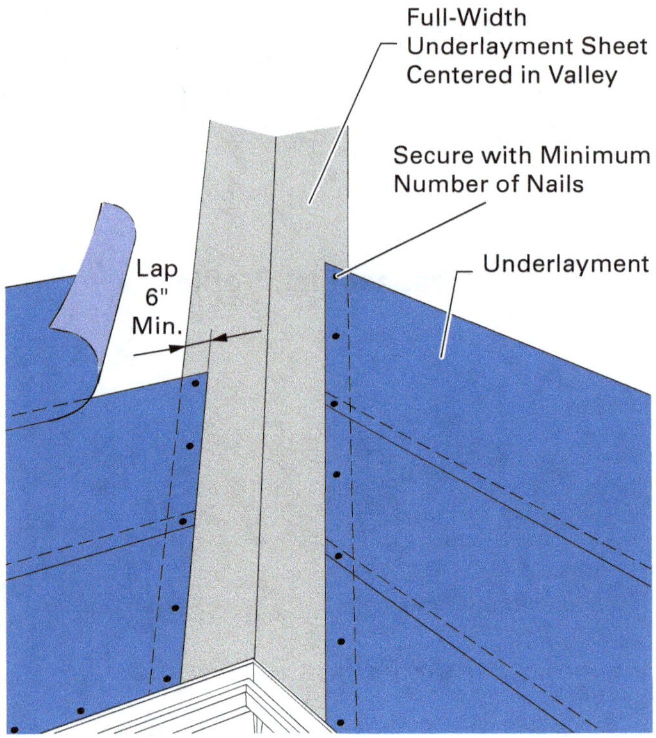

Figure 31 Underlayment centered in valley.

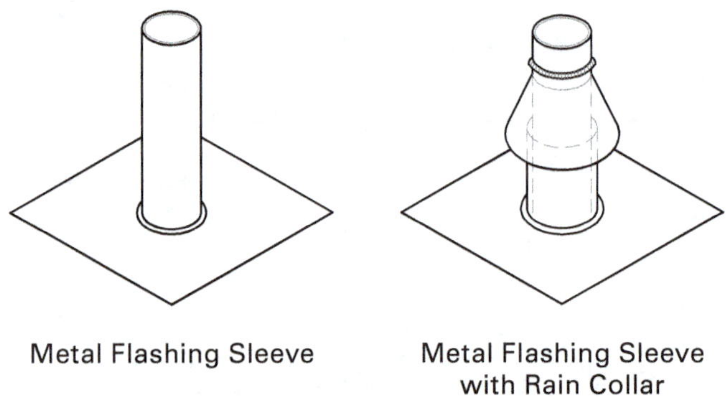

Figure 32 Penetration flashing.

- *Backer flashings* (*Figure 34*) *or cricket flashings* (*Figure 35*) — Should be installed where a metal shingle roof intersects the upslope side of a vertical roof penetration. Backer flashings are used on penetrations that are 24" (61 cm) wide or less, and cricket flashings are used on penetrations greater than 24".
- *Counterflashing* — Should be applied to base flashings, such as apron, step, channel, cricket, and backer flashings, to protect the base flashings' top edges from water intrusion (*Figure 36*).

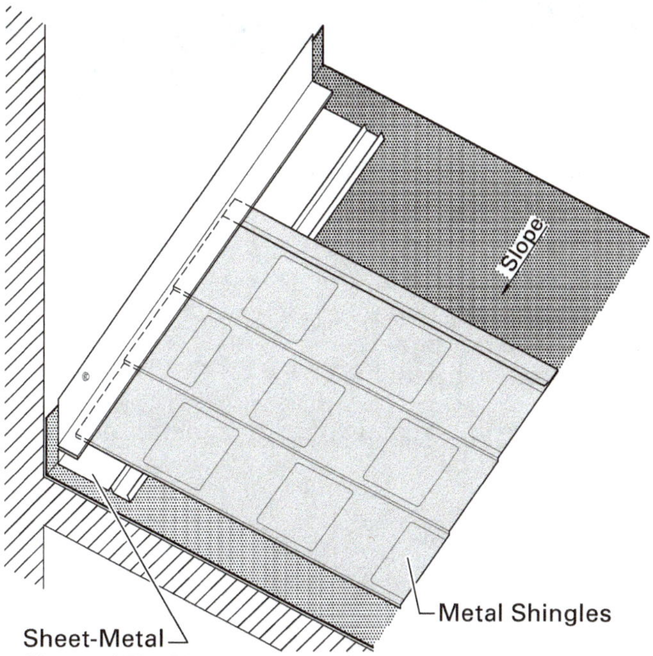

Figure 33 Metal-shingle channel flashing.

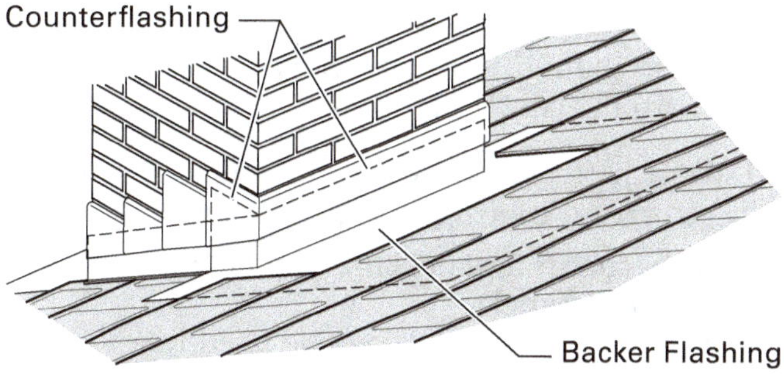

Figure 34 Backer flashing.

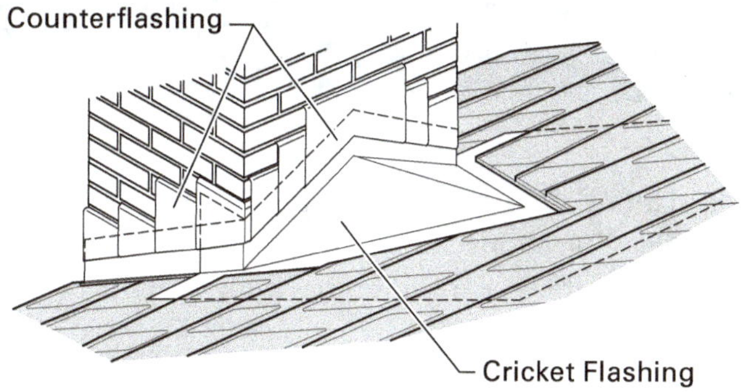

Figure 35 Cricket flashing.

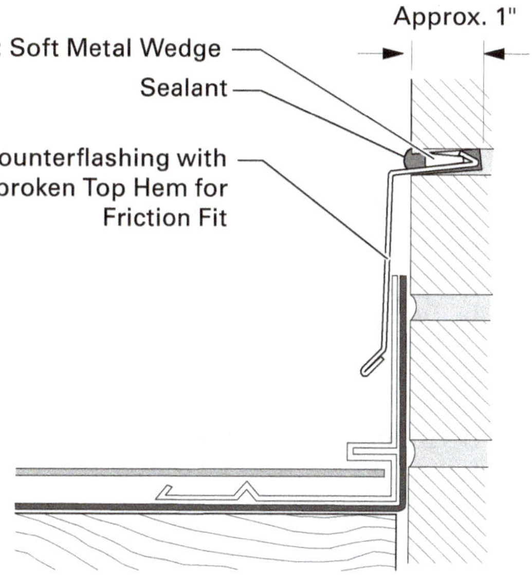

Figure 36 Counterflashing.

2.0.0 Section Review

1. The gauge of flashing metal should _____.
   a. always be exactly the same as the roof panels
   b. be equal to or lighter than the roof panels
   c. be equal to or heavier than the roof panels
   d. be one gauge lighter than the roof panels

2. Structural metal roofs can only be installed over a continuous or closely spaced roof deck.
   a. True
   b. False

3. Metal shingles can sometimes mimic the color and texture of asphalt shingles.
   a. True
   b. False

Review Questions

1. Underlayments used with metal roofing systems should _____.
   a. be thicker than required for other types of roofing
   b. be placed on top of flashings instead of under them
   c. meet the requirements as a high-temperature product
   d. tailor each panel with snips

2. Bonding metal joints with a molten mixture of metals such as lead and tin is called _____.
   a. soldering
   b. interlocking
   c. welding
   d. adhering

3. Which of the following is a category of snap-together seams?
   a. Mechanical seams
   b. Snap-back seams
   c. Snap-lock seams
   d. Hooked seams

4. Bermuda horizontal panels use which of the following seam types?
   a. Mechanical seams
   b. Snap-on seams
   c. Snap-together seams
   d. Hooked seams

5. Interlocking folds and edges include the use of continuous hook strips and cleats.
   a. True
   b. False

6. Fixed points that help accommodate thermal expansion and loads in metal panels should be limited to _____.
   a. six per panel
   b. four per panel
   c. two per panel
   d. one per panel

7. The various metals used in a metal roof system *must* be _____.
   a. identical
   b. compatible
   c. dissimilar
   d. painted

8. In architectural and structural metal roof systems, small penetration flashings such as vent pipes are *best* located _____.
   a. in panel seams
   b. to the left side of a panel
   c. to the right side of a panel
   d. in the center of a panel

9. Which of the following statements about structural roofing panels is *correct*?
   a. Structural panels are identical to architectural panels.
   b. Structural panels can be used in low-slope systems.
   c. Structural panels can only be installed over a continuous substrate.
   d. Structural panels are available in only one profile.

10. Metal shingle roofs *must* always be _____.
    a. water-shedding
    b. flexible
    c. insulated
    d. low-slope

Trade Terms Introduced in This Module

Architectural metal roofs: A metal panel system constructed of 26-gauge or thinner metals used in steep-slope applications that require a continuous or closely spaced substrate underneath the system for support.

Fixed points: Points of attachment on steep-slope metal roof systems that secure the system and represent an obstacle to the natural expansion and contraction of metal. A fixed point for a panel is usually located at a roof's ridges or hips, or at the eaves or valleys.

Flux: A chemical substance that helps remove and prevents the formation of oxides on metal surfaces as they are heated for soldering, and acts as a vehicle to help solder flow more readily.

Galvanic corrosion: A form of corrosion (rusting) resulting from the creation of an electric current flowing between dissimilar metals in contact with each other.

Metal shingle roofs: Steep-slope metal roof systems installed in overlapping courses like traditional shingles.

Solder: A metal alloy often containing tin and lead that is melted using a hot iron to seal seams and join metals.

Soldering: A non-fusion process used to bond metal objects together with solder. See *solder.*

Standing-seam roofs: Metal roofs with a panel seam made by turning up the edges of adjacent metal panels and interlocking them in a variety of ways. Common to both architectural and structural metal roofs.

Structural metal roofs: A metal panel system constructed of 24-gauge or thicker metals used primarily in low-slope applications but also for steep-slope applications.

Additional Resources

This module presents thorough resources for task training. The following reference material is suggested for further study.

National Roofing Contractors Association (NRCA), **www.nrca.net**.

NCCER Module 00103, *Introduction to Hand Tools*.

NCCER Module 00104, *Introduction to Power Tools*.

NCCER Module 16105, *Introduction to Steep-Slope Roofing*.

NCCER Module 16106, *Introduction to Low-Slope Roofing*.

NCCER Module 16107, *Sheet Metal in Roofing*.

NCCER Module 75901, *Fall Protection Orientation*.

Occupational Safety and Health Administration (OSHA), US Department of Labor, **www.osha.gov**.

Sheet Metal and Air Conditioning Contractors' National Association (SMACNA), **www.smacna.org**.

Figure Credits

iStock@titine974, Module Opener

iStock@Nicola Forenza, Figure 1

iStock@nejdetduzen, Figure 2

Youngstown Glove Company, Figure 3

Courtesy of the National Roofing Contractors Assoc., Figures 4–11, 14, 21–24, 26, 28–29, 31–36

iStock@Jesse James metal roofing, Figure 12

iStock@StockSolutions, Figure 13

Courtesy of Malco Tools, Figure 15

iStock@Mo Semsem, Figure 16 (A)

iStock@Steven Giles, Figure 16 (B)

Sievert Industries, Inc., Figure 19

iStock@Bilanol, Figure 20

iStock@SUMITH NUNKHAM, Figure 27

iStock@Vitaliy Halenov, Figure 30

Section Review Answer Key

Section 1.0.0

Answer	Section Reference	Objective
1. a	1.1.2	1a

Section 2.0.0

Answer	Section Reference	Objective
1. c	2.1.1	2a
2. b	2.2.0	2b
3. a	2.3.0	2c

User Update

NCCER makes every effort to keep its textbooks up-to-date and free of technical errors. We appreciate your help in this process. If you find an error, a typographical mistake, or an inaccuracy in NCCER's curricula, please submit a User Update form by visiting **https://www.nccer.org/olf**. You can also scan the QR code using the camera on your phone or mobile device to access the form.

Thermoplastic Roof Systems

Overview

Thermoplastic roof systems are single-ply systems commonly found on commercial, low-slope roofs. Installation of thermoplastic roof systems is different for every job. Roofing professionals must be familiar with the variety of common thermoplastic field membranes, flashing accessories, and attachment methods—and understand how those components come together during installation or repair—to create effective and long-lasting weatherproofing for the building below.

Module 16206

Trainees with successful module completions may be eligible for credentialing through the NCCER Registry. To learn more, go to **www.nccer.org** or contact us at 1.888.622.3720. Our website, **www.nccer.org**, has information on the latest product releases and training.

Your feedback is welcome. You may email your comments to **curriculum@nccer.org**, send general comments and inquiries to **info@nccer.org**, or fill in the User Update form at the back of this module.

This information is general in nature and intended for training purposes only. Actual performance of activities described in this manual requires compliance with all applicable operating, service, maintenance, and safety procedures under the direction of qualified personnel. References in this manual to patented or proprietary devices do not constitute a recommendation of their use.

16206 V1.0

16206
Thermoplastic Roof Systems

Objectives

Successful completion of this module prepares you to do the following:

1. Identify types of thermoplastic roof systems and their basic components.
   a. Identify types of thermoplastic field membranes.
   b. List and describe pre-manufactured flashing accessories.
   c. Identify components used for attachment.
2. Explain how thermoplastic roof systems are installed and repaired.
   a. Identify tools and equipment used to install thermoplastic roof systems.
   b. Describe how thermoplastic roof systems are installed.
   c. Explain basic repair methods for damaged thermoplastic roof systems.

Performance Tasks

Under supervision, you should be able to do the following:

1. Install a fastener and membrane plate into a substrate.
2. Clean a membrane lap to prepare for seaming.
3. Use a hand welder to create a seam.

Trade Terms

Application rate
Ballast
Fasteners
Fishmouth
Flashings
Fleece-backing
Hot-air welding
Probing
PVC
Seam
Self-adhering membranes
Thermoplastic
T-joint
TPO
Weighted roller

Industry Recognized Credentials

If you are training through an NCCER-accredited sponsor, you may be eligible for credentials from NCCER's Registry. The ID number for this module is 16206. Note that this module may have been used in other NCCER curricula and may apply to other level completions. Contact NCCER's Registry at 1.888.622.3720 or go to **www.nccer.org** for more information.

You can also show off your industry-recognized credentials online with NCCER's digital badges. Transform your knowledge, skills, and achievements into badges that you can share across social media platforms, send to your network, and add to your resume. For more information, visit **www.nccer.org**.

Videos and Digital Resources

How to Access Resources

This craft has additional videos and resources to enhance your learning experience. To view these resources, scan the QR below. The videos and resources are separated by module.

You can scan this code using the camera on your phone or mobile device to view these videos and resources.

Contents

1.0.0 Thermoplastic System Types and Components 1
1.1.0 Membrane Types 1
1.1.1 Standard Membranes 1
1.1.2 Fleece-Back Membranes 2
1.1.3 Self-Adhered Membranes 2
1.2.0 Flashing Accessories 3
1.3.0 Attachment Components 3
1.3.1 Adhesives 3
1.3.2 Fasteners 4
1.3.3 Induction Plates 4
1.3.4 Ballast 4
2.0.0 Installation and Repair 7
2.1.0 Tools and Equipment 7
2.1.1 Hand Tools 7
2.1.2 Power Tools 8
2.2.0 Installation Procedures 10
2.2.1 Adhered Installation 10
2.2.2 Mechanically Attached Installation 12
2.2.3 Induction Welded Installation 13
2.2.4 Ballasted Installation 15
2.3.0 Basic Repair Methods 16
2.3.1 Repairing a Puncture 16
2.3.2 Repairing a Cut 16
2.3.3 Repairing a Fishmouth 17
2.3.4 Unadhered Areas 17

Figures

Figure 1 Thermoplastic roof system 2
Figure 2 Membrane with fleece-backing 2
Figure 3 T-joint cover 3
Figure 4 Adhesive applied to a TPO roof system by spraying 4
Figure 5 Fastener and membrane plate for mechanical attachment 5
Figure 6 Common induction plates 5
Figure 7 Ballasted roof system with river rock and vegetation 5
Figure 8 Weighted roller 8
Figure 9 Push broom 8
Figure 10 Seam probes 9
Figure 11 Heat welding a seam 9
Figure 12 Robotic welder 10
Figure 13 General procedure for adhering a thermoplastic roof membrane (excludes flashings and penetrations) 11
Figure 14 Mechanical attachment of roof membrane 12
Figure 15 Induction welded installation 14
Figure 16 Membrane layout 14
Figure 17 Ballasted roof system 15
Figure 18 Fishmouth 17

Section One

1.0.0 Thermoplastic System Types and Components

Objective

Identify types of thermoplastic roof systems and their basic components.

a. Identify types of thermoplastic field membranes.
b. List and describe pre-manufactured flashing accessories.
c. Identify components used for attachment.

Trade Terms

Ballast: A material—such as smooth river stone, crushed stone, standard precast concrete pavers or interlocking, beveled, doweled, or contoured fit lightweight concrete pavers—that weighs down a membrane to hold it in place.

Fasteners: Any of a wide variety of mechanical securement devices, including nails, staples, screws, cleats, clips, and bolts.

Flashings: Components used to weatherproof or seal edges of a roof system at perimeters, penetrations, walls, expansion joints, valleys, drains, and other places where the roof covering is interrupted or terminated. For example, membrane base flashing covers the edge of the field membrane, and cap flashings or counterflashings shield the upper edges of the base flashing.

Fleece-backing: A nonwoven polyester material adhered to the underside of a thermoplastic sheet for reinforcement.

PVC: Polyvinyl chloride, a common type of thermoplastic used in roofing systems.

Self-adhering membranes: Membranes that can bond to both a substrate and to themselves at overlaps without the use of an additional adhesive.

Thermoplastic: A roofing material designed to be heated and fused to itself. When cooled, it remains flexible, waterproof, and strong.

T-joint: An area where three field membrane sheets overlap.

TPO: Thermoplastic polyolefin, a common type of thermoplastic used in roofing systems.

Weighted roller: A heavy object pushed over adhered or self-adhered membrane to smooth it out and ensure sufficient adhesion to the substrate.

Thermoplastic roof systems (*Figure 1*) are one of two standard types of single-ply systems, the other being thermoset. They are categorized as single-ply systems because they consist of one layer of weatherproofing. The name *thermoplastic* refers to the material's ability to repeatedly soften when heated and harden when cooled. This trait makes thermoplastic roof systems energy efficient and durable. They are also aesthetically pleasing and easy to maintain and repair.

> **NOTE**
> Differences between thermoplastics and thermosets are described in NCCER Module 16106, *Introduction to Low-Slope Roofing.*

Two common types of thermoplastics encountered in roofing systems are polyvinyl chloride (PVC) and thermoplastic polyolefin (TPO). Since its introduction as a commercial roofing membrane in the 1990s, TPO has become the fastest growing commercial roof membrane on the market.

Thermoplastic roofing systems consist of a field membrane that is attached to the substrate by adhesives, mechanical fasteners, ballast, or induction welding plates. Flashing accessories are used to weatherproof or seal the edges and penetrations of a roof system.

1.1.0 Membrane Types

There are three common membrane types for TPO and PVC: smooth membranes, membranes with fleece-backing, and self-adhering membranes. It is important to recognize these different membranes because they allow for different methods of attachment to the substrate.

1.1.1 Standard Membranes

Standard (or smooth-back) thermoplastic membranes do not come with an additional material attached, unlike fleece-back and self-adhered membranes. They are versatile when it comes to installation because they allow for several different methods of attachment to the substrate. The membrane shown in *Figure 1* is a smooth membrane.

Figure 1 Thermoplastic roof system.

1.1.2 Fleece-Back Membranes

Fleece-backing is a nonwoven polyester material adhered to the underside of a thermoplastic sheet for reinforcement (*Figure* 2). Fleece-back membranes can be more resistant to punctures, hail, and damage from foot traffic than smooth membranes.

While smooth membranes can attach to the substrate with a variety of methods, fleece-back membranes typically require an adhesive, such as hot asphalt or low-rise foam.

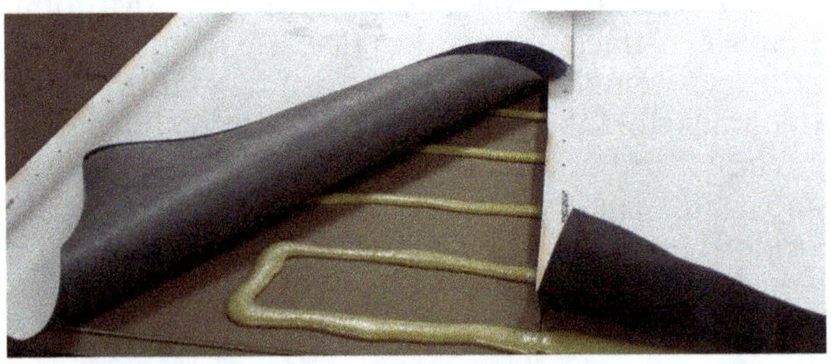

Figure 2 Membrane with fleece-backing.

WARNING!

Always check the safety data sheet (SDS) when using adhesives. Long sleeves and gloves may be needed to protect the skin from chemicals.

1.1.3 Self-Adhered Membranes

Self-adhered membranes, also called *peel-and-stick membranes,* come with adhesive already on the back of the sheet. They can bond to a substrate without the use of an additional attachment method. This makes them cleaner and easier to install than other types of membranes.

NOTE

Self-adhered membranes have temperature limitations for installation. If the temperature is too cold, they may not adhere.

1.2.0 Flashing Accessories

Flashings are components used to weatherproof or seal edges of a roof system at perimeters, penetrations, walls, expansion joints, valleys, drains, and other places where the roof covering is interrupted or terminated. In low-slope systems, flashings are often made of the same material as the roof membrane.

These are some of the most common prefabricated flashing accessories encountered in thermoplastic roof systems:

- *Boots* — A covering made of flexible material or a combination of flexible and rigid material that may be pre-formed to a particular shape, used to exclude dust, dirt, moisture, etc., from around a penetration. Boots are commonly placed around vent pipes.
- *Corners* — Used to weatherproof any location where there is a 90-degree bend, such as where the roof meets a parapet wall.
- *Self-adhering cover tape* — Used to strip in metal edging. Self-adhering cover tape typically comes in widths of 6", 9", and 12", with adhesive on the bottom and membrane on the top. Also known as *peel-and-stick cover tape*.
- *Sealant pocket* — A sheet metal sleeve placed around a penetration and filled with grout to provide a watertight seal.
- T-joint *cover* (*Figure 3*) — An accessory used to weatherproof T-joints, which may be round or square with rounded edges.
- *Thermoplastic coated metals* — A piece of long, thin metal half-coated with TPO or PVC, providing an additional service to weld field membrane to where needed, such as along a roof edge.

Roofers can fabricate flashings from detail membrane and wall flashing membrane on the jobsite. Detail membrane, also known as *unreinforced flashing membrane*, can be shaped and stretched to ensure a tight seal around a penetration. Detail membrane must be compatible with the field membrane, but it cannot be used as field membrane. It comes in a roll and is measured and cut on-site.

Wall flashing membrane runs along a vertical wall, such as where the edge of the roof meets a parapet wall. Generally made of the same materials as the field membrane, it is typically reinforced and comes in a variety of thicknesses.

1.3.0 Attachment Components

Thermoplastic roof systems may be attached to a building in three general ways: mechanically attached with fasteners, adhered, and weighed down with ballast. They may also be attached by induction welding, an increasingly popular method. Induction welding is an alternative way to mechanically attach a thermoplastic membrane.

NOTE

Attachment methods are introduced in NCCER Module 16106, *Introduction to Low-Slope Roofing.*

1.3.1 Adhesives

Adhesives are glues that secure the thermoplastic roof membrane to the substrate. The type of thermoplastic membrane used, such as TPO or PVC, will dictate the type of adhesive needed.

CAUTION

Using the wrong adhesive may cause a chemical reaction that can cause damage to the membrane. Check Section 10 of the SDS for this information.

WARNING!

Adhesives contain chemicals that may be flammable. Ensure a fire extinguisher is within reach when working with adhesives. Consult the SDS to determine the correct type of fire extinguisher and other related safety guidelines.

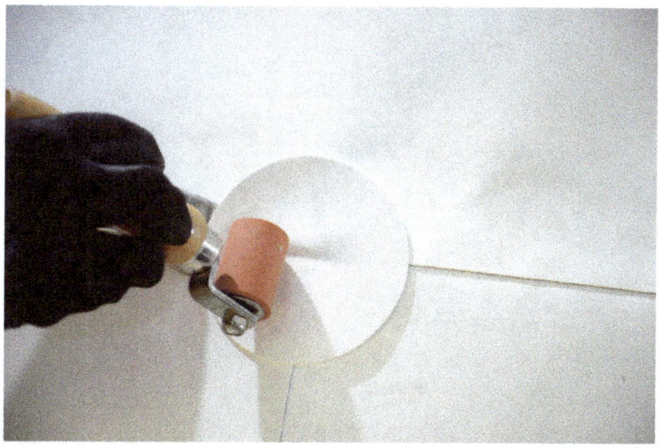

Figure 3 T-joint cover.

Some adhesives are applied to both the membrane and the substrate. This is called a *two-sided application*. Other adhesives are applied to the substrate or membrane only. This is called a *one-sided application*. The adhesives are typically installed using a roller or sprayer per the manufacturer's guidelines (*Figure 4*). These systems are then typically rolled using a **weighted roller** to ensure proper adhesion to the substrate. Self-adhered membranes come from the manufacturer with adhesive already attached.

1.3.2 Fasteners

Fasteners mechanically secure the thermoplastic roof membrane to the substrate. They are typically comprised of a screw driven through a round metal plate (*Figure 5*). The type of fastener used is dependent on factors such as the substrate or deck and system type. Plates used for fastening insulation are typically larger than membrane plates.

1.3.3 Induction Plates

Induction plates (*Figure 6*) are specially coated metal plates that are placed under the membrane. When heated from above with an induction welder, the coating on the plates melts and fuses with the thermoplastic roof membrane.

1.3.4 Ballast

Ballasted roof systems (*Figure 7*) are attached by being weighed down with ballast to hold the membrane in place. Ballast materials also protect thermoplastic membranes from the sun's UV light, which can cause the membrane to break down.

Some common ballast materials used on thermoplastic roof systems include:

- *River rock (aggregate)* — Loose-laid pieces of rock that have been weathered. River rock is larger and smoother than gravel.
- *Pavers* — Precast slabs, usually made of concrete, used to weigh down (or ballast) a roof system and provide a functional finish and walking surface.
- *Vegetation* — A planter on top of the field membrane that weighs down the roof system. In addition to being eco-friendly, this ballast type can cut stormwater management and energy expenses for building owners.

> **NOTE**
> Vegetation is not considered a roofing product and falls under different fall protection standards than roof systems.

Figure 4 Adhesive applied to a TPO roof system by spraying.

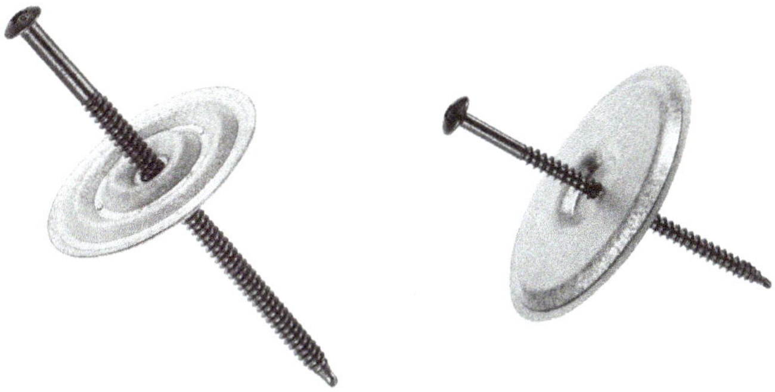

(A) Membrane Plate (B) Insulation Plate

Figure 5 Fastener and membrane plate for mechanical attachment.

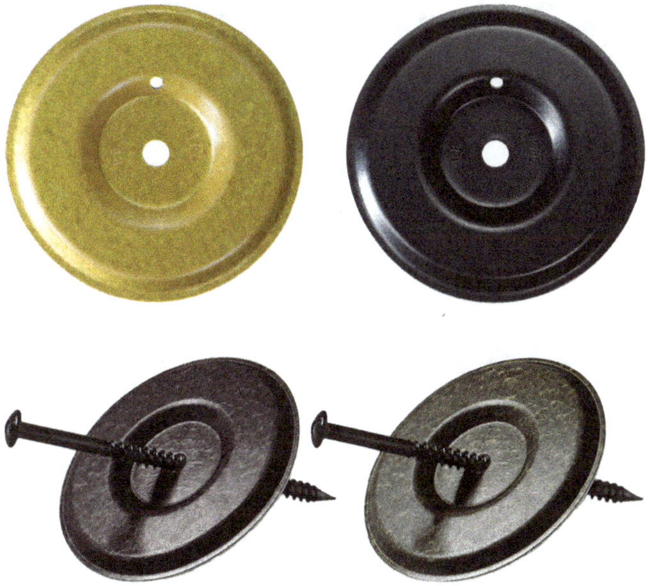

Figure 6 Common induction plates.

Figure 7 Ballasted roof system with river rock and vegetation.

Going Green

Vegetation

When used as a ballast, vegetation provides a range of environmental benefits, such as managing runoff and absorbing heat to naturally cool a building. Although vegetative roof systems date back to at least the 1930s, they have grown increasingly popular with climate change awareness. Green Roofs for Healthy Cities reported a 15 percent increase in the number of these types of roof systems between 2013 and 2019.

1.0.0 Section Review

1. Which type of flashing accessory can be used to weatherproof a location where there is a 90-degree bend?
   a. T-joint cover
   b. Corners
   c. Self-adhering cover tape
   d. Pitch pocket
2. Which of the following types of thermoplastic membrane can bond to the substrate and to itself at overlaps without the use of an additional attachment method?
   a. Smooth
   b. Fleece-backed
   c. Standard
   d. Self-adhered
3. Which of the following attachment components holds a roof membrane in place by weighing it down?
   a. Ballast
   b. Induction plates
   c. Fasteners
   d. Adhesives

SECTION TWO

2.0.0 INSTALLATION AND REPAIR

Objective

Explain how thermoplastic roof systems are installed and repaired.

a. Identify tools and equipment used to install thermoplastic roof systems.
b. Describe how thermoplastic roof systems are installed.
c. Explain basic repair methods for damaged thermoplastic roof systems.

Performance Tasks

1. Install a fastener and membrane plate into a substrate.
2. Clean a membrane lap to prepare for seaming.
3. Use a hand welder to create a seam.

Trade Terms

Application rate: The average quantity (mass, volume, or thickness) of material applied per unit area.

Fishmouth: An opening in the edge of a membrane caused by wrinkling or shifting of sheets during installation.

Hot-air welding: A method of melting and fusing together the overlapping edges of separate sheets of roof membrane by the application of heat (in the form of hot air) and pressure. Also called *heat welding*.

Probing: Checking for voids or open areas in a seam using a seam probe hand tool.

Seam: A joint formed by mating two separate, overlapping pieces of roof membrane.

This section will cover the essential tools and installation methods for adhered, mechanical, induction welded, and ballasted attachment. Workers on thermoplastic roof systems should be familiar with the basic steps for each of these methods, although in some cases separate crews may handle specific steps such as flashing.

When installing thermoplastic roof membrane, it is essential to check your work as you go to ensure adequate weatherproofing. This section will cover how to check your work by probing.

Workers on thermoplastic roof systems should also be prepared to repair damage caused during or after installation. This section will cover repair methods for three basic types of membrane damage.

2.1.0 Tools and Equipment

Workers on thermoplastic roof systems rely on special hand tools and power tools to install or repair the field membrane. Some of these tools are used in conjunction with one another, such as two-inch silicone hand rollers and hot-air hand welders.

> **NOTE**
> Basic tools for low-slope roofing are covered in NCCER Module 00103, *Introduction to Hand Tools*, and NCCER Module 00104, *Introduction to Power Tools*.

2.1.1 Hand Tools

Roofers use a variety of basic hand tools when installing thermoplastic roof systems. The following are some hand tools specific to thermoplastic systems:

- *Scissors/shears* — Used to create a flashing accessory from detail or wall membrane, or to cut rounded edges on flashing accessories to assist in hand welding.
- *Crayon* — Used to mark components hidden underneath field membrane, such as drains, for other workers' awareness.

A variety of rollers can be used when attaching the membrane to the substrate.

- *Adhesive roller* — Used to roll the adhesive onto a surface.
- *Weighted roller* — A heavy object pushed over adhered or self-adhered membrane to smooth it out and ensure sufficient adhesion to the substrate (*Figure 8*).
- *Push broom* — A wide broom used to remove debris from substrates and to smooth out a membrane to promote adhesion (*Figure 9*).
- *Two-inch silicone hand roller* — Used in conjunction with a hot-air hand welder to press on a seam during hot-air welding and ensure a tight bond.
- *Seam probe* (*Figure 10*) — A small piece of metal, usually bent at a 90-degree angle, that can slide along a seam to test for open welds or seams. Seam probes typically come from the manufacturer with a sharp point and tend to sharpen with use. They should be dulled with a file before the first use and regularly afterwards to prevent punctures.

Figure 8 Weighted roller.

2.1.2 Power Tools

Thermoplastic roof installers should be familiar with power tools used for hot-air welding, adhesive attachment, and induction welding.

Hot-air welding is a method of melting and fusing together the overlapping edges of separate sheets of roof membrane by the application of heat (in the form of hot air) and pressure.

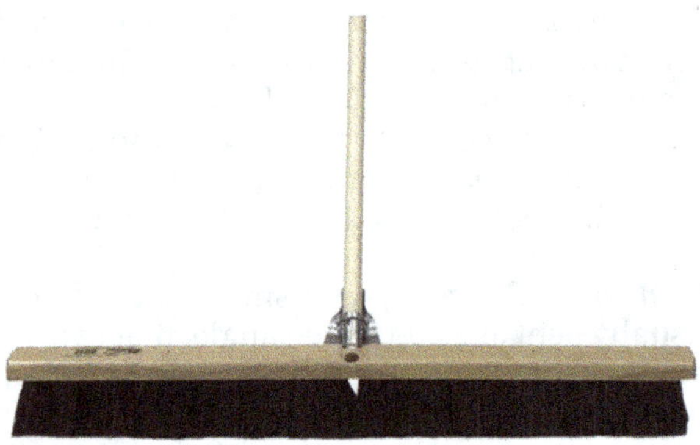

Figure 9 Push broom.

Hot-air welder and hand welders (*Figure 11*) blow extremely hot air, typically over 500°F (260°C), into the seam between two sheets of thermoplastic membrane, melting and bonding them together. Hand welders are used in conjunction with a two-inch silicone hand roller to press on the seam and ensure a tight bond (*Figure 11* [*A*]). They are also typically used for flashings and detailing.

> **WARNING!**
>
> High temperatures associated with hot-air welders may lead to damage, fires, or burns if they are not properly used.

The following steps outline a general procedure for operating hot-air hand welders:

Step 1 Adjust the heat and airflow controls following manufacturer's guidelines.

Step 2 Position the nozzle between the membrane rolls, perpendicular (at a 90-degree angle) to the lap.

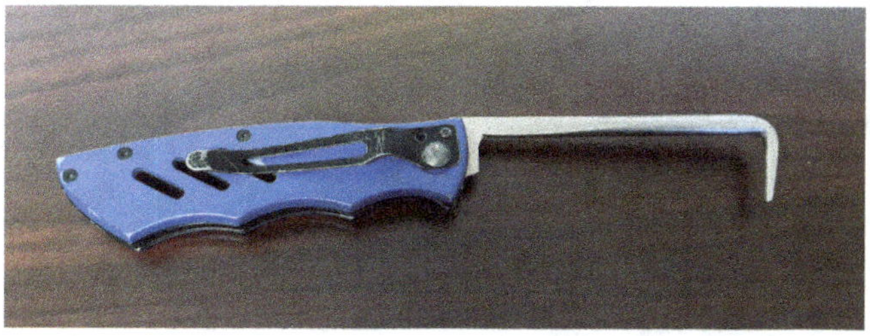

(A) Seam Probe

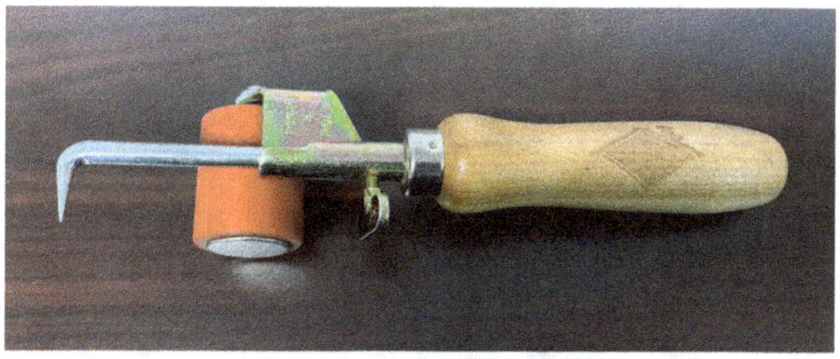

(B) Seam Roller and Probe

Figure 10 Seam probes.

Step 3 Pre-weld the seam by pulling the nozzle steadily along the lap while rolling the hand roller back and forth over the top membrane.

Step 4 Reposition the nozzle between the membrane rolls, at a 45-degree angle relative to the lap.

Step 5 Finish the weld by pulling the nozzle steadily along the lap while rolling the hand roller back and forth over the top membrane.

Hot-air robotic welders (*Figure 12*) automate and speed up the above process, making them especially useful for commercial roofing jobs. It is a push tool that has a welder and rollers attached. It has adjustment controls for speed, heat, and airflow.

Hot-air welders typically connect to a generator via an extension cord for power. It is important that the generator and cord suit the type of welder being used. For hand welders, the generator should have a minimum power supply of 100 volts, 15 amps, and 2,500 watts continuous. Be sure to use an extension cord that is rated for the current and voltage. For robotic welders, the generator should have a minimum power supply of 220 volts, 30 amps, and 10,000 watts continuous. The extension cord should be a #10 wire with a maximum length of 100'.

(A) Heat Welding a Seam with Hand Seamer

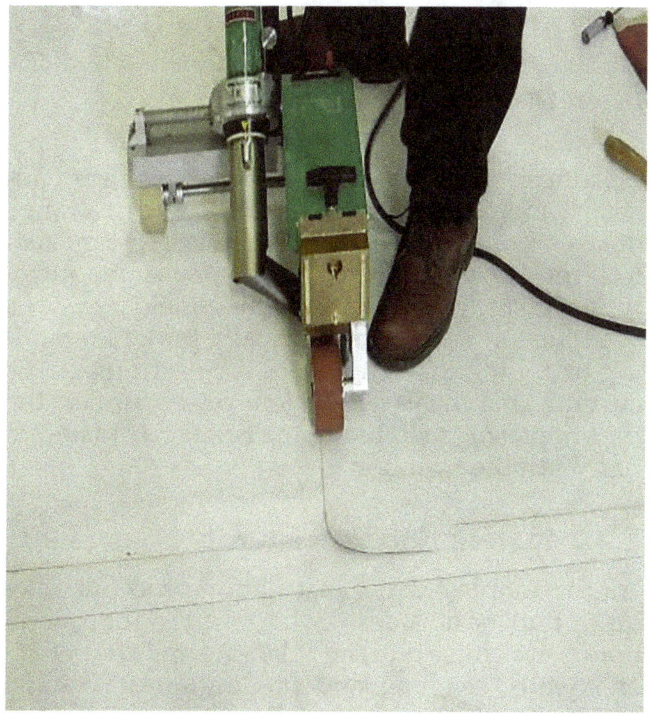

(B) Heat Welding a Seam with Hot-Air Welder

Figure 11 Heat welding a seam.

> **NOTE**
> GFCIs should be used with the power source to ensure the load associated with the generator is supported. Consult manufacturer recommendations for power requirements.

Figure 12 Robotic welder.

Adhesive sprayers come with a container of adhesive and a spraying hose. They come in a range of sizes from small boxes, containers, or canisters to large hoppers. Look to the sprayer manufacturer for spraying rates and guidelines.

Induction welders are typically portable standing or rolling tools. It creates an electromagnetic current that heats induction plates under the thermoplastic membrane, causing the plates to fuse with the membrane.

2.2.0 Installation Procedures

All thermoplastic roof installation procedures require four basic steps: laying out the roof membrane, attaching it to the substrate and/or perimeter walls, seaming and probing, and flashing. The order of these steps will vary depending on the installation method.

NOTE

The procedures in this module assume insulation has already been installed over the deck. These procedures are examples to help you gain an understanding of the process. The exact steps and order will vary depending on the specific roof design and materials specified. Always consult applicable guidelines from your employer, including project specifications, building codes, and manufacturer instructions.

2.2.1 Adhered Installation

Adhering a thermoplastic roof membrane involves some basic steps, including laying out sheets, applying adhesive, brooming and rolling, and heat-welding seams. Each step must be done carefully to ensure adequate adhesion to the substrate. *Figure 13* shows the steps for laying out and adhering each sheet. (Note that *Figure 13* is a simplified procedure and does not cover flashings and penetrations.)

The following steps are an example of a general procedure for adhering thermoplastic roof membrane using bonding adhesive:

Step 1 Unroll the roof membrane so that wrinkles and buckles are not formed. Any wrinkles or buckles must be removed from the sheet prior to permanent securement.

Step 2 Lay out and orient the membrane roll so that the side laps run perpendicular to the slope of the roof, so water runs towards the drainage points. Allow the membrane to relax in place and come to ambient temperature.

Step 3 Cut the membrane roll to the desired length. Full-width rolls can be installed throughout the field and perimeter of the roof. Half sheets are not necessary.

1 First Sheet

- ▶ **Unroll the membrane and allow it to relax.**
- ▶ **Size and cut the membrane to fit the area.**
  - *Fold edges up all vertical surfaces.*
  - *Position membrane to establish water-shedding sequence.*
  - *Lay membrane into position free of wrinkles and buckles.*
- ▶ **Adhere approximately 50% of the membrane.**
  - *Fold half of the membrane back on itself.*
  - *Apply solvent-based adhesive on the exposed substrate and the exposed membrane.*
  - *Allow adhesive sufficient time to flash or dry to the touch.*
- ▶ **Unfold the membrane back into its original position, avoiding wrinkles or buckles.**
- ▶ **Broom or roll the membrane immediately.**

2 Second Sheet

- ▶ **Unroll the membrane and allow it to relax.**
- ▶ **Size and cut the membrane to fit the area.**
  - *Fold edges up all vertical surfaces.*
  - *Position membrane to shed water.*
  - *Establish consistent end and side laps per manufacturer specifications.*
- ▶ **Adhere approximately 50% of the membrane.**
  - *Fold half of the membrane back on itself.*
  - *Apply solvent-based adhesive on the exposed substrate and the exposed membrane.*
  - *Keep lap areas of top and bottom sheet free of adhesive.*
- ▶ **Unfold the membrane back into its original position, avoiding wrinkles or buckles.**
- ▶ **Broom or roll the membrane immediately.**
- ▶ **Hot-air weld all seams.**
  - *Clean and prep each lap area before hot-air welding and probe all seams when cooled.*

Repeat Step 2 for all subsequent sheets.

Figure 13 General procedure for adhering a thermoplastic roof membrane (excludes flashings and penetrations).

Step 4 Unroll the membrane sheet in place and fold back the sheet in the long dimension (to expose the underside) to allow adhering of the membrane, one half of a sheet at a time. Do not crease.

Step 5 Apply the bonding adhesive to the substrate and/or membrane using a roller, brush, or squeegee, depending on the adhesive type. Follow the manufacturer's guidelines for application temperature, rate, and coverage for the adhesive used.

Step 6 Wait for the adhesive to dry. Solvent-based adhesive must be allowed to dry until tacky to the touch before mating the membrane. Water-based adhesive must be allowed to flash off completely.

Step 7 Bond the membrane to the substrate and apply pressure by means of a roller or push broom to ensure complete bonding. Avoid trapping air between the membrane and the substrate.

Step 8 Lay down the next membrane roll. Follow manufacturer's guidelines for how many inches of overlap there should be between each roll for the side and end laps.

Step 9 Use a hot-air welder to bond the membrane sheets together at the seams, to create a watertight seal.

Step 10 Repeat *Steps 1–9*, laying down each roll successively and overlapping each seam.

Step 11 After the membrane has cooled, run a probe along the seam to check for adequate bonding. If the probe slips between the membranes, mark the spot with a crayon to be repaired with a T-joint patch.

Step 12 Secure flashing components to areas where the roof system is interrupted or terminated. Roof membrane must be mechanically secured at the perimeter, at the base of internal walls and curbs, and at all penetrations.

NOTE

Always probe seams when they cool, and probe again when the job is completed.

Self-adhering membranes come with the adhesive already included so that it does not need to be applied to the membrane. The undersurface of a self-adhering membrane is protected by a release paper or film, which prevents the membrane from bonding to itself during shipping and handling. During installation, the release paper is peeled off while the membrane is pressed onto the substrate. Follow manufacturer's instructions for specific installation guidelines.

2.2.2 Mechanically Attached Installation

Mechanical attachment (*Figure 14*) is a common attachment method for thermoplastic roof installation. As with adhered installation, this method involves layout of the membrane, then attachment, then seaming, probing, and flashing.

The following steps outline a general procedure for mechanically attaching thermoplastic roof membrane:

Step 1 Unroll the roof membrane so that wrinkles and buckles are not formed. Any wrinkles or buckles must be removed from the sheet prior to permanent securement.

Step 2 Lay out and orient the membrane roll so that the side laps run perpendicular to the slope of the roof, so water runs towards the drainage points. Allow the membrane to relax in place and come to ambient temperature.

Figure 14 Mechanical attachment of roof membrane.

Step 3 Cut the membrane roll to the desired length. Full-width rolls can be installed throughout the field and perimeter of the roof. Half sheets are not necessary.

Step 4 Lay out the membrane plates along the seam area, following manufacturer's guidelines for how the plates should be spaced.

> **NOTE**
> Location and determination of half sheets will vary depending on wind uplift resistance requirements in the region.

Step 5 Drill the fasteners through the plates and into the deck below at the edge of the membrane roll. Space fasteners as recommended by the manufacturer. You should feel the fastener touch the top flute of the deck; follow manufacturer's guidelines for how many inches the fastener should penetrate. When installing over a metal deck, the fasteners should penetrate the top of the flutes.

Step 6 Lay down the next membrane roll. Follow manufacturer's guidelines and code requirements for how many inches of overlap there should be between each roll.

Step 7 Repeat *Steps 1–5,* laying down each roll successively and overlapping each seam.

Step 8 Use a hot-air welder to bond the membrane sheets together at the seams, to create a watertight seal.

Step 9 When the membrane has cooled, run your probe along the seam to check for adequate bonding. If your probe slips in between the membranes, mark the spot to be repaired with hot-air welding.

Step 10 Secure flashing components to areas where the roof system is interrupted or terminated. Roof membrane must be mechanically secured at the perimeter, at the base of internal walls and curbs, and at all penetrations.

> **NOTE**
> Always probe seams when they cool, and probe again when the job is completed.

2.2.3 Induction Welded Installation

Induction welding is becoming an increasingly popular method of thermoplastic roof system installation (*Figure 15*). Note that unlike other attachment methods, induction welding does not begin with layout of the membrane. The induction plates must first be put down on top of the boards in a specific pattern.

The following steps outline a general procedure for induction welding thermoplastic roof membrane:

Step 1 Lay out the induction plates over the insulation boards, following manufacturer's guidelines for how the plates should be patterned and spaced.

Step 2 Fasten the induction plates to the substrate.

Step 3 Lay out and orient the membrane roll so that the side laps run perpendicular to the slope of the roof, so water runs towards the drainage points (*Figure 16*). Allow the membrane to relax in place and come to ambient temperature.

Step 4 Cut the membrane roll to the desired length and make sure it is laid out correctly.

Step 5 Lay down the next membrane roll. Follow manufacturer's guidelines for how many inches of overlap there should be between each roll.

Step 6 Repeat *Steps 1–5,* laying down each roll successively and overlapping each seam.

Step 7 Use a hot-air welder to bond the membrane sheets together at the seams, to create a watertight seal.

Step 8 After the membrane has cooled, run your probe along the seam to check for adequate bonding. If your probe slips in between the membranes, mark the spot to be repaired with hot-air welding.

Step 9 Place the induction welder over an induction plate. It is important to place the magnet with precision over the plate and avoid moving, wrinkling, or twisting the membrane. Follow manufacturer's guidelines for operating the welder.

> **CAUTION**
> Before using an induction welder (prior to beginning *Step 9*), ensure the calibration settings are appropriate for the conditions and time of day.

Figure 15 Induction welded installation.

Figure 16 Membrane layout.

Step 10 Place a weighted magnet over the plate for the length of time specified by the manufacturer. Ensure the magnet has been wiped off so no metal shavings puncture the membrane.

Step 11 Repeat *Steps 9–10* for all induction plates.

Step 12 Secure flashing components to areas where the roof system is interrupted or terminated. Roof membrane must be mechanically secured at the perimeter, at the base of internal walls and curbs, and at all penetrations.

> **NOTE**
> Always probe seams when they cool, and probe again when the job is completed.

2.2.4 Ballasted Installation

Ballasted or loose-laid roof systems (*Figure 17*) are often used on buildings that have walkways or high-traffic areas on the roof. Ballasted roof systems are heavy, so the roof deck must be designed to withstand the weight of the system. Membrane laps are heat welded, and the system is weighed down with ballast.

The following steps outline a general procedure installing a ballasted thermoplastic roof membrane:

Step 1 Unroll the roof membrane so that wrinkles and buckles are not formed. Any wrinkles or buckles must be removed from the sheet prior to permanent securement.

Step 2 Lay out and orient the membrane roll so that the side laps run perpendicular to the slope of the roof, so water runs towards the drainage points. Allow the membrane to relax in place and come to ambient temperature.

Step 3 Cut the membrane roll to the desired length. Full-width rolls can be installed throughout the field and perimeter of the roof. Half sheets are not necessary.

Step 4 Use a hot-air welder to bond the membrane sheets together at the seams, to create a watertight seal.

Step 5 When the membrane has cooled, run your probe along the seam to check for adequate bonding. If your probe slips in between the membranes, mark the spot to be repaired with hot-air welding.

Figure 17 Ballasted roof system.

Step 6 Secure flashing components to areas where the roof system is interrupted or terminated. Roof membrane must be mechanically secured at the perimeter, at the base of internal walls and curbs, and at all penetrations.

Step 7 Install a loose-laid gravel mat or separator sheet (sometimes called a *poly mat*) over the membrane to protect it from puncturing under the ballast.

Step 8 Apply the specified ballast material over the membrane, following manufacturer's guidelines for application rate.

> **NOTE**
> Always probe seams when they cool, and probe again when the job is completed.

2.3.0 Basic Repair Methods

Mistakes happen even under the best conditions for roof installation. To ensure a quality and efficient job, workers on thermoplastic roof systems should know how to repair any damage created during the installation process.

This section will cover basic repair methods for three common types of membrane damage: a puncture, a cut, and a fishmouth.

Workers should be generally aware of wrinkles in the field membrane. Wrinkles do not necessarily void warranties or pose a risk to weatherproofing, but over time they could reduce the long-term performance of the roof system. They may also raise aesthetic concerns for building owners.

2.3.1 Repairing a Puncture

A puncture is a small hole in the field membrane. Punctures pose a risk to weatherproofing because they can allow moisture to penetrate the roofing system.

The following steps outline a general procedure for repairing punctures during thermoplastic roof installation:

Step 1 Pull back the cut membrane to examine underneath for damage to the substrate or the presence of dirt or moisture.

Step 2 Remove dirt and residue from the membrane using a manufacturer-approved, low-residue cleaner and a mildly abrasive pad to extend at least $1/4$" around the perimeters of the patch.

Step 3 Wipe clean with a cloth moistened with a manufacturer-approved membrane cleaner until it returns to factory white.

Step 4 If required, apply a manufacturer-approved primer evenly around the puncture.

Step 5 Wait for the primer to flash off and dry.

Step 6 Cut a piece of field membrane to create a patch, or use a patch prefabricated by the manufacturer.

Step 7 Round all corners of the patch.

Step 8 Apply the patch over the puncture using a hot-air welder and hand roller.

Step 9 Seal edges as required by manufacturer's guidelines.

2.3.2 Repairing a Cut

A cut is essentially an elongated puncture. The process for repairing a cut is similar to that for a puncture, but there are some additional steps.

The following steps outline a general procedure for repairing cuts during thermoplastic roof installation:

Step 1 Pull back the cut membrane to examine underneath for damage to the substrate or the presence of dirt or moisture.

Step 2 Remove dirt and residue from the membrane using a manufacturer-approved, low-residue cleaner and a mildly abrasive pad.

Step 3 Wipe clean with a cloth moistened with a manufacturer-approved membrane cleaner until it returns to factory white.

Step 4 Mechanically fasten the membrane on each edge of the cut.

Step 5 If required, apply a manufacturer-approved primer evenly around the cut.

Step 6 Wait for the primer to flash off and dry.

Step 7 Cut a piece of field membrane to create a patch, or use a patch prefabricated by the manufacturer.

Step 8 Round all corners of the patch.

Step 9 Apply the patch over the cut using a hot-air welder and hand roller.

Step 10 Seal edges as required by manufacturer's guidelines.

2.3.3 Repairing a Fishmouth

A fishmouth (*Figure 18*) is an opening in the edge of a membrane caused by wrinkling or shifting of sheets during installation. Fishmouths pose a risk to weatherproofing as they can cause seams to open and allow moisture into the system.

The following steps outline a general procedure for repairing a fishmouth during thermoplastic roof installation:

Step 1 Pull back the cut membrane to examine underneath for damage to the substrate or the presence of dirt or moisture.

Step 2 Remove the fishmouth by cutting or slicing.

Step 3 Remove dirt and residue from the membrane using a manufacturer-approved, low-residue cleaner and a mildly abrasive pad.

Step 4 Wipe clean with a cloth moistened with a manufacturer-approved membrane cleaner until it returns to factory white.

Step 5 If required, apply a manufacturer-approved primer evenly around the damaged area.

Step 6 Wait for the primer to flash off and dry.

Step 7 Cut a piece of field membrane to create a patch, or use a patch prefabricated by the manufacturer.

Step 8 Round all corners of the patch.

Step 9 Apply the patch over the area using a hot-air welder and hand roller.

Step 10 Seal edges as required by manufacturer's guidelines.

2.3.4 Unadhered Areas

Unadhered areas compromise the performance of the roof system. Roofers may locate unadhered areas simply by pushing on the field membrane or flashings. Inadequately adhered areas of the membrane may also show movement by billowing or fluttering in the wind. If you suspect you have found an unadhered area, alert your foreman so that they can determine the appropriate action.

Figure 18 Fishmouth.

2.0.0 Section Review

1. What is the difference between a puncture and a cut?
   a. A puncture is a cut that hasn't been cleaned.
   b. Punctures occur on flashing membranes and cuts occur on field membranes.
   c. A cut is an elongated puncture.
   d. There is no difference; punctures and cuts refer to the same type of membrane damage.

2. What should a roofing worker do before using a seam probe for the first time?
   a. Sharpen it with a file
   b. Dull it with a file
   c. Bend it to the preferred angle
   d. Nothing

3. In which of the following thermoplastic roof attachment methods does attachment to the substrate come last?
   a. Adhered
   b. Mechanically attached
   c. Induction welded
   d. Ballasted

Review Questions

1. Which of the following statements about detail membrane is *true*?
   a. It cannot be used in lieu of field membrane.
   b. It is primarily used to weatherproof vertical walls.
   c. It is reinforced.
   d. It can be used anywhere on the roof.

2. Which of the following is a benefit of self-adhered membranes?
   a. They are more effective in cold climates.
   b. They are less labor-intensive to install than other types of membranes.
   c. They are generally less expensive than other types of membranes.
   d. They do not require flashing accessories.

3. What should you do to a patch before applying it over a puncture, cut, or fishmouth?
   a. Coat it in manufacturer-approved primer.
   b. Sharpen all edges with a file.
   c. Round all corners.
   d. Roll it with a hand roller.

4. Which of the following is a difference between gravel and river rock ballast materials?
   a. Gravel is loose-laid over the field membrane and river rock is not.
   b. Gravel is larger than river rock.
   c. River rock provides protection from UV light and gravel does not.
   d. River rock is smoother than gravel.

5. Which of the following is the *correct* order of steps for mechanically attached installation?
   a. Laying out the membrane, attachment, seaming, flashing, and probing.
   b. Laying out the membrane, seaming, probing, flashing, and attachment.
   c. Attachment, seaming, probing, flashing, and laying out the membrane.
   d. Flashing, laying out the membrane, seaming, probing, and attachment.

6. Which of the following is a possible shape for a T-joint cover?
   a. T-shaped
   b. Triangle
   c. Round
   d. The roofing worker cuts it to the preferred shape on-site.

7. How does hot-air welding work?
   a. An electromagnetic current heats plates under the membrane, causing them to melt and fuse with the membrane.
   b. Adhesive is allowed to dry on the membrane until it is tacky enough to create a bond.
   c. A heavy object is pushed over adhered or self-adhered membrane.
   d. Heat is applied between overlapping edges of separate roof membrane sheets so that they melt and fuse together.

8. Which of the following statements about induction welding is *true*?
   a. Installation of an induction welded roof system begins with layout of the membrane.
   b. Induction plates must be fastened on top of the membrane.
   c. Induction welding does not require hot-air welding at seams.
   d. A stand-up magnet must be placed over the induction plates after the induction welder is removed.

9. Which of the following is a type of fastener?
   a. T-joint cover
   b. Paver
   c. Induction plate
   d. Screw

10. Which of the following tools is useful for checking your work when installing thermoplastic field membrane?
    a. Seam probe hand tool
    b. Scissors
    c. Weighted roller
    d. Hot-air robotic welder

Trade Terms Introduced in This Module

Application rate: The average quantity (mass, volume, or thickness) of material applied per unit area.

Ballast: A material, such as smooth river stone, crushed stone, standard precast concrete pavers or interlocking, beveled, doweled, or contoured fit lightweight concrete pavers, that weighs down a membrane to hold it in place.

Fasteners: Any of a wide variety of mechanical securement devices, including nails, staples, screws, cleats, clips, and bolts.

Fishmouth: An opening in the edge of a membrane caused by wrinkling or shifting of sheets during installation.

Flashings: Components used to weatherproof or seal edges of a roof system at perimeters, penetrations, walls, expansion joints, valleys, drains, and other places where the roof covering is interrupted or terminated. For example, membrane base flashing covers the edge of the field membrane, and cap flashings or counterflashings shield the upper edges of the base flashing.

Fleece-backing: A nonwoven polyester material adhered to the underside of a thermoplastic sheet for reinforcement.

Hot-air welding: A method of melting and fusing together the overlapping edges of separate sheets of roof membrane by the application of heat (in the form of hot air) and pressure. Also called *heat welding*.

Probing: Checking for voids or open areas in a seam using a seam probe hand tool.

PVC: Polyvinyl chloride, a common type of thermoplastic used in roofing systems.

Seam: A joint formed by mating two separate, overlapping pieces of roof membrane.

Self-adhering membranes: Membranes that can bond to both a substrate and to themselves at overlaps without the use of an additional adhesive.

Thermoplastic: A roofing material designed to be heated and fused to itself. When cooled, it remains flexible, waterproof, and strong.

T-joint: An area where three field membrane sheets overlap.

TPO: Thermoplastic polyolefin, a common type of thermoplastic used in roofing systems.

Weighted roller: A heavy object pushed over adhered or self-adhered membrane to smooth it out and ensure sufficient adhesion to the substrate.

Additional Resources

This module presents thorough resources for task training. The following reference material is suggested for further study.

GAF Roofing, YouTube channel. **www.youtube.com**.

National Roofing Contractors Association (NRCA), **www.nrca.net**.

NCCER Module 00103, *Introduction to Hand Tools*.

NCCER Module 00104, *Introduction to Power Tools*.

NCCER Module 16106, *Introduction to Low-Slope Roofing*.

NCCER Module 16207, *EPDM Roof Systems*.

Figure Credits

Courtesy of GAF, Module Opener, Figures 1–6, 8, 11 (B), 12, 15–16

Kraft Tool Co.®, Figure 9

Patrick D Murphy Co., Inc., Architects, Figures 10, 17–18

Courtesy of the National Roofing Contractors Assoc., Figures 11 (A), 14

Section Review Answer Key

SECTION 1.0.0

Answer	Section Reference	Objective
1. b	1.2.0	1b
2. d	1.1.3	1a
3. a	1.3.4	1c

SECTION 2.0.0

Answer	Section Reference	Objective
1. c	2.3.2	2c
2. b	2.1.1	2a
3. c	2.2.3	2b

User Update

NCCER makes every effort to keep its textbooks up-to-date and free of technical errors. We appreciate your help in this process. If you find an error, a typographical mistake, or an inaccuracy in NCCER's curricula, please submit a User Update form by visiting **https://www.nccer.org/olf**. You can also scan the QR code using the camera on your phone or mobile device to access the form.

EPDM Roof Systems

Overview

EPDM roof systems are single-ply systems that have been commonly used on low-slope roofs since the 1970s. Roofing professionals must be familiar with common types of EPDM membrane, as well as the components and attachment methods used to secure them to the substrate for effective weatherproofing.

Module 16207

Trainees with successful module completions may be eligible for credentialing through the NCCER Registry. To learn more, go to **www.nccer.org** or contact us at 1.888.622.3720. Our website, **www.nccer.org**, has information on the latest product releases and training.

Your feedback is welcome. You may email your comments to **curriculum@nccer.org**, send general comments and inquiries to **info@nccer.org**, or fill in the User Update form at the back of this module.

This information is general in nature and intended for training purposes only. Actual performance of activities described in this manual requires compliance with all applicable operating, service, maintenance, and safety procedures under the direction of qualified personnel. References in this manual to patented or proprietary devices do not constitute a recommendation of their use.

16207 V1.0

16207

EPDM Roof Systems

Objectives

Successful completion of this module prepares you to do the following:

1. Identify types of EPDM roof systems and their basic components.
   a. Identify types of EPDM field membranes.
   b. List and describe pre-manufactured flashing accessories.
   c. Identify attachment methods and the related attachment components.
2. Explain how EPDM roof systems are installed and repaired.
   a. Identify tools and equipment used to install EPDM roof systems.
   b. Describe how EPDM roof systems are installed.
   c. Explain basic repair methods for damaged EPDM roof systems.

Performance Tasks

Under supervision, you should be able to do the following:

1. Create a seam from a lap.
   - Position two pieces of membrane to create a properly sized 6" end lap.
   - Position a third piece of membrane over the first two with a 6" lap, creating a T-joint.
   - Install a T-joint patch.
   - Seal the edges of the seam.
2. Install EPDM membrane on an inside and an outside corner.

Trade Terms

Alignment marks
Application rate
Ballast
EPDM
Fishmouth
Flashings
Laps
Mil
Point loading
Seam
Seam tape
Self-adhered membrane
Thermoplastic
Thermoset
T-joint
Volatile organic compounds (VOCs)
Weighted roller

Industry Recognized Credentials

If you are training through an NCCER-accredited sponsor, you may be eligible for credentials from NCCER's Registry. The ID number for this module is 16207. Note that this module may have been used in other NCCER curricula and may apply to other level completions. Contact NCCER's Registry at 1.888.622.3720 or go to **www.nccer.org** for more information.

You can also show off your industry-recognized credentials online with NCCER's digital badges. Transform your knowledge, skills, and achievements into badges that you can share across social media platforms, send to your network, and add to your resume. For more information, visit **www.nccer.org**.

Videos and Digital Resources

How to Access Resources

This craft has additional videos and resources to enhance your learning experience. To view these resources, scan the QR below. The videos and resources are separated by module.

You can scan this code using the camera on your phone or mobile device to view these videos and resources.

Contents

1.0.0 EPDM System Types and Components ... 1
1.1.0 Membrane Types ... 2
1.1.1 Reinforced and Nonreinforced Membranes ... 2
1.1.2 Self-Adhered Membranes ... 2
1.2.0 Flashing and Seaming Accessories ... 3
1.3.0 Attachment Methods and Components ... 4
1.3.1 Adhesives and Primers ... 4
1.3.2 Fasteners ... 5
1.3.3 Ballast ... 6
2.0.0 Installation and Repair ... 8
2.1.0 Tools and Equipment ... 8
2.2.0 Installation Procedures ... 10
2.2.1 Adhered Installation ... 11
2.2.2 Mechanically Attached Installation ... 13
2.2.3 Ballasted Installation ... 15
2.3.0 Basic Repair Techniques ... 15
2.3.1 Repairing a Puncture ... 16
2.3.2 Repairing a Cut ... 16
2.3.3 Repairing a Fishmouth ... 17

Figures

Figure 1 EPDM roof system 1
Figure 2 EPDM membranes 3
Figure 3 EPDM pipe boot 4
Figure 4 Example of a corner 4
Figure 5 T-joint patches 4
Figure 6 Seam tape 4
Figure 7 EPDM adhesive 5
Figure 8 EPDM membrane with fasteners and plates 6
Figure 9 Ballasted roof system with river rock and vegetation 6
Figure 10 Scrub pad for cleaning EPDM membrane 9
Figure 11 Weighted roller 9
Figure 12 Seam rollers 10
Figure 13 Spraying an adhesive 10
Figure 14 Discharge the cleaning solution into an appropriate container until all the adhesive has cleared the spray wand 10
Figure 15 Orientation of EPDM membrane sheet to the path of water drainage 12
Figure 16 Priming the lap 12
Figure 17 Setting the seam tape 12
Figure 18 Removing the release film from the seam tape 13
Figure 19 EPDM boot installed on a penetration 13
Figure 20 Sealed pipe boot 14
Figure 21 Edge of membrane sheet with fasteners and alignment marks 14
Figure 22 Applying primer to the lapped area 14
Figure 23 T-joint cover at a curb corner 15
Figure 24 Fishmouth 18

SECTION ONE

1.0.0 EPDM SYSTEM TYPES AND COMPONENTS

Objective

Identify types of EPDM roof systems and their basic components.

a. Identify types of EPDM field membranes.
b. List and describe pre-manufactured flashing accessories.
c. Identify attachment methods and the related attachment components.

Trade Terms

Ballast: Material that weighs down a membrane to hold it in place. Ballast includes smooth river stone; standard precast concrete pavers; or interlocking, beveled, doweled, or contoured fit lightweight concrete pavers.

EPDM: Acronym for *ethylene propylene diene monomer*; a common type of thermoset used in single-ply roof systems.

Flashings: Components used to weatherproof or seal edges of a roof system at perimeters, penetrations, walls, expansion joints, valleys, drains, and other places where the roof covering is interrupted or terminated. For example, membrane base flashing covers the edge of the field membrane, and cap flashings or counterflashings shield the upper edges of the base flashing.

Mil: A unit of measurement equal to one-thousandth of an inch, or 0.001" (0.0254 mm). The thickness of a membrane sheet is described in mils (for example, 45-mil or 60-mil membrane).

Point loading: The act of spacing out roofing materials in different areas on the roof so that all the weight of the roofing materials is not in one place.

Seam: The weatherproof bonding of two overlapping pieces of roof membrane.

Seam tape: Self-adhering cured or uncured synthetic-rubber tape used for creating seams between membrane sheets.

Self-adhered membrane: A membrane with factory-applied adhesive that is able to bond itself without the use of an additional adhesive.

Thermoplastic: A roofing material designed to be heated and fused to itself. When cooled, it remains flexible, waterproof, and strong.

Thermoset: A class of polymers that when cured using heat, chemical, or other means, change into a substantially infusible and insoluble material.

T-joint: An area where multiple field membrane sheets overlap.

Volatile organic compounds (VOCs): Compounds and chemicals that easily evaporate at room temperature. They are found in products such as solvents, paints, and adhesives.

An **EPDM** roof system (*Figure 1*) is a single-ply system consisting of a single layer of weatherproofing. Introduced to the US market in the early 1960s, EPDM gained popularity during the Middle East oil embargo of the early 1970s. The embargo caused shortages of petroleum products and price increases that negatively affected competing asphalt-based products. While EPDM shares the single-ply market with **thermoplastic** membranes such as TPO and PVC, it remains a cost-effective option for commercial single-ply roofing.

EPDM stands for its chemical name, *ethylene propylene diene monomer*, which is a synthetic rubber classified as a **thermoset**. Thermosets maintain their shape and form after being heated during the manufacturing process. Unlike thermoplastics, which can be reheated and reshaped during installation, thermosets are not heated again during the installation process. While there are other types of thermosets used for low-slope roofing, EPDM overwhelmingly dominates the thermoset market. Since EPDM is not heated during installation, heat-related safety concerns are eliminated.

Figure 1 EPDM roof system.

Did You Know?

Low-Slope Market Share

The chart shown here illustrates roofing sales in 2015 by roof system type. Thermoset roof systems accounted for 24 percent of roofing sales.

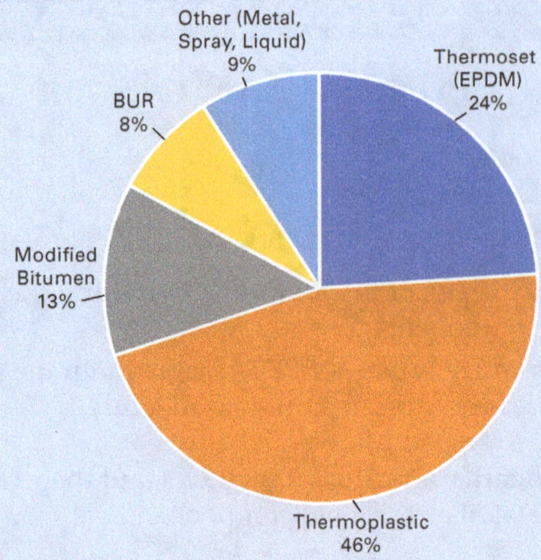

EPDM roof systems consist of a field membrane that is attached to the substrate by adhesives, mechanical fasteners, or ballast. Flashing accessories are used to seal the edges of the roof system as well as other interruptions and penetrations.

1.1.0 Membrane Types

EPDM membrane (*Figure* 2) is known for its durability, along with being particularly resistant to damage from ultraviolet (UV) light and extreme weather. It comes in three standard thicknesses: 0.045" thick, 0.060" thick, and 0.090" thick. You will commonly see these measurements written as 45 mil, 60 mil, and 90 mil. Most EPDM membranes are black, but they may also be surfaced with other colors.

EPDM is commonly sold in large rolls. A typical roll of 45-mil material is 10' wide and 100' long. Nonreinforced material weighs roughly 0.32 lb/ft^2 (1.5 kg/m^2), while reinforced material generally weighs 30–35 percent more. A 10' × 100' roll of nonreinforced 45-mil membrane weighs about 330 lb (150 kg). All other types and thicknesses will be significantly heavier.

WARNING!

Think carefully before you try to handle and maneuver a roll of material on your own. Seek help and practice good lifting techniques when maneuvering heavy materials.

When placing materials on the roof, ensure the roof is point loaded. Point loading means spacing out roofing materials in different areas on the roof so that all the weight of the roofing materials is not in one place. Putting too much material in one area can lead to roof collapse.

1.1.1 Reinforced and Nonreinforced Membranes

As its name implies, reinforced roof membrane has been strengthened with a thin layer of material such as fiberglass, polyester, nylon, or polyethylene. These materials are added in the form of a mat between layers of EPDM. Reinforced membranes were introduced to the market in the late 1980s and are more resistant to shrinkage and punctures than nonreinforced membranes. Reinforcement may be inside the membrane, or the membrane may have a backing to reinforce it, as shown in *Figure* 2.

Fleece-back membranes are a type of reinforced membrane. The reinforcement comes from fleece-backing, a type of nonwoven polyester material adhered to the underside of an EPDM sheet. The fleece-backing, like a reinforcement layer, adds strength and puncture resistance.

1.1.2 Self-Adhered Membranes

Self-adhered membrane, also called *peel-and-stick membrane,* comes with a factory-applied adhesive on the back of the sheet. It can bond directly to a substrate, without additional adhesive. This makes it cleaner and easier to install than other EPDM types, although the product is more expensive. A seam, however, is created in much the same way as standard EPDM, using primer and seam tape.

Self-adhered membrane must be applied in specific conditions. Manufacturer instructions will specify the appropriate temperature range for installation. The factory adhesive is initially less effective in cold weather, although the adhesive bond improves as the weather warms. Once applied, it is an effective option in any climate.

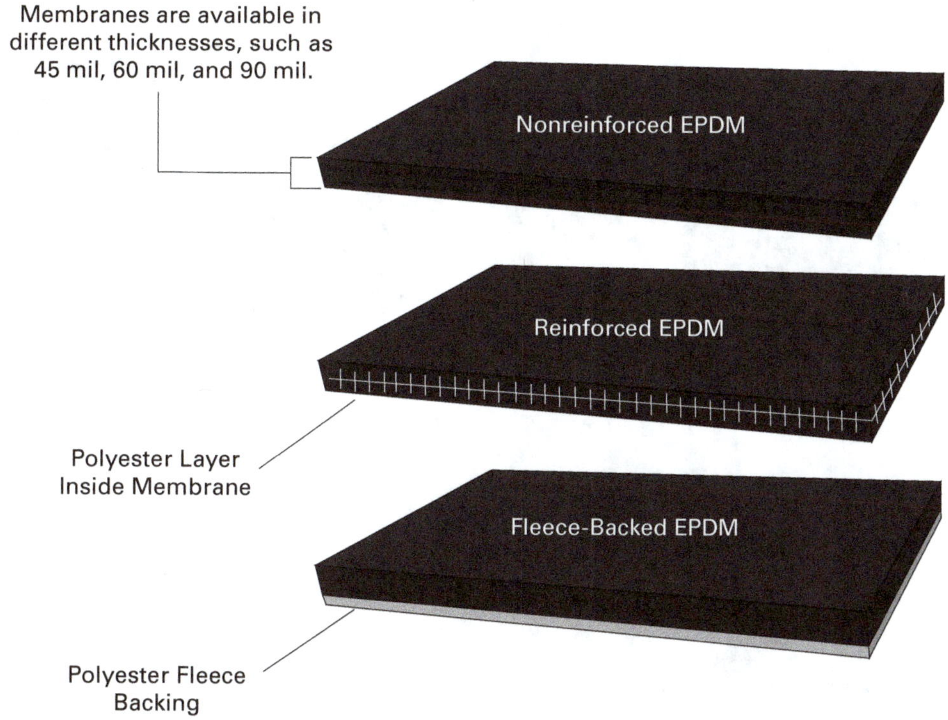

Figure 2 EPDM membranes.

> **CAUTION**
>
> Ensuring the substrate is clean, dry, and in good condition is particularly important for self-adhered membranes. Dirt or debris could prevent the system from adhering, leading to system failure.

1.2.0 Flashing and Seaming Accessories

Flashings are used to weatherproof or seal edges of a roof system at perimeters, penetrations, walls, expansion joints, valleys, drains, and similar places where the roof covering is interrupted or terminated. In low-slope systems like EPDM, flashings are generally made of the same material as the roof membrane.

These are some of the most common prefabricated flashing accessories encountered in EPDM roof systems, although there are others:

- *Boots* — Coverings made of flexible material or a combination of flexible and rigid material pre-formed to a particular shape, used to exclude dust, dirt, moisture, etc., from around a penetration (*Figure 3*). Boots are commonly placed around pipes and similar round objects. Larger penetrations and rectangular shapes usually require flashing that is field-fabricated and applied.
- *Corners* — Used to weatherproof any location where three intersecting planes come together, as shown in *Figure 4*.
- *Pitch pocket* — A sheet metal sleeve placed around a penetration and filled with grout to provide a watertight seal.
- **T-joint** *patch (Figure 5)* — An accessory used to weatherproof T-joints.
- **Seam tape** *(Figure 6)* — Used to create a seam and adhere flashing for all EPDM membrane types. It is an aggressive adhesive material that bonds the upper and lower surfaces of a lap together. Seam tape is usually 3" wide. Each roll includes a paper or plastic liner to prevent the tape from sticking to itself. Also known as *peel-and-stick tape*.
- *Cover tape* — Used where seams need to be spliced. Self-adhering cover tape usually comes in widths of 6", 9", and 12", with adhesive on the bottom and membrane on the top. Also known as *splice tape*.

Roofers can fabricate flashings from detail membrane and wall flashing membrane on the jobsite. Detail membrane, also known as *unreinforced flashing membrane*, can be shaped and stretched to ensure a tight seal around a penetration. Detail membrane must be compatible with the field membrane, but it cannot be used as field membrane. It comes in a roll and is measured and cut on-site.

Figure 3 EPDM pipe boot.

Figure 5 T-joint patches.

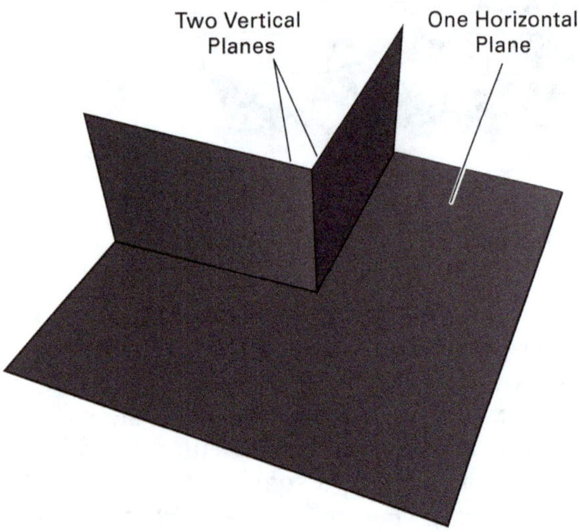

Figure 4 Example of a corner.

Figure 6 Seam tape.

Wall flashing membrane runs along a vertical wall, such as where the edge of the roof meets a parapet wall. Generally made of the same materials as the field membrane, it is typically reinforced and comes in a variety of thicknesses.

Uncured rubber is another option that can be used to flash around irregular shapes. Uncured rubber is thermoset material that has not been "set" yet and self-cures in the sun.

1.3.0 Attachment Methods and Components

EPDM roof systems may be attached in three primary ways: adhered, mechanically attached with fasteners, or weighed down with ballast. Both fully adhered and ballasted EPDM systems are more resistant to uplift.

> **NOTE**
> Attachment methods are introduced in NCCER Module 16106, *Introduction to Low-Slope Roofing.*

1.3.1 Adhesives and Primers

Adhesives (*Figure 7*) are used to secure EPDM roof membrane to the substrate. They can be applied by spraying, rolling, or brushing. The contents should be well stirred before use.

Adhesives and primers contain **volatile organic compounds (VOCs)**. VOCs are gases that are released into the air from products or processes. Different adhesives and primers contain different levels of VOCs. In some cases, it may be necessary to temporarily shut down ventilation units while applying adhesives and primers until the fumes settle.

WARNING!

Before applying an adhesive or primer, review the Safety Data Sheet (SDS) and take the necessary precautions. Adhesive and primer products may be highly flammable and/or an inhalation hazard. Ensure that any electrically powered equipment is properly grounded and keep the equipment as far from the product as possible.

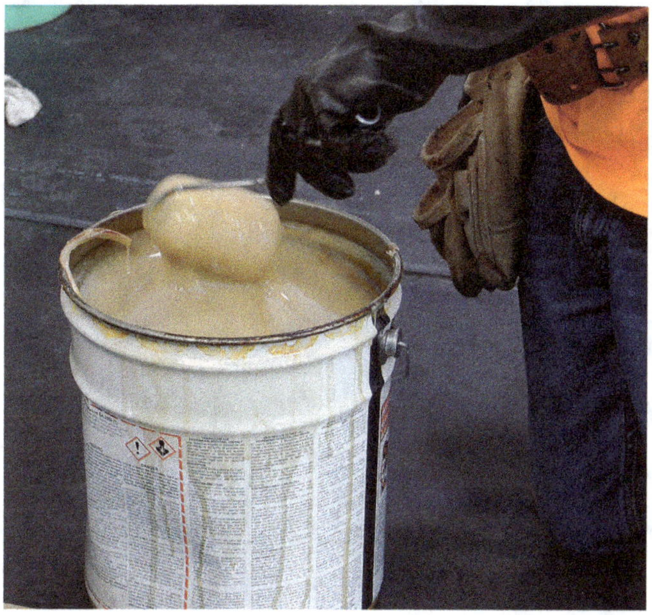

Figure 7 EPDM adhesive.

Adhesives

There are two classes of adhesives used for EPDM attachment. Note that there is a difference in how the two products are used:

- *Water-based adhesives* are applied only to the substrate and not to the membrane itself. The membrane is then placed while the adhesive is still wet.
- *Solvent-based adhesives* are applied to both the substrate and the membrane. The adhesive is then allowed to dry to a point where it does not come off when you touch it with a gloved knuckle or finger. The amount of time required can vary significantly due to the weather conditions and the current temperature of the substrate and membrane. If the materials are hot, the time required for the adhesive to set up may be short.

CAUTION

Keep containers of adhesive tightly closed when not in use. The solvent, regardless of type, will consistently evaporate as long as the container is open, spoiling product consistency.

Water-based adhesives may be required for buildings that are occupied (such as hospitals) to minimize odors. However, water-based adhesives are more temperature sensitive and don't perform as well as solvent-based adhesives. Water-based adhesives usually take longer to dry than solvent-based ones, since the water evaporates more slowly than most petroleum-based solvents. Solvent-based adhesives are typically required for fleece-backed EPDM.

Primers

Primers are products that appear similar to adhesives, but primers are definitely not adhesives. They are used in locations where two pieces of EPDM membrane are being joined to each other, such as flashings and seams. Primers are cleaning agents that promote and improve adhesion. As they dry, they also become tacky, much like an adhesive. This is why those new to the craft easily confuse the two products.

Like a solvent-based adhesive, the primer must be allowed to dry properly before the seam or flashing is closed, and it is tested in the same manner as an adhesive. The material should not be stringy or remain attached to a gloved fingertip or knuckle when touched. Note that no primer is used between the EPDM membrane and the substrate throughout the field. It is only used to prepare seams and flashing materials, typically before seam tape is applied.

1.3.2 Fasteners

Fasteners mechanically secure the EPDM roof membrane through the insulation and into the deck. A common technique is a screw driven through a round metal plate. The type and length of fastener needed depends on the roof deck material. On a mechanically attached EPDM roof, fasteners are applied to the underlying membrane at a lap joint, and then the upper layer of the lap is adhered to the lower membrane over the fasteners (*Figure 8*), forming a seam. Due to the potential for leaks, mechanical fasteners must not be left exposed on the roof.

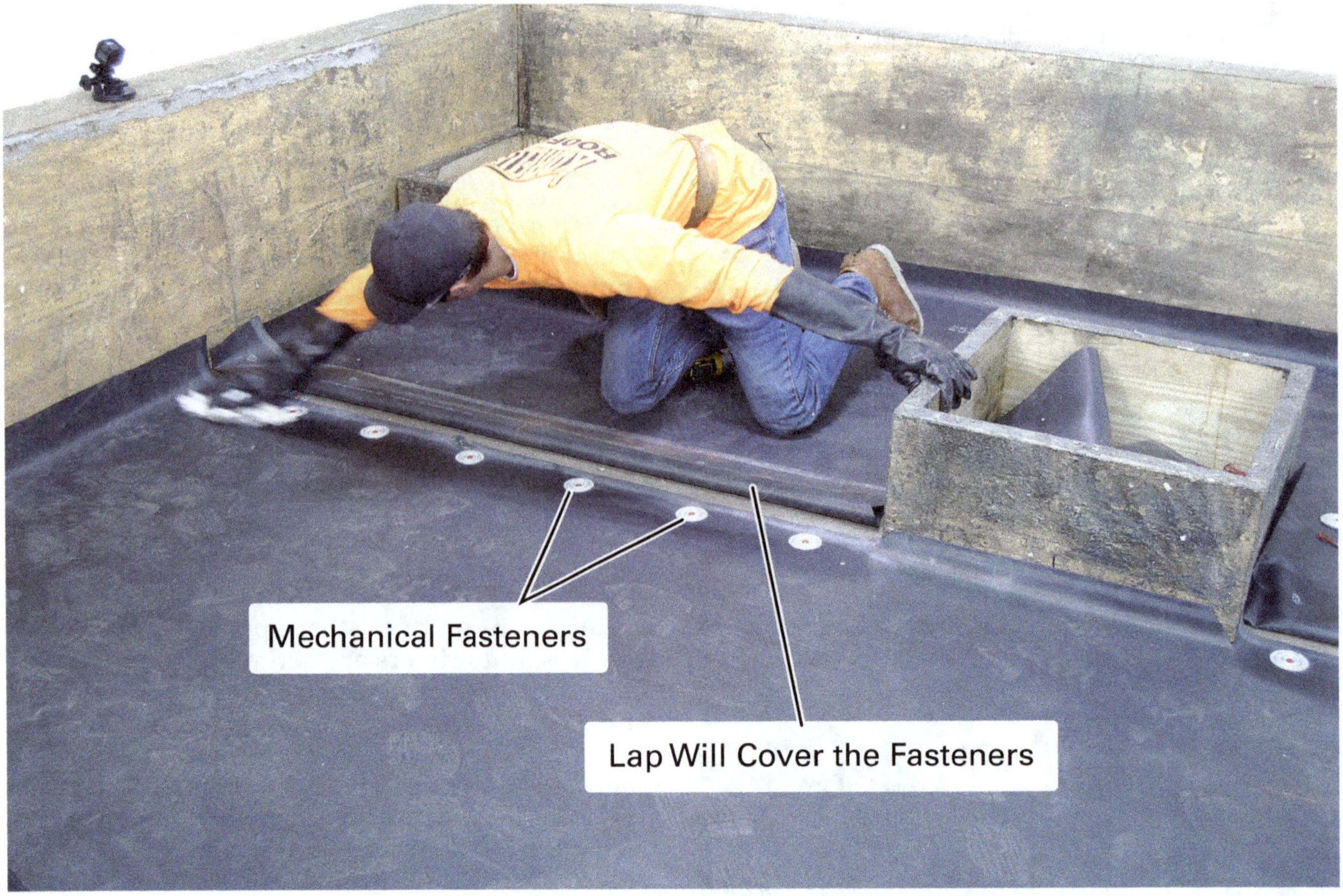

Figure 8 EPDM membrane with fasteners and plates.

1.3.3 Ballast

Ballast (*Figure 9*) holds a roof membrane system in place by simply weighing it down. In this type of installation, a slip sheet is required underneath the ballast to protect the membrane. Some common ballast materials used on EPDM roof systems include the following:

- *River rock* — Rock that has a relatively smooth surface and lacks sharp edges. River rock is larger and smoother than gravel.
- *Pavers* — Precast slabs of a consistent size and shape, usually made of concrete, that provide ballast. Pavers may be used to create walk paths across the roof, while the rest of the roof is covered by rock. This arrangement is common where HVAC equipment is installed on the rooftop and HVAC technicians will be consistent visitors.

WARNING!

Pavers are very heavy. Always use safe lifting techniques and get help before attempting to move something heavy.

Figure 9 Ballasted roof system with river rock and vegetaion.

Going Green

Energy Efficient Roofing

Adding ballast materials to EPDM roofs can make them nearly as energy saving as lighter-colored roofs. A study from Oak Ridge National Laboratory found that ballasted EPDM remains in a surface temperature range comparable to an exposed white membrane. This makes it an effective alternative for keeping the building cooler at a reduced cost. The study also noted that the performance of white roofs degraded over time due to the accumulation of dirt, debris, and stains, while ballasted roofs maintained their effectiveness.

1.0.0 Section Review

1. Which of the following is a characteristic of self-adhered EPDM membrane?
   a. It is the least expensive form of EPDM membrane.
   b. Self-adhered EPDM is best installed in warm weather.
   c. The adhesive is especially effective when it is installed in cold weather.
   d. The adhesive adds flexibility and strength to the membrane.

2. It is common to use a variety of materials, such as TPO and PVC, when flashing EPDM roof systems.
   a. True
   b. False

3. To reduce the potential for leaks, _____.
   a. a primer should always be applied before an adhesive is applied
   b. add adhesive to the underside of the metal plate before installing a fastener
   c. mechanical fasteners must not be left exposed on the roof
   d. prime the immediate area before installing a mechanical fastener

Section Two

2.0.0 Installation and Repair

Objective

Explain how EPDM roof systems are installed and repaired.

a. Identify tools and equipment used to install EPDM roof systems.
b. Describe how EPDM roof systems are installed.
c. Explain basic repair methods for damaged EPDM roof systems.

Performance Tasks

1. Create a seam from a lap.
   - Position two pieces of membrane to create a properly sized 6" end lap.
   - Position a third piece of membrane over the first two with a 6" lap, creating a T-joint.
   - Install a T-joint patch.
   - Seal the edges of the seam.
2. Install EPDM membrane on an inside and an outside corner.

Trade Terms

Alignment marks: Marked or chalked lines extending the length of membrane and used to align the lap of the adjacent sheet accurately and consistently.

Application rate: The average quantity (mass, volume, or thickness) of material applied per unit area.

Fishmouth: An opening in the edge of a membrane caused by wrinkling or shifting of sheets during installation.

Laps: Areas where membrane sheets or flashings overlap the membrane below. Laps are sealed to create seams.

Weighted roller: A heavy object pushed over adhered or self-adhered membrane to smooth it out and eliminate bubbles, wrinkles, fishmouths, and other deficiencies.

There are a variety of tools and methods used to install EPDM roof systems. EPDM may be adhered, mechanically attached, or ballasted. Regardless of which approach is used, you will create seams by adhering membranes together at laps. This is commonly done with primer and seam tape.

When installing EPDM, you must be prepared to repair damage or resolve mistakes that occur during installation. This section also introduces simple repair techniques for three common types of membrane damage.

2.1.0 Tools and Equipment

EPDM roof systems are not especially demanding in terms of tools and equipment. Basic tools are generally all that is required to install and repair EPDM.

> **NOTE**
> Basic tools for low-slope roofing are covered in NCCER Module 00103, *Introduction to Hand Tools,* and NCCER Module 00104, *Introduction to Power Tools.*

Tools required to install EPDM roof systems include the following:

- *Long scissors/shears* — Used to cut membrane. Long scissors are preferred, as they make better cuts.
- *White rags* — Used with EPDM cleaners. White rags are preferred since they have no dye that could contaminate the surface.
- *Chalk/crayons/pencils* — Used to mark guidelines for primer and seam tape. Also used to mark components hidden underneath field membrane, such as drains, for other workers' awareness.
- *Cordless drill* — Used to drive fasteners for mechanical attachment.
- *Stand-up drill* — An ergonomic alternative to a handheld drill that allows roofers to drive fasteners without bending down.
- *Mildly abrasive scrub pads* (*Figure 10*) — Used with cleaning agents to scrub membrane laps. Scrub pads may also be used to apply primer.
- *Rollers* — Dipped in adhesive or primer and used to roll the material onto a surface.

- *Paint brushes* — Dipped in adhesive or primer and used to spread the material onto a surface.
- *Caulking gun* — Used to apply sealants.
- **Weighted roller** — A heavy roller used over the field membrane and seams to smooth it out and ensure consistent adhesion (*Figure 11*).
- *Seam roller (Figure 12)* — Small, handheld rollers are also used along seams and flashing joints. Also called a *hand roller.*
- *Push broom* — Sometimes used instead of a roller to smooth out wrinkles and apply slight pressure to the membrane, ensuring consistent adhesion. Of course, it is also used to sweep the substrate clean. Common leaf blowers may also be used to blow debris from the installation area.

EPDM adhesives can also be sprayed using a spray wand. The adhesive is provided by the manufacturer in a pressurized tank, which is then connected to a spray wand and hose (*Figure 13*). Although the largest projects may require more sophisticated and powerful spray equipment that remains on the ground, these self-contained spray systems are convenient and effective for many projects.

Using spray equipment typically requires the following steps:

WARNING!

Always follow the manufacturer's instructions for the product in use. Be sure to wear the appropriate PPE and review the SDS for the adhesive before beginning.

It is important to follow applicable safety procedures for putting away a pressurized tool. If you are not sure what to do, ask your supervisor.

Step 1 Connect one end of the hose to the spray wand. You will need a wrench to secure the connection.

Step 2 Attach the other end of the hose to the pressurized adhesive tank and tighten the fitting.

Step 3 Follow the manufacturer instructions to ensure the contents are well mixed.

Step 4 Ensure the trigger of the spray wand is in a safe position and open the valve on the tank.

Step 5 Place the end of the spray wand in a box to contain the spray. Pull the trigger to purge air out of the hose and wand, and to ensure the spray pattern is consistent.

Figure 10 Scrub pad for cleaning EPDM membrane.

Figure 11 Weighted roller.

Step 6 Apply the adhesive using overlapping strokes as shown in *Figure 13*.

Step 7 Once the adhesive has been applied, close the valve on the tank. Place the end of the spray wand in a box again, to capture the spray, and pull the trigger to relieve the remaining pressure in the hose and wand.

Step 8 Disconnect the hose from the adhesive tank. The adhesive tank can be used again, until it is empty.

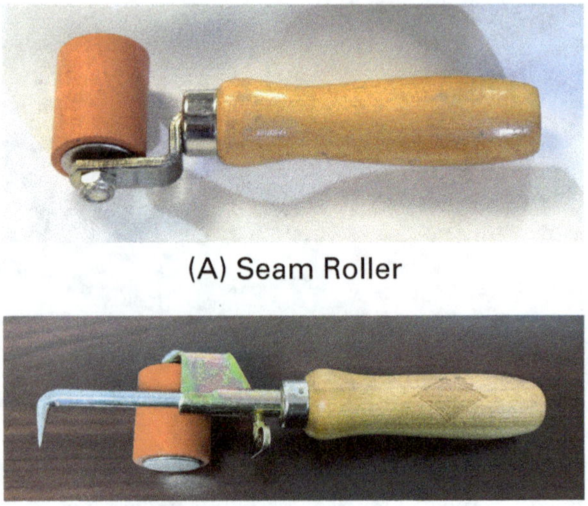

(A) Seam Roller

(B) Seam Roller with Probe

Figure 12 Seam rollers.

Figure 13 Spraying an adhesive.

One of the most important steps when working with an adhesive is cleanup. If the equipment is not properly cleaned, adhesive will cure inside the hose and spray wand and will destroy them. To ensure the equipment remains in good condition and ready for the next application, manufacturers provide adhesive cleaning products in pressurized tanks. After disconnecting the hose from the adhesive tank, connect it to the tank of adhesive cleaner. Since the discharge will be a liquid solvent instead of adhesive, it is best to discharge it into an appropriate container. Open the tank valve and pull the trigger to discharge the adhesive and cleaning solution (*Figure 14*). Then close the tank valve. Pull the trigger again to discharge any remaining pressure and disassemble the equipment.

CAUTION

Failure to clean the hose and spray wand properly and thoroughly will likely cause permanent damage to the wand and hose. Check with your supervisor to find out if the hose is disposable.

2.2.0 Installation Procedures

With teamwork, EPDM is relatively easy and quick to install compared to some other roof systems. Many of the steps for adhered, mechanically fastened, and ballasted installation are the same as those for thermoplastic roof systems, described in NCCER Module 16206, *Thermoplastic Roof Systems*. However, one important difference is found in the seaming process. While thermoplastic membranes are heated and welded to create a seam, EPDM membranes are adhered at seams.

NOTE

The procedures in this section are examples only and assume the deck has been properly prepared. Always comply with company procedures and manufacturer's instructions when installing a roof membrane.

WARNING!

Before beginning an installation, don the required PPE and ensure you are familiar with the precautions in the SDS for each product to be used on the job.

Figure 14 Discharge the cleaning solution into an appropriate container until all the adhesive has cleared the spray wand.

2.2.1 Adhered Installation

An adhered installation process is often used for EPDM membranes today. The approach is somewhat slower than mechanically attaching the membrane, but the material is more consistently attached and secured to the substrate when it is adhered. Adhesives may be liquid applied or spray applied, and some membranes may come with adhesive already on the back of the membrane (self-adhered).

The following steps outline a general procedure for adhering EPDM roof membrane. However, installation methods can vary due to weather conditions, product characteristics, and contractor preferences:

Step 1 Unroll the roof membrane so that wrinkles and buckles are not formed. Any wrinkles or buckles must be removed from the sheet prior to permanent securement.

WARNING!

Rolling out an EPDM roll creates static electricity, possibly resulting in a spark. These sparks can ignite fumes that settle in low spots on the roof and start a fire.

Step 2 Lay out and orient the membrane roll so that the water is routed over the lip of the seam and not against it (*Figure 15*). Allow the membrane to relax in place and reach the ambient temperature before adhering it.

Step 3 Cut the membrane roll to the desired length and confirm it is oriented correctly.

Step 4 Fold the membrane sheets back to expose the underside and the substrate.

Step 5 Apply the adhesive to the membrane and/or substrate by spraying, rolling, or brushing. Once applied, the layer of adhesive should be smooth, thin, and free of voids or puddles. When spraying adhesive, follow the manufacturer's guidelines for temperature, humidity, and application rate. Only a single coat is applied.

- Apply water-based adhesives only to the substrate.
- Apply solvent-based adhesives to both the membrane underside and the substrate.

NOTE

For solvent-based adhesives, wait for the adhesive to dry. Confirm that the adhesive is ready to form a bond by touching a finger or knuckle to the surface while wearing a nitrile glove. The adhesive should be tacky but not stringy, leaving no residue on the glove. When using a water-based adhesive, the membrane is generally placed while the adhesive is still wet.

Step 6 Fold the membrane back over the adhesive and substrate.

Step 7 Roll the first membrane sheet with a weighted roller to ensure consistent contact with the substrate and adhesive. Some professionals may prefer a push broom for this purpose.

Step 8 Using a white marker, draw alignment marks on the edge of the sheet to guide seaming. Alignment marks show where to lay the second sheet to ensure adequate overlap. They also provide a guideline for where to apply primer. Place the alignment marks about ¼" (~6 mm) from where the edge of the overlapping sheet will actually align, and about 2' (~61 cm) apart.

Step 9 Lay down the next membrane roll parallel to the first one, ensuring that it overlaps the previous section to create a reliable seam. The overlap is typically 6" (~15 cm). Do not apply adhesive to the lapped areas.

Step 10 Repeat *Steps 1–9*, laying down each roll successively, overlapping each previous section consistently and in a manner that will shed water over the side laps as shown in *Figure 15*.

Step 11 Reopen the lap and apply a primer to both EPDM mating surfaces. Once applied, the primer should be smooth, thin, and free of voids or puddles. The primer should extend just beyond the alignment marks (*Figure 16*) on the lower membrane, and at least the width of the lap on the upper membrane.

Step 12 Wait for the primer to dry according to the manufacturer's instructions. Confirm that the primer is ready by touching it with a gloved finger or knuckle. It should be tacky but not stringy, leaving no residue on the glove.

Step 13 Position the outside edge of the seam tape along the alignment marks and set it (*Figure 17*). With the release liner still on the top side of the tape, apply pressure with a roller to eliminate bubbles and wrinkles in the tape.

Step 14 Close the lap by laying the top sheet back over the bottom sheet. Just the edge of the seam tape should be visible.

Step 15 Pull the release film off the tape (*Figure 18*), pressing or hand rolling over the new seam as you go. Note the 45-degree angle at which the release film is being held as it is pulled, and that it is kept low and close to the roof.

Step 16 At every T-joint, apply a T-joint cover and seal the edges. Apply a manufacturer-recommended sealant around the perimeter of the T-joint cover.

Step 17 Anchor any flashing components to areas where the roof system is interrupted or terminated.

Self-adhered membranes do not require additional adhesive to be applied. The underside of a self-adhered membrane is protected by a release film, which prevents the membrane from bonding to itself during shipping and handling. During installation, the release paper is peeled off while the membrane is pressed onto the substrate. Primer may or may not be required for self-adhered membranes. Follow the manufacturer's instructions for specific installation guidelines.

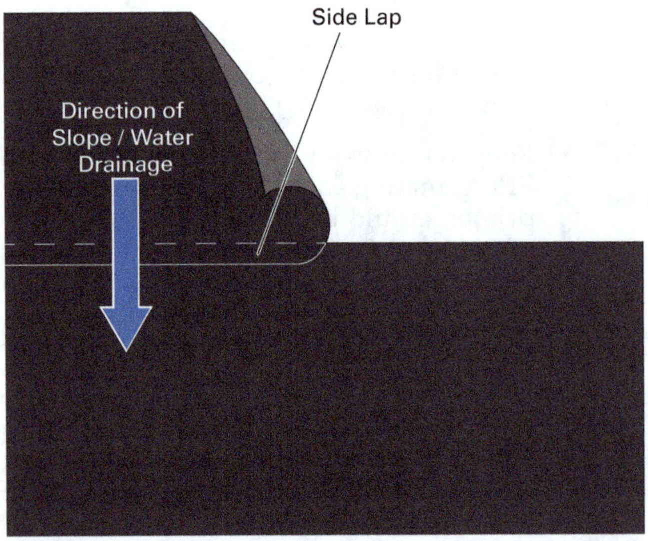

Figure 15 Orientation of EPDM membrane sheet to the path of water drainage.

Flashing a Roof Penetration

These same techniques are used to flash penetrations, curbs, and similar areas. In *Figure 18*, note the 4" PVC pipe penetrating the roof. The membrane has been cut and feathered to fit around the pipe as the membrane went down. A prefabricated 4" EPDM boot will be placed around the penetration. The boot extends away from the pipe several inches.

The membrane is usually mechanically fastened to the substrate around a penetration like this one, before the boot is placed and within its perimeter. The surfaces are then cleaned (if needed), primed, taped, and rolled just like a seam. A clamp is placed around the neck of the boot to ensure a good seal is formed against the pipe. Sealant is added to the lip of the boot around the pipe, and around the perimeter of the boot. The completed installation is shown in *Figure 19*. Note the two T-joint patches installed adjacent to it.

Figure 16 Priming the lap.

Figure 17 Setting the seam tape.

Applying Sealants

Some seams may require sealant—for example, membrane edges that have been hand-cut need sealant. For every project, seams must be inspected as a quality assurance measure. This must be done before any sealant is applied.

Beyond looking for visible signs that the membrane isn't properly sealed along the edge, there are two primary things to look for when inspecting a seam:

- *Evidence of the primer* — A small amount of primer should still be visible outside the seam. If there is no primer visible, then part of the edge will not likely adhere as it should.
- *Visible seam tape* — Just a small line of seam tape should be visible beyond the edge of the top layer of membrane. If the edge of it disappears beneath the edge of the top layer, water and debris can also get beneath the membrane and work to weaken the joint further over time.

Figure 18 Removing the release film from the seam tape.

Figure 19 EPDM boot installed on a penetration.

Since the sealant is clearly visible, it needs to look neat and professional. It also needs to be done in a way that doesn't interfere with water drainage. A rough, bumpy bead will hold small pockets of water back, and it doesn't look very good either. A smooth, steady, consistent bead does the best job of sealing the seam and has a professional look (*Figure 20*). The sealant bead should cover the visible portion of the seam tape as well as the edge of the top layer of membrane.

2.2.2 Mechanically Attached Installation

Mechanical attachment is another common installation method for EPDM roof systems. However, since the membrane is not secured across its entire surface, pressure changes within the structure can cause the membrane to lift away from the substrate and appear inflated. Due to the stress that uplift places on the membrane, mechanical attachment generally requires reinforced EPDM membrane to reduce the potential for tearing to occur at fastener locations.

The following steps outline a general procedure for mechanically attaching EPDM roof membrane:

Step 1 Unroll the roof membrane so that wrinkles and buckles are not formed. Any wrinkles or buckles must be removed from the sheet prior to permanent securement.

Step 2 Lay out and orient the membrane roll so that the water is routed over the lip of the seam and not against it, as shown in *Figure 15*. Allow the membrane to relax in place and reach the ambient temperature before adhering it.

Step 3 Cut the membrane roll to the desired length and make sure it is oriented correctly.

Step 4 Draw alignment marks after laying out each sheet to guide the placement of the next sheet. Place the alignment marks about $\frac{1}{4}$" (~6 mm) from where the edge of the overlapping sheet will actually align, and about 2' (~61 cm) apart. When working on a short seam, shorter spacing is better.

Step 5 Lay out the membrane plates, following the manufacturer's guidelines for how the plates should be patterned and spaced.

Step 6 Drill the fasteners directly through the insulation boards and into the deck below at the edge of the membrane roll. Follow the manufacturer's guidelines for fastener length to ensure they penetrate to the proper depth. For EPDM, fasteners along the seams are generally spaced at 12" (~30 cm) intervals.

Step 7 Lay down the next membrane roll, ensuring that there is enough overlap to accommodate the seam tape (typically 6" or about 15 cm). The mechanical fasteners should be placed within the lap area (*Figure 21*).

Step 8 Repeat *Steps 1–7*, laying down each roll successively and overlapping each seam.

Step 9 After aligning the sheets, reopen the lap and apply primer equally to both mating surfaces (*Figure 22*). Once applied, the primer should be smooth, thin, and free of voids or puddles. The primer should extend just beyond the alignment marks on the lower membrane, and at least the width of the lap on the upper membrane.

Step 10 Wait for the primer to dry according to the manufacturer's instructions. Confirm that the primer is ready by touching it with a gloved finger or knuckle. It should be tacky but not stringy, leaving no residue on the glove. (Liquid-applied adhesives may also be used, although they are uncommon.)

Step 11 Position the outside edge of the seam tape along the alignment marks and set it. With the release liner still on top of the tape, press with a hand roller to eliminate bubbles and wrinkles in the tape. Roll diagonally at a 45-degree angle from the edge of the seam.

Step 12 Close the lap by laying the top sheet back over the bottom sheet. Just the edge of the seam tape should be visible.

Step 13 Pull the release film off the tape, pressing or hand rolling over the seam as you go.

Step 14 At every intersection of horizontal and vertical seams, apply a T-joint cover and seal the edges (*Figure 23*). Apply a manufacturer-recommended sealant around the perimeter of the T-joint cover.

Step 15 Anchor any flashing components to areas where the roof system is interrupted or terminated.

Figure 20 Sealed pipe boot.

Figure 21 Edge of membrane sheet with fasteners and alignment marks.

Figure 22 Applying primer to the lapped area.

Figure 23 T-joint cover at a curb corner.

2.2.3 Ballasted Installation

Although ballasted EPDM systems are less common than adhered and mechanically attached installations, you may encounter them and should be familiar with the installation process.

The following steps outline a general procedure for installing a ballasted EPDM roof:

Step 1 Unroll the roof membrane so that wrinkles and buckles are not formed. Any wrinkles or buckles must be removed from the sheet prior to permanent securement.

Step 2 Lay out and orient the membrane roll so that the water is routed over the lip of the seam and not against it, as shown in *Figure 15*. Allow the membrane to relax in place and reach the ambient temperature before adhering it.

Step 3 Cut the membrane roll to the desired length and make sure it is oriented correctly.

Step 4 Lay down the next membrane roll, ensuring that there is enough overlap to accommodate the seam tape, typically 6" (~15 cm). Do not apply adhesive to the lapped areas.

Step 5 Repeat *Steps 1–4*, laying down each roll successively and overlapping each seam.

Step 6 After aligning the sheets, place alignment marks on the bottom sheet along the edges of each run of membrane. Place the alignment marks about 1/4" (~6 mm) from where the edge of the overlapping sheet will actually align, and about 2' (~61 cm) apart.

Step 7 Reopen the lap and apply a primer to both EPDM mating surfaces. Once applied, the primer should be smooth, thin, and free of voids or puddles. The primer should extend just beyond the alignment marks on the lower membrane, and at least the width of the lap on the upper membrane.

Step 8 Wait for the primer to dry according to the manufacturer's instructions. Confirm that the primer is ready by touching it with a gloved finger or knuckle. It should be tacky but not stringy, leaving no residue on the glove.

Step 9 Position the outside edge of the seam tape along the alignment marks and set it. With the release liner still on top of the tape, press with a hand roller to eliminate bubbles and wrinkles in the tape.

Step 10 Close the lap by laying the top sheet back over the bottom sheet. Just the edge of the seam tape should be visible.

Step 11 Pull the release film off the tape, pressing or hand rolling over the seam as you go.

Step 12 At every intersection of horizontal and vertical seams, apply a T-joint cover and seal the edges. Apply a manufacturer-recommended sealant around the perimeter of the T-joint cover.

Step 13 Anchor flashing components to areas where the roof system is interrupted or terminated. The membrane will be mechanically attached at the perimeter and similar flashing areas.

Step 14 Apply the ballast material over the membrane carefully, following the manufacturer's guidelines for application rate.

> **NOTE**
> Some ballasted installations require a separator material to protect the EPDM membrane from the ballast.

2.3.0 Basic Repair Techniques

While EPDM is a durable and long-lasting single-ply membrane, it can be damaged or misapplied during its installation. Therefore, workers should be able to fix their own common mistakes. This section will cover basic repair techniques for three types of membrane damage: a puncture, a cut, and what is known as a fishmouth.

Workers should be generally aware of wrinkles in the field membrane. Wrinkles do not necessarily void warranties or pose a risk to weatherproofing, but over time cracks can form in these areas. They may also raise aesthetic concerns for building owners.

Repairs are sometimes required during the installation process, when the membrane is still fresh and relatively clean. However, repairs may also be required after the membrane has weathered for months or years. In this case, thorough cleaning of the surface is required before primers or adhesives are applied. Always use products specifically designed for this purpose, and for the type of membrane currently in use. Never attempt to use gasoline or similar products for this purpose.

CAUTION

Gasoline, certain oils, roof cement, and oil-based paint can severely damage or destroy EPDM membrane. Avoid using these and similar materials on all EPDM products. As a best practice to avoid accidents, these products should not be present on the roof during the installation process.

WARNING!

Membrane cleaners are made of dangerous chemicals. Consult the SDS for safe use of membrane cleaner.

2.3.1 Repairing a Puncture

A *puncture* is a small hole in the field membrane. Punctures pose a risk because they can allow moisture to penetrate the roof system. Whenever you are working on this or another type of membrane roof, closely examine any areas where a tool or other sharp object is dropped. It is better to make a repair as soon as a deficiency is discovered than to wait for a leak to develop. Use a marker or crayon to mark deficiencies immediately when you notice them so they are not overlooked or forgotten.

The following steps outline a general procedure for repairing punctures in the EPDM roof membrane:

Step 1 Identify the puncture and make sure it is marked.

Step 2 Examine the area and prepare a patch. Self-adhering membrane is often used as it is more convenient, but a patch can also be cut from standard membrane. The patch should extend at least 3" (~8 cm) away from the edge of the puncture in all directions. This means that a patch should never be smaller than 6" × 6" (~15 cm × 15 cm) or a 6" circle, but it may be larger. Round all the corners of the patch—square corners are more likely to be disturbed and broken loose by foot traffic.

Step 3 Remove dirt and residue from the membrane using a manufacturer-approved, low-residue cleaner and a mildly abrasive scrub pad. Clean an area that is several inches larger than the patch on all sides.

Step 4 Wipe the surface clean with a cloth moistened with a manufacturer-approved membrane cleaner.

WARNING!

Consult the SDS for safe use of membrane cleaner.

Step 5 Lay the patch over the repair area, with the patch centered on the puncture. Mark the edges of the patch. Remove it and then apply a manufacturer-approved primer evenly over the area. The primer should extend at least $\frac{1}{4}$" beyond the perimeter of the patch.

Step 6 Wait for the primer to dry according to the manufacturer's instructions.

Step 7 Adhere the patch over the puncture following the manufacturer's guidelines.

Step 8 Seal the edges as directed by the manufacturer's guidelines. This typically requires the application of an approved elastomeric sealant around the edges.

2.3.2 Repairing a Cut

A cut is essentially an elongated puncture. Workers must frequently use the freshly covered roof as a surface on which to lay out additional sheets and flashing pieces. It is very easy to accidentally slice the new roof below the piece being cut. To prevent this from happening, protect the new roof by laying something over it (such as lumber, layers of cardboard, metal, etc.) when cutting the membrane.

If the membrane is cut during installation procedures, it will need to be repaired. The repair process for a cut is similar to that for a puncture, but cuts do require mechanical fasteners to be installed on each side before the patch is applied. The following steps outline a general procedure for repairing cuts in the EPDM membrane:

Step 1 Identify the cut and make sure it is marked.

Step 2 Pull back the cut membrane to examine underneath for significant damage to the substrate or the presence of dirt or moisture. Remove any debris or dry any moisture. If the damage beneath the membrane is significant, report the problem to your supervisor before patching the cut.

Step 3 Examine the area and prepare a patch. For cuts, the patch is more likely to be a rectangle that is 6" (~15 cm) wide and as long as necessary to cover the cut from end to end, plus 6". This allows the patch to extend 3" (~7.6 cm) beyond each end of the cut. Round all the corners of the patch.

Step 4 Remove dirt and residue from the membrane using a manufacturer-approved, low-residue cleaner and a scrub pad. Clean an area that is several inches larger than the patch on all sides.

Step 5 Wipe the area clean with a cloth moistened with a manufacturer-approved membrane cleaner.

Step 6 Mechanically fasten the membrane along each edge of the cut. Ensure that the fasteners are of the proper length to penetrate the deck below.

Step 7 Lay the patch over the repair area, with the patch centered on the cut. Mark the edges of the patch. Remove it and then apply a manufacturer-approved primer evenly over the area. The primer should extend at least ¼" beyond the perimeter of the patch.

Step 8 Wait for the primer to dry according to the manufacturer's instructions. Confirm that the primer is ready by touching it with a gloved finger or knuckle. It should be tacky but not stringy, leaving no residue on the glove.

Step 9 Adhere the patch over the puncture following the manufacturer's guidelines.

Step 10 Seal the edges as directed by the manufacturer's guidelines. This typically requires the application of an approved elastomeric sealant around the edges.

2.3.3 Repairing a Fishmouth

A fishmouth (*Figure 24*) is a raised opening in a lapped edge or seam, usually caused by wrinkling or shifting of membrane during installation. Fishmouths are a risk as they collect dirt and water and threaten the integrity of the seam. The one shown in *Figure 24* is rather large; they are often much smaller.

The following steps outline a general procedure for repairing fishmouths in the EPDM membrane:

Step 1 Identify the fishmouth and make sure it is marked.

Step 2 Remove the fishmouth by cutting or slicing the loose portion away.

Step 3 Examine the area and prepare a patch. The patch may be a square, rectangle, or circle, and large enough to extend at least 3" (~7.6 cm) away from the edge of the removed section in all directions. Round all the corners of the patch.

Step 4 Remove dirt and residue from the membrane using a manufacturer-approved, low-residue cleaner and a scrub pad. Clean an area that is several inches larger than the patch on all sides.

Step 5 Wipe clean with a manufacturer-approved membrane cleaner.

Step 6 Lay the patch over the repair area, with the patch centered on the removed fishmouth. Mark the edges of the patch. Remove it and then apply a manufacturer-approved primer evenly over the area. The primer should extend at least ¼" beyond the perimeter of the patch.

Step 7 Wait for the primer to dry according to the manufacturer's instructions. Confirm that the primer is ready by touching it with a gloved finger or knuckle. It should be tacky but not stringy, leaving no residue on the glove.

Step 8 Adhere the patch over the removed fishmouth following the manufacturer's guidelines.

Step 9 Seal the edges as directed by the manufacturer's guidelines. This typically requires the application of an approved elastomeric sealant around the edges.

Figure 24 Fishmouth.

Membrane Field-Uplift Testing

Over the years, uplift testing of membrane roofs has sometimes been a post-installation requirement to test the quality of the bond between the membrane and substrate. One procedure was developed by ASTM International, formerly known as American Society forTesting and Materials International, while the other was developed by the Factory Mutual (FM) Global Insurance Company.

The two procedures are similar and require the use of a 5' × 5' (~1.5 m × 1.5 m) chamber equipped with a vacuum pump. By creating a negative pressure inside the chamber, which is placed on the membrane, the membrane's adhesion to the substrate is tested.

It is important to note that the validity of the testing has been discussed for many years.There are many factors that can affect the results and lead to an invalid evaluation. For a complete discussion on the process and the factors of concern, access an Industry Issue Update prepared by the NRCA titled *Field-UpliftTesting,* located at **https://nrca.net/technical/guidelines-resources/bulletins-reports-archive**. If you see a device similar to this one being lifted to a roof, you will understand how it will be used.

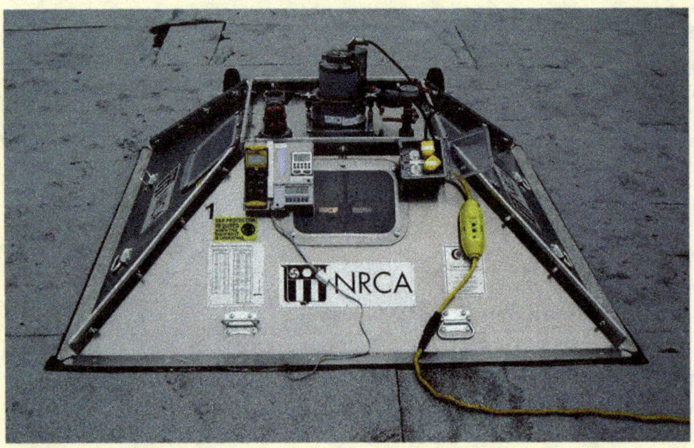

Figure Credit: Courtesy of the National Roofing Contractors Assoc.

2.0.0 Section Review

1. Which of the following may be used to clean laps?
   a. Steel wool
   b. Scrub pad
   c. Gasoline
   d. Pressure washer

2. When preparing the laps of a seam with primer, where should the primer be applied?
   a. Apply it to the upper membrane only, about half the width of the lap.
   b. Apply it to the lower membrane only, up to but not past the alignment marks.
   c. Apply it to both the upper and lower membrane, extending just beyond the alignment marks on the lower membrane.
   d. Apply it to both the upper and lower membrane, up to but not past the alignment marks on the lower membrane.

3. Gasoline is highly recommended for cleaning a weathered EPDM membrane.
   a. True
   b. False

Review Questions

1. A measurement of 45 mil is equal to _____.
   a. 4,500"
   b. 45"
   c. 4.5"
   d. 0.045"

2. Which of the following is another name for *self-adhered membrane*?
   a. Peel-and-stick membrane
   b. Reinforced membrane
   c. Nonreinforced membrane
   d. Mechanically attached membrane

3. Water-based adhesives usually take longer to dry than solvent-based adhesives.
   a. True
   b. False

4. Solvent-based adhesives _____.
   a. can be mixed with water-based adhesives
   b. are only applied to the membrane
   c. are applied to both the substrate and the membrane
   d. can be used instead of a primer on EPDM seams

5. To create an EPDM side-lap seam, alignment marks should be placed on the lower membrane about how far from where the edge of the lap will be?
   a. ¼" (~6.4 mm)
   b. ⅞" (~2.2 cm)
   c. 6" (~15 cm)
   d. 12" (~30 cm)

6. When taping an EPDM seam, where should the tape be placed?
   a. On the underside of the upper membrane sheet, flush with the edge.
   b. On the bottom membrane sheet of the lap, aligned with the edge of the sheet where it meets the substrate.
   c. On the bottom membrane sheet, set back at least 1" (~2.5 cm) from the edge of the upper sheet.
   d. On the bottom membrane sheet, aligned with the alignment marks that will ensure just the edge of the tape will be visible after the lap is closed.

7. Regarding water movement across the roof, the EPDM membrane *must* be placed so that _____.
   a. the longest seams travel in the same direction as the water
   b. the water never encounters a seam as it moves down the roof
   c. the water is routed over the lip of the seam and not against it
   d. flashing around penetrations will be optional instead of required

8. When using a hand roller to smooth out a seam, you should roll _____.
   a. straight along the edge of the seam
   b. at a 15-degree angle from the edge seam
   c. at a 45-degree angle from the edge seam
   d. at a 90-degree angle from the edge seam

9. Which of the following may be required to protect the membrane in a ballasted EPDM roof system?
   a. Adhesive
   b. Roof cement
   c. A concrete substrate
   d. A separator sheet

10. In the context of single-ply roofing, what is a *fishmouth*?
    a. A long cut
    b. A raised opening or void along a seam edge
    c. A bubble in the field membrane
    d. A hole on the membrane that is too large to call a puncture

Trade Terms Introduced in This Module

Alignment marks: Marked or chalked lines extending the length of membrane and used to align the lap of the adjacent sheet accurately and consistently.

Application rate: The average quantity (mass, volume, or thickness) of material applied per unit area.

Ballast: Material that weighs down a membrane to hold it in place. Ballast includes smooth river stone; standard precast concrete pavers; or interlocking, beveled, doweled, or contoured fit lightweight concrete pavers.

EPDM: Acronym for *ethylene propylene diene monomer*; a common type of thermoset used in single-ply roof systems.

Fishmouth: An opening in the edge of a membrane caused by wrinkling or shifting of sheets during installation.

Flashings: Components used to weatherproof or seal edges of a roof system at perimeters, penetrations, walls, expansion joints, valleys, drains, and other places where the roof covering is interrupted or terminated. For example, membrane base flashing covers the edge of the field membrane, and cap flashings or counterflashings shield the upper edges of the base flashing.

Laps: Areas where membrane sheets or flashings overlap the membrane below. Laps are sealed to create seams.

Mil: A unit of measurement equal to one-thousandth of an inch, or 0.001" (0.0254 mm). The thickness of a membrane sheet is described in mils (for example, 45-mil or 60-mil membrane).

Point loading: The act of spacing out roofing materials in different areas on the roof so that all the weight of the roofing materials is not in one place.

Seam: The weatherproof bonding of two overlapping pieces of roof membrane.

Seam tape: Self-adhering cured or uncured synthetic-rubber tape used for creating seams between membrane sheets.

Self-adhered membrane: A membrane with factory-applied adhesive that is able to bond itself without the use of an additional adhesive.

Thermoplastic: A roofing material designed to be heated and fused to itself. When cooled, it remains flexible, waterproof, and strong.

Thermoset: A class of polymers that when cured using heat, chemical, or other means, change into a substantially infusible and insoluble material.

T-joint: An area where multiple field membrane sheets overlap.

Volatile organic compounds (VOCs): Compounds and chemicals that easily evaporate at room temperature. They are found in products such as solvents, paints, and adhesives.

Weighted roller: A heavy object pushed over adhered or self-adhered membrane to smooth it out and eliminate bubbles, wrinkles, fishmouths, and other deficiencies.

Additional Resources

This module presents thorough resources for task training. The following reference material is suggested for further study.

American Lung Association, **www.lung.org**.

National Roofing Contractors Association (NRCA), **www.nrca.net**.

NCCER Module 00103, *Introduction to Hand Tools*.

NCCER Module 00104, *Introduction to Power Tools*.

NCCER Module 16106, *Introduction to Low-Slope Roofing*.

NCCER Module 16206, *Thermoplastic Roof Systems*.

Figure Credits

Patrick D Murphy Co., Inc., Architects, Module Opener, Figure 12 (B)

Courtesy of the National Roofing Contractors Assoc., Figures 1, 6–9, 16–24

Versico Roofing Systems, Figures 3, 5

iStock@chaundthar v, Figure 10

Courtesy of GAF, Figures 11, 12 (A)

HB Fuller, Figures 13–14

Section Review Answer Key

Section 1.0.0

Answer	Section Reference	Objective
1. b	1.1.2	1a
2. b	1.2.0	1b
3. c	1.3.2	1c

Section 2.0.0

Answer	Section Reference	Objective
1. b	2.1.0	2a
2. c	2.2.1; 2.2.2; 2.2.3	2b
3. b	2.3.0	2c

User Update

NCCER makes every effort to keep its textbooks up-to-date and free of technical errors. We appreciate your help in this process. If you find an error, a typographical mistake, or an inaccuracy in NCCER's curricula, please submit a User Update form by visiting **https://www.nccer.org/olf**. You can also scan the QR code using the camera on your phone or mobile device to access the form.

Built-Up Roof Systems

Overview

Built-up roof (BUR) systems are one of the oldest and most proven systems in use today. They were the first low-slope roof system widely used in the United States, dating back to the nineteenth century. Installing BUR systems requires teamwork, expertise, and knowledge of a variety of components and tools.

Module 16208

Trainees with successful module completions may be eligible for credentialing through the NCCER Registry. To learn more, go to **www.nccer.org** or contact us at 1.888.622.3720. Our website, **www.nccer.org**, has information on the latest product releases and training.

Your feedback is welcome. You may email your comments to **curriculum@nccer.org**, send general comments and inquiries to **info@nccer.org**, or fill in the User Update form at the back of this module.

This information is general in nature and intended for training purposes only. Actual performance of activities described in this manual requires compliance with all applicable operating, service, maintenance, and safety procedures under the direction of qualified personnel. References in this manual to patented or proprietary devices do not constitute a recommendation of their use.

16208 V1.0

16208
Built-Up Roof Systems

Objectives

Successful completion of this module prepares you to do the following:

1. Identify types of built-up roof (BUR) systems and their basic components.
   a. List and describe components of a BUR membrane.
2. Explain how BUR systems are installed and identify common repair methods.
   a. Identify tools and equipment used to install BUR systems.
   b. Describe how BUR systems are installed.
   c. Explain basic repair methods for damaged BUR systems.

Performance Tasks

Under supervision, you should be able to do the following:

1. Miter a cant strip to a vertical projection on an inside or outside corner.
2. Using cold or hot bitumen, install a three-ply built-up roof on a mock-up or jobsite.
3. Cut in BUR plies around a curb and over a cant strip and seal at top edge.
4. Install a plumbing vent flashing.
5. Cut and install a drip edge corner.
6. Install and strip in a drip or metal edge.
7. Install a cap sheet surface using cold or hot bitumen.

Trade Terms

Bill of lading
Bitumen
Cant strip
Fishmouth
Flash point (FP)
Mole run
Ply
Redundant
Scrim
Spudding
Tapered edge strip
Viscosity

Industry Recognized Credentials

If you are training through an NCCER-accredited sponsor, you may be eligible for credentials from NCCER's Registry. The ID number for this module is 16208. Note that this module may have been used in other NCCER curricula and may apply to other level completions. Contact NCCER's Registry at 1.888.622.3720 or go to **www.nccer.org** for more information.

You can also show off your industry-recognized credentials online with NCCER's digital badges. Transform your knowledge, skills, and achievements into badges that you can share across social media platforms, send to your network, and add to your resume. For more information, visit **www.nccer.org**.

Videos and Digital Resources

How to Access Resources

This craft has additional videos and resources to enhance your learning experience. To view these resources, scan the QR below. The videos and resources are separated by module.

You can scan this code using the camera on your phone or mobile device to view these videos and resources.

Contents

1.0.0 Built-Up Roofing 1
  1.1.0 BUR Components 1
    1.1.1 Sheets and Plies 2
    1.1.2 Bitumen 2
    1.1.3 Surfacing 3
    1.1.4 Flashings 3
    1.1.5 Accessories 5
2.0.0 Installation and Repair 6
  2.1.0 Tools and Equipment 6
    2.1.1 Hand Tools 6
    2.1.2 Power Tools 7
  2.2.0 Installation Procedures 7
    2.2.1 Installing a Base Sheet 8
    2.2.2 Installing Plies 9
    2.2.3 Installing Surfacing 9
  2.3.0 Basic Repair Methods 9
    2.3.1 Three-Course Repair 9
    2.3.2 Fishmouths and Mole Runs 10

Figures

Figure 1 BUR membrane installation 1
Figure 2 Components of a BUR system 1
Figure 3 Example of a BUR flashing detail drawing 4
Figure 4 Example of a membrane flashing in a BUR system 5
Figure 5 Mop and mop cart 6
Figure 6 Spud bar 7
Figure 7 Kettle at a jobsite 7
Figure 8 Gravel conveyor 8
Figure 9 Shingle-fashion arrangement used in BUR membranes 9
Figure 10 Hot-mopped and cold-applied bitumen 10

SECTION ONE

1.0.0 BUILT-UP ROOFING

Objective

Identify types of built-up roof (BUR) systems and their basic components.

a. List and describe components of a BUR membrane.

Trade Terms

Bill of lading: A detailed list of items shipped; it is delivered by the carrier with the shipment.

Bitumen: An asphalt-based product that provides waterproofing and adhesion.

Cant strip: A long, triangular object placed in 90-degree bends before applying the membrane to reduce the stress on the membrane.

Flash point (FP): The temperature at which a material can ignite.

Ply: A layer of reinforcement in a built-up or mod-bit roof system.

Redundant: Having multiple layers for extra protection.

Scrim: Reinforcement fibers in modified bitumen sheets.

Tapered edge strip: A rigid board shaped with built-in slope and used in penetration flashings to provide slope toward drains.

Viscosity: The consistency of a liquid. Liquids with high viscosity are thicker, while liquids with low viscosity are thinner.

Built-up roof (BUR) systems have been used in the United States since the 1840s. They are made of multiple layers of sheets, felts, fabrics, or mats assembled on the roof in shingle fashion with alternating layers of bitumen. A roofing crew works together to cold-apply or hot-mop the bitumen onto the substrate and then roll the sheets into it (*Figure 1*).

BUR systems have sheet-applied (membrane) flashings or liquid-applied flashings and are surfaced with a coating or cap sheet. A traditional surfacing is a flood coat of bitumen into which gravel can be embedded.

These systems are called *built-up* because the components are assembled in place on the roof.

Figure 1 BUR membrane installation.

BUR systems are very effective at weatherproofing because they are redundant. This means each repetitive layer makes it less likely for water to get through. If one layer breaks down or gets damaged, the remaining layers provide protection. A layer of sheets in a BUR system is referred to as a ply. BUR systems are sometimes informally referred to as *tar and gravel* roofs.

1.1.0 BUR Components

BUR systems include the following primary components (*Figure 2*):

- Plies, or reinforcement sheets
- Bitumen
- Surfacing
- Flashings (membrane or liquid-applied)
- Accessories

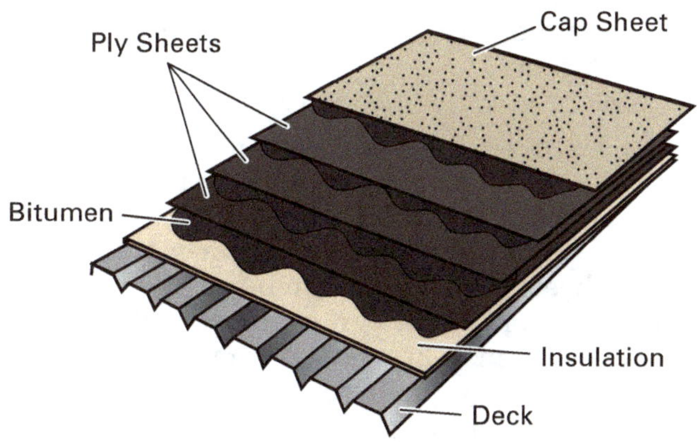

Figure 2 Components of a BUR system.

Teamwork in Tragedy

When American Airlines Flight 77 hit the US Pentagon on September 11, 2001, jet fuel ignited on the roof and damaged the building's BUR and slate roof systems. Firefighters worked to extinguish the fire, but the slate shingles performed so well at keeping water out that it hindered their efforts.

Within weeks, the National Roofing Contractors Association (NRCA) rallied its resources and organized a plan for replacing the 40,000-square-foot roof system as a gift from the industry. The repairs included more consideration for fire protection. In all, 124 contractors, distributors, and manufacturers made financial contributions; 16 manufacturers made material donations; and 15 contracting companies donated labor. Total donations exceeded $500,000.

Figure Credit: "Aerial view of the Pentagon. Original image from Carol M. Highsmith's America, Library of Congress collection. Digitally enhanced by rawpixel." by Carol M. Highsmith is marked with CC0 1.0

1.1.1 Sheets and Plies

Roll-roofing materials serve as reinforcing layers in BUR membranes. These layers fall into three general categories in a roof system:

- *Base sheets* — The first sheet, which is usually mechanically attached to a nailable substrate. Base sheets serve several purposes, including the following:
  - Separate a roof system from a substrate and prevent bitumen from dripping into the structure
  - Provide support over slightly rough or irregular substrates
  - Serve as an attachment or base layer for adhering insulation boards on nailable roof decks
  - Serve as a first layer in BUR membranes, temporary roof systems, and built-up vapor retarders
  - Serve as a single-layer vapor retarder
- *Ply sheets* — Ply sheets are installed directly over base sheets or insulation boards as layers in a BUR system. Multiple plies are installed in BUR systems by mopping bitumen and rolling the sheets into it. The number of plies needed depends on the design of the specific system. Ply sheets may be made of the following materials:
  - *Glass ply* (fiberglass membrane coated with asphalt)
  - *Polyester* (heat-resistant polyester)
  - *Asphalt-saturated organic felt* (organic fibers coated with asphalt)
- *Mineral-surfaced cap sheets* — The final sheet installed in a BUR system as a surfacing. (Note that BUR systems may also be surfaced with gravel, coatings, or other materials.) Cap sheets may contain the following reinforcement materials:
  - *Glass ply* (fiberglass membrane coated with asphalt)
  - *Polyester* (heat-resistant polyester)
  - *Composite ply* (polyester, glass, or polyester trilaminate reinforcement coated with asphalt)

Cap sheets contain reinforcement and bitumen. The reinforcement in cap sheets is called **scrim** and may be made of a variety of synthetic fibers formed into fabric mats.

1.1.2 Bitumen

Bitumen is a versatile material with weatherproofing and adhesive properties. The most common bitumens used for BUR membranes are as follows:

- Cold-applied bitumen adhesives
- Cold-applied polymer-modified bitumen adhesives
- Hot-applied bitumen
- Hot-applied polymer-modified bitumen

NOTE

The terms *bitumen* and *asphalt* may be used interchangeably when referring to roofing products.

Bitumen is derived from petroleum. Hot-applied bitumen is thermoplastic in nature, which means it becomes fluid with heat and reverts to a solid state when cool. Historically, coal tar bitumen has also been used for BUR membranes. However, coal tar is no longer widely used in BUR construction. Different types of bitumen vary based on viscosity and melting temperature.

Hot bitumen is heated in a kettle and pumped to the roof through an insulated supply pipe. Roofers then pour it into luggers or buckets and apply it to the substrate using mops. It must be applied to the roof at a specified application rate, such as 2.5 gallons per 100 square feet (25 pounds per square).

WARNING!

Operating a kettle requires special training and special attention to safety. Consult product packaging, bill of lading, or Safety Data Sheet (SDS) to ensure that the flash point (FP) of bitumen is not reached. When overheated, kettles can explode. (As a general rule, the limit is the flash point minus 50°F [~28°C].)

1.1.3 Surfacing

Mineral-surfaced cap sheets are often used as the top layer of a BUR membrane. This material is typically made of a fiberglass mat. The mat is filled and coated on both sides with bitumen and surfaced on the side that will be exposed to weather with mineral granules. Mineral-surfaced cap sheets are available in a variety of colors. Other surfacing materials may be used, including liquid-applied coatings, or a flood coat of bitumen with embedded gravel.

1.1.4 Flashings

Membrane flashings are used to terminate a BUR membrane at a roof's perimeters, penetrations, and interruptions. Membrane flashings typically include a base sheet (or multiple base layers) and a cap sheet. Flashings may also include additional components, such as a cant strip to alleviate the stress on the membrane against walls, or tapered edge strip to provide slope toward a drain.

Did You Know?

BUR Beginnings

Like many industrial cities, Cincinnati adopted gas streetlamps in the nineteenth century. Gas for the lamps was provided by the Cincinnati Gas Light and Coke Co., which was established in 1843 to convert coal to gas and pipe it throughout the city. An unfortunate byproduct of this process was a sticky, black substance known as *coal tar*. The industry policy at the time was to dump this waste into the nearest stream. This was not unusual, as the Ohio River served as the waste disposal system for many other Cincinnati industries.

Samuel Warren, who established one of the earliest roofing companies in the United States and pioneered BUR technology, asked the Cincinnati Gas Light and Coke Co. for permission to take all the waste tar off its hands, and they accepted his offer. The coal tar made a low-cost, abundant substitute for the pine tar originally used by roofers.

Figure Credit: iStock@alexeys

Metal flashing components include drip edges, copings, and counterflashings. All these components work together to keep water out of perimeters, penetrations, and interruptions. An example of a flashing detail in a BUR system is shown in *Figure 3*.

NOTE

Metal roofing components and flashings are described in NCCER Module 16107, *Sheet Metal in Roofing*.

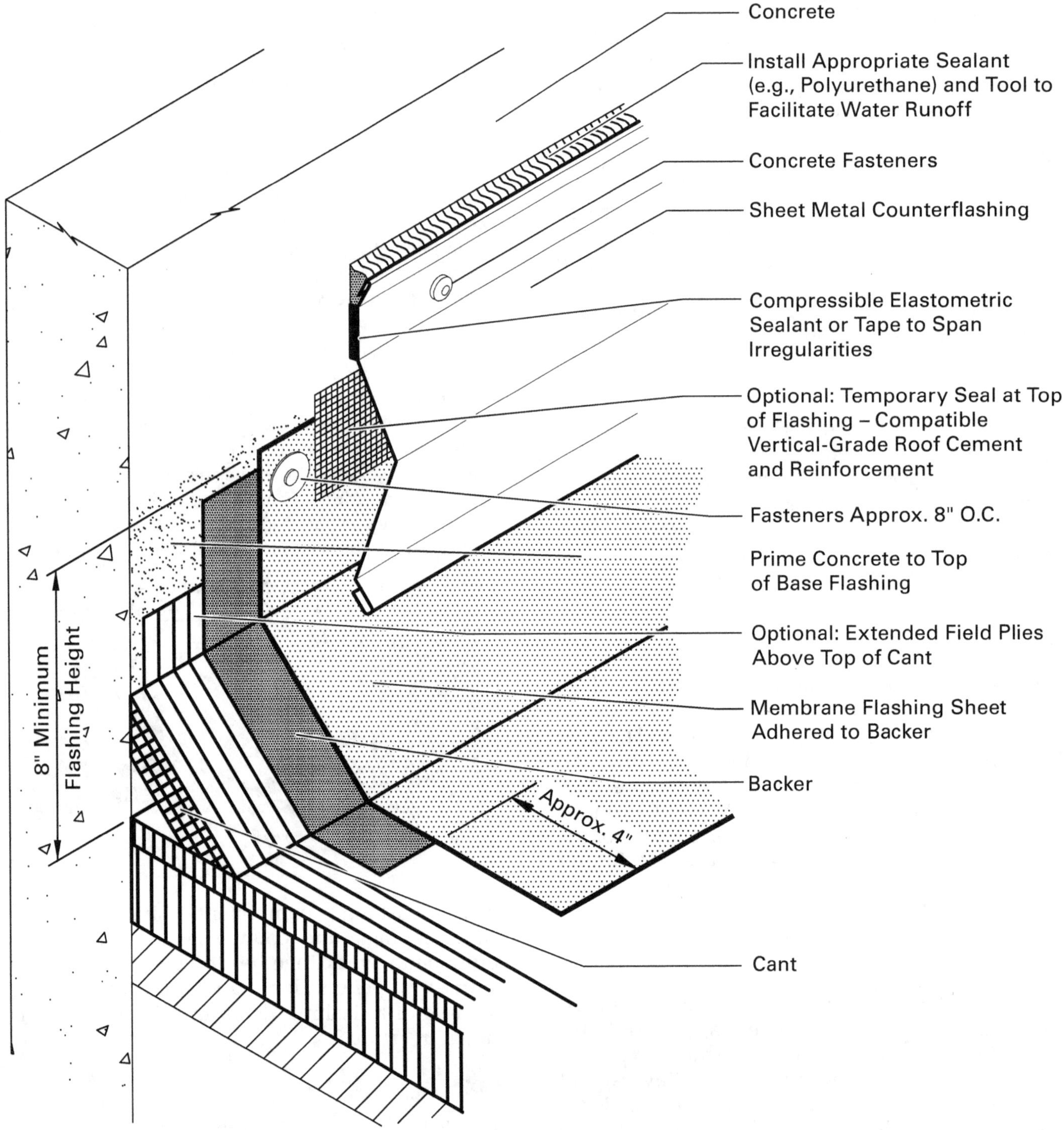

Notes:

1. This detail is applicable for hot-, cold- or torch-applied flashing systems.
2. This detail does not allow for differential movement between the deck and wall. Counterflashing detail may be a two-piece reglet and counterflashing.

Figure 3 Example of a BUR flashing detail drawing.

Membrane Flashings

The following types of sheets are used for membrane flashings (*Figure 4*) in BUR systems:

- *APP modified-bitumen sheets* — Smooth- or granule-surfaced sheets available with fiberglass and/or polyester reinforcements. The bitumen in APP sheets is modified with plastic.
- *SBS modified-bitumen base sheets* — Among the most common bituminous roofing sheets used for BUR. The bitumen in SBS sheets is modified with rubber.
- *SBS modified-bitumen membrane* — Granule-surfaced or metal-foil-surfaced sheets available in different weights with fiberglass or polyester reinforcements.

Liquid-Applied Flashings

Liquid-applied flashings typically contain a one- or two-component resin. They generally consist of a base coat, polyester fabric reinforcement, and top coat. In some systems, they may include a surfacing, such as a coating or granules.

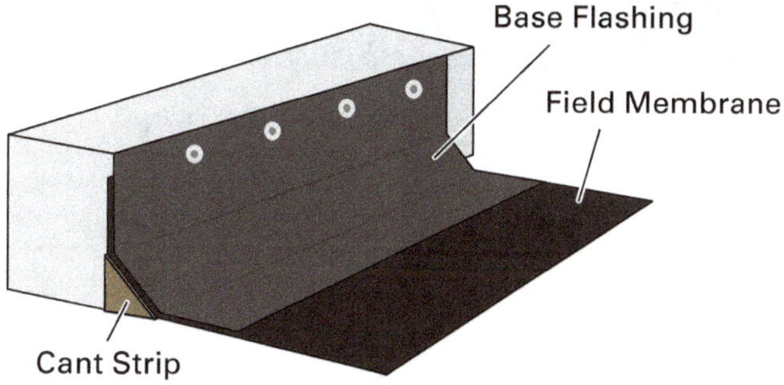

Figure 4 Example of a membrane flashing in a BUR system.

Liquid-applied flashings are commonly used in situations where penetrations or roof surfaces are irregularly shaped and difficult to flash using a membrane flashing.

1.1.5 Accessories

Accessory products commonly used in constructing BUR membranes include rosin-sized sheathing paper, reinforcing fabrics (such as fiberglass, cotton, or polyester), roof cements, and adhesives.

1.0.0 Section Review

1. BUR stands for _____.
   a. bitumen underlayment roofing
   b. butadiene urethane roofing
   c. built-up roofing
   d. built-in uplift resistance

2. Where in a BUR system are base sheets installed?
   a. Under the ply sheets
   b. Between the ply sheets
   c. Over the ply sheets
   d. Only at perimeter edges

Section Two

2.0.0 Installation and Repair

Objective

Explain how BUR systems are installed and identify common repair methods.

a. Identify tools and equipment used to install BUR systems.
b. Describe how BUR systems are installed.
c. Explain basic repair methods for damaged BUR systems.

Performance Tasks

1. Miter a cant strip to a vertical projection on an inside or outside corner.
2. Using cold or hot bitumen, install a three-ply built-up roof on a mock-up or jobsite.
3. Cut in BUR plies around a curb and over a cant strip and seal at top edge.
4. Install a plumbing vent flashing.
5. Cut and install a drip edge corner.
6. Install and strip in a drip or metal edge.
7. Install a cap sheet surface using cold or hot bitumen.

Trade Terms

Fishmouth: An opening in the edge of a membrane caused by wrinkling or shifting of sheets during installation.

Mole run: A long wrinkle in a roof membrane.

Spudding: Removing a gravel surface to expose the roof membrane.

A wide variety of tools are required to install built-up roofs. Each must be used carefully and safely. To install quality roof systems, roofers must be familiar with each of these tools and the steps involved in using them. They must also be able to fix mistakes that occur during installation, such as wrinkling and openings in membrane sheets.

2.1.0 Tools and Equipment

Roofers use many tools to install BUR systems. Some are used specifically for hot bitumen or cold-applied bitumen. It is important to be familiar with each of these tools and be able to use them safely.

> **NOTE**
> Basic tools used for construction work are covered in NCCER Module 00103, *Introduction to Hand Tools*, and NCCER Module 00104, *Introduction to Power Tools*.

2.1.1 Hand Tools

Roofers must be familiar with the following hand tools for BUR construction:

- *Mop* — Used to apply hot bitumen (*Figure 5*).
- *Mop cart* — A rolling bucket filled with bitumen into which the mop is placed and transported (*Figure 5*).
- *Bucket* — Used to transport bitumen.
- *Lugger* — A secondary container on the roof for hot bitumen that has a spigot to pour material into smaller containers for installation.
- *Pouring can* — Used to spread bitumen.
- *Felt slitter* — Used to cut sheets.
- *Utility knife* — Used to cut sheets.
- *Gravel spreader* — Used to spread gravel surfacing.
- *Hot machine* — Used to spread bitumen (also called a *chain mop* or *johnny mop*).
- *Torch* — Used to heat APP or SBS sheets and make them pliable or to heat bitumen to a liquid state.
- *Hot-air gun* — Used to heat APP or SBS cap sheet laps.
- *Trowel* — Used to press or smooth down sheets around flashings or tight areas or to spread roof cement. Types of trowels include the following:
  - Margin trowel (squared shape)
  - Linoleum trowel (used for cold-applied systems)

Figure 5 Mop and mop cart.

- *Notched squeegee* — Used for spreading cold-applied bitumen.
- *Broom* — Used to smooth down sheets and ensure tight adhesion to the bitumen underneath.
- *Roller* — Used to smooth down sheets and ensure tight adhesion to the bitumen underneath.
- *Paint roller and pan* — Used to apply primer.
- Spudding *tools (Figure 6)* — Used for repair and reroofing projects, as well as for cleanup during new installations.
  - *Ripping hammer* (used to loosen gravel)
  - *Spud bar* (used to loosen gravel)
  - *Shovel* (used to relocate or remove gravel)
- *Chalk line* — Used to mark lines on the substrate.
- *Wheelbarrow* — Used to transport materials, such as gravel.

2.1.2 Power Tools

Roofers must be familiar with the following power tools for BUR construction:

- *Kettle* — Used to heat bitumen to a molten liquid and pump it to the roof through a pipe (*Figure 7*).
- *Sprayer* — Used for cold process installations, adhesives, and coatings.
- *Power buggy* — Used to spread gravel.
- *Powered felt slitter* — Used to cut sheets.
- *Tearoff machine* — A large, powered, chisel-like tool used to remove sections of roof assembly from the deck.

Figure 6 Spud bar.

- *Conveyor* — Used to transport roofing materials, such as gravel, from the ground to the roof (*Figure 8*).
- *Roof hoist* — Used to lift roofing materials to the roof.

2.2.0 Installation Procedures

Steps for installing a BUR system include mechanically attaching a base sheet, adhering multiple plies with bitumen, and adding a surfacing to the top of the system. The following is a very general procedure for installing a BUR system:

> **NOTE**
> Always refer to all applicable documentation when installing a roof system, including building codes, manufacturer instructions for equipment and materials, construction drawings, and project specifications.

Step 1 Make sure the substrate is clean, dry, and in good condition.

Step 2 Mechanically attach a base sheet, if applicable.

Step 3 Install insulation and/or cover boards as applicable.

Step 4 Install the specified number of ply sheets in shingle fashion (*Figure 9*).
- Mop the hot bitumen or use a notched rubber broom to spread cold bitumen.
- Roll the sheets into the bitumen.

Figure 7 Kettle at a jobsite.

Figure 8 Gravel conveyor.

WARNING!

Handling hot bitumen is extremely dangerous as it may reach temperatures of up to 500°F (260°C). Always be aware of the following safety considerations when working with hot bitumen:

- If bitumen is heated above the flash point, it may ignite and cause fires or explosions. Consult the bitumen packaging or SDS to ensure that the temperature of bitumen is always at least 50°F (~28°C) below the flash point.
- To protect yourself from burns, use extreme care and wear personal protective equipment (PPE) when working with hot bitumen. Necessary PPE may include work boots, long pants, long sleeves, heat-resistant gloves with tight cuffs, safety glasses and/or a face shield, and other protective clothing. Clothing must be made of a material that doesn't melt, such as cotton. If hot bitumen gets on your skin, seek medical attention immediately.
- Always stay hydrated by drinking enough water, especially when working with hot bitumen on a hot day.
- Mops are subject to spontaneous combustion if not periodically spun out and laid on a non-combustible surface. As the outer layer of bitumen on the mop cools, the heat of the bitumen underneath may cause the outer layer to crack and explode. To prevent this, spin out mops regularly.

Step 5 Seal gaps by applying adhesive and spreading it smooth with a trowel.

Step 6 Flash any penetrations.

Step 7 Apply the specified surfacing, such as a coating or cap sheet.

2.2.1 Installing a Base Sheet

A base sheet/ply is used as the first layer over a nailable deck or over insulation. Roof designers specify the type of base sheet and attachment method that should be used.

Base sheets are often mechanically attached to nailable substrates so the ply sheets can be adhered on top of them.

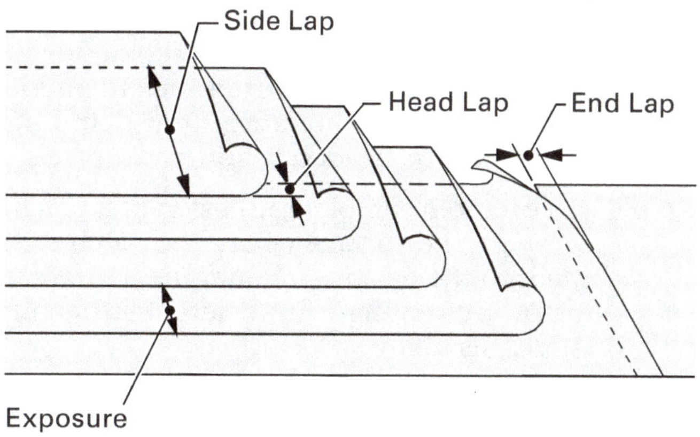

Figure 9 Shingle-fashion arrangement used in BUR membranes.

2.2.2 Installing Plies

Plies may be adhered with hot-mopped or cold-applied bitumen (*Figure 10*). These two processes use different tools and come with unique safety considerations.

Hot-mopped bitumen must be just the right temperature and viscosity. If it is too hot, it will be too thin to provide enough coverage. If it is not hot enough, it may not spread or adhere effectively. Hot-mopped bitumen is transported on the roof in buckets. A roofer will dip a mop in the mop cart or bucket and spread it on the substrate while other crew members roll in the sheets and cut them to fit as needed. Roofers will snap chalk lines or use lines applied by manufacturer as guides to install sheets in a straight line.

WARNING!

Mops may seem heavier throughout a day's work. Be aware of dehydration and overexertion by drinking water, using good posture, and taking breaks when needed.

For cold-applied installations, roofers typically use a notched rubber squeegee or brush to evenly coat the substrate. The sheets are then rolled into the modified bitumen and pressed into it with a broom or roller to ensure even adhesion. Using cold-applied bitumen is less dangerous than using hot bitumen, but the process takes longer.

2.2.3 Installing Surfacing

Before applying mineral-surfaced cap sheets, the cap sheet should be unrolled, cut into workable lengths, and allowed to lie flat. This allows it to relax and reduces the possibility of it curling during installation. Then, they are applied to the roof with bitumen adhesive just like the ply sheets.

What's wrong with this picture?

Figure Credit: Courtesy of the National Roofing Contractors Assoc.

Other membrane surfacing options include a flood coat of hot bitumen and aggregate or liquid-applied coatings.

NOTE

Coatings are described in detail in NCCER Module 16210, *Liquid-Applied Roofing.*

2.3.0 Basic Repair Methods

Mistakes happen even under the best conditions for roof installation. To ensure a quality and efficient job, workers on BUR roof systems should know how to repair any damage created during the installation process.

2.3.1 Three-Course Repair

Three-coursing is commonly used to repair BUR and polymer-modified bitumen flashings. This repair involves successive applications of a base course of asphalt roof cement, reinforcing fabric, and a second course of asphalt roof cement.

(A) Hot-Mopped Bitumen

(B) Cold-Applied Bitumen

Figure 10 Hot-mopped and cold-applied bitumen.

Where a three-course repair will remain exposed, it should additionally include an application of a surfacing material.

2.3.2 Fishmouths and Mole Runs

A **fishmouth** is an opening in the lapped edge of a membrane sheet or flashing. A **mole run** is a long ridge or wrinkle in a roof membrane. Fishmouths and mole runs pose a risk to weatherproofing and should be cut and patched. The following is a general procedure for repairing a fishmouth or mole run:

Step 1 Remove rough spots, debris, contaminants, and loose surfacing from the membrane or flashing surface to be repaired. Granule-surfaced membranes should be washed, brushed, broomed, or blown clean.

Step 2 Cut the fishmouth or mole run down the center to lay the material flat. Remove any debris from under the fishmouth and ensure it is dry.

Step 3 Use a self-adhering, polymer-modified bitumen patch, and ensure it extends at least 3" beyond any part of the cut and is well adhered. A hot-air gun or detailing torch may be needed to soften the patch.

Step 4 To promote complete adhesion of a patch, prepare the surface where the patch will tie into the existing membrane. Different steps are required for different surfacing types, as follows:

- *Gravel surfacing* — Spud gravel surfacing to expose the bitumen flood coat or plies. Apply heat to the bitumen with a roofing torch or hot-air gun. Smooth any irregularities and allow it to cool.
- *Granule-surfaced cap sheets* — Apply heat and press granule surfacing into the bitumen. Where a patch will be installed in cold-applied adhesive, prime the surface of the membrane with asphalt primer and allow it to dry.
- *Aluminum coatings* — Use a stiff-bristle broom to release and sweep away loose coating. Use a hot-air gun or torch to heath the patch area only until the surface turns black.
- *Elastomeric coatings* — Use a hot-air gun or torch to heat the surface sufficiently to scrape the coating off.

Step 5 Where the membrane was not degranulated or heated, such as when installing a patch in cold-applied adhesive, lightly prime the surface of the membrane with asphalt primer and allow it to dry.

CAUTION

Primer contains solvents and is used to enhance adhesion; however, overuse of primer can harm the membrane.

WARNING!

Additional PPE may be needed to safely handle primer. Primer is also highly flammable, so be aware of any heat sources and do not place primer containers close to them. Fumes from some primers may collect in low-lying areas and be ignited by static sparks, causing a fire. Always consult the SDS for proper PPE and related hazards.

Step 6 Where plies are removed from the membrane, install the same number of plies of reinforcing material in cold-applied or hot bitumen. Extend the bottom ply at least 6" beyond the area to be repaired and each succeeding ply at least 3" beyond the previous ply.

Step 7 Check the edges of the patch for voids by probing it with a trowel. If voids are present, use the trowel to lift the unattached edge of the patch, apply heat to the underside, and press the trowel down over the patch until bleed-out is visible.

NOTE

If patching over a vertical substrate, fasten the top of the patch to the wall or curb approximately 8" on center using fasteners suitable for the substrate. Consult manufacturer instructions for exact measurements and fastening patterns.

Step 8 Reapply the surfacing as required.

Step 9 Replace any flashing metal or other components removed when preparing the repair area.

2.0.0 Section Review

1. Which of the following tools is *best* for applying hot bitumen?
   a. Squeegee
   b. Mop
   c. Gravel spreader
   d. Trowel

2. Which of the following tools is used to transport roofing materials from the ground to the roof?
   a. Roller
   b. Power buggy
   c. Felt stretcher
   d. Conveyor

3. Which of the following tools is usually used for cold-applied installations?
   a. Mop
   b. Kettle
   c. Squeegee
   d. Broom

Review Questions

1. A layer of sheets in a BUR system is called a _____.
   a. flood coat
   b. ply
   c. cap
   d. mat

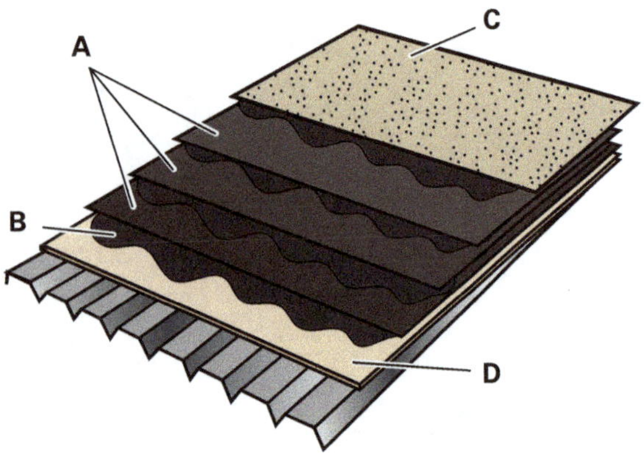

Figure RQ01

2. Refer to *Figure RQ01*. Which component is the bitumen?
   a. A
   b. B
   c. C
   d. D

3. Refer to *Figure RQ01*. Which component is the insulation?
   a. A
   b. B
   c. C
   d. D

4. Refer to *Figure RQ01*. Which component is the plies?
   a. A
   b. B
   c. C
   d. D

5. Refer to *Figure RQ01*. Which component is the cap sheet?
   a. A
   b. B
   c. C
   d. D

6. Which of the following statements about bitumen is *true*?
   a. It is made of pine tar.
   b. It becomes fluid when heated and solid when cool.
   c. It is not used on BUR systems.
   d. It is an insulator.

7. Liquid-applied flashings typically consist of _____.
   a. one coat of resin
   b. three coats of resin and a granule-surfaced cap sheet
   c. a base coat, polyester fabric reinforcement, and top coat
   d. a layer of bitumen

8. Which of the following tools is used to apply roof cement?
   a. Trowel
   b. Mop
   c. Kettle
   d. Sprayer

9. What is the *first* thing you should do when installing a BUR system?
   a. Seal gaps by applying adhesive and spreading it smooth.
   b. Apply a flood coat of bitumen.
   c. Make sure the substrate is clean, dry, and in good condition.
   d. Apply the specified surfacing, such as a coating or cap sheet.

10. Prior to installing them, mineral-surfaced cap sheets should be _____.
    a. coated with two-component adhesive
    b. washed with water
    c. unrolled and allowed to lie flat
    d. placed in a humid area

11. Which of the following is an opening in the edge of a membrane sheet?
    a. Tunnel
    b. Fishmouth
    c. Puncture
    d. Ridge

12. How far should a patch extend beyond any part of the membrane damage?
    a. At least 1"
    b. At least 2"
    c. At least 3"
    d. At least 5"

Trade Terms Introduced in This Module

Bill of lading: A detailed list of items shipped, delivered by the carrier with the shipment.

Bitumen: An asphalt-based product that provides waterproofing and adhesion.

Cant strip: A long, triangular object placed in 90-degree bends before applying the membrane to reduce the stress on the membrane.

Fishmouth: An opening in the edge of a membrane caused by wrinkling or shifting of sheets during installation.

Flash point (FP): The temperature at which a material can ignite.

Mole run: A long wrinkle in a roof membrane.

Ply: A layer of reinforcement in a built-up or mod-bit roof system.

Redundant: Having multiple layers for extra protection.

Scrim: Reinforcement fibers in modified bitumen sheets.

Spudding: Removing a gravel surface to expose the roof membrane.

Tapered edge strip: A rigid board shaped with built-in slope and used in penetration flashings to provide slope toward drains.

Viscosity: The consistency of a liquid. Liquids with high viscosity are thicker, while liquids with low viscosity are thinner.

Additional Resources

This module presents thorough resources for task training. The following reference material is suggested for further study.

National Roofing Contractors Association (NRCA), **www.nrca.net**.

NCCER Module 00103, *Introduction to Hand Tools*.

NCCER Module 00104, *Introduction to Power Tools*.

NCCER Module 16107, *Sheet Metal in Roofing*.

NCCER Module 16209, *Modified Bitumen Roof Systems*.

NCCER Module 16210, *Liquid-Applied Roofing*.

Figure Credits

Patrick D Murphy Co., Inc., Architects, Module Opener

Courtesy of the National Roofing Contractors Assoc., Figures 1, 3, 5–7, 9–10

Silver Systems Incorporated, Figure 8

Courtesy of GAF, Figure 10 (A)

iStock@Niko_Cingaryu, Figure 10 (B)

Section Review Answer Key

Section 1.0.0

Answer	Section Reference	Objective
1. c	1.0.0	1
2. a	1.1.1	1a

Section 2.0.0

Answer	Section Reference	Objective
1. b	2.1.1	2a
2. d	2.1.2	2a
3. c	2.2.2	2b

User Update

NCCER makes every effort to keep its textbooks up-to-date and free of technical errors. We appreciate your help in this process. If you find an error, a typographical mistake, or an inaccuracy in NCCER's curricula, please submit a User Update form by visiting **https://www.nccer.org/olf**. You can also scan the QR code using the camera on your phone or mobile device to access the form.

Modified Bitumen Roof Systems

Overview

Modified bitumen (mod-bit) roofing is an evolution of built-up roofing (BUR). It was first introduced in Europe in the mid-1960s and made its way to the United States and Canada around 1975. The bitumen in mod-bit roof systems is modified with polymers to add desirable properties for roofing. Unlike BUR systems, mod-bit membranes are assembled in sheets and rolls at a factory before they are applied to the roof. Installing mod-bit systems requires teamwork, expertise, and knowledge of a variety of components and tools.

Module 16209

Trainees with successful module completions may be eligible for credentialing through the NCCER Registry. To learn more, go to **www.nccer.org** or contact us at 1.888.622.3720. Our website, **www.nccer.org**, has information on the latest product releases and training.

Your feedback is welcome. You may email your comments to **curriculum@nccer.org**, send general comments and inquiries to **info@nccer.org**, or fill in the User Update form at the back of this module.

This information is general in nature and intended for training purposes only. Actual performance of activities described in this manual requires compliance with all applicable operating, service, maintenance, and safety procedures under the direction of qualified personnel. References in this manual to patented or proprietary devices do not constitute a recommendation of their use.

16209 V1.0

16209

Modified Bitumen Roof Systems

Objectives

Successful completion of this module prepares you to do the following:

1. Identify types of modified bitumen (mod-bit) roof systems and their basic components.
   a. List and describe components of a mod-bit membrane.
2. Explain how mod-bit roof systems are installed and identify common repair methods.
   a. Identify tools and equipment used to install mod-bit roof systems.
   b. Describe how mod-bit roof systems are installed.
   c. Explain basic repair methods for damaged mod-bit roof systems.

Performance Tasks

Under supervision, you should be able to do the following:

1. Miter a cant strip to a vertical projection on an inside or outside corner.
2. Using cold or hot bitumen, install a mod-bit roof on a mock-up or jobsite.
3. Install a plumbing vent flashing.
4. Cut and install a drip edge corner.
5. Install and strip in a drip or metal edge.
6. Install a cap sheet surface using cold or hot bitumen.
7. Cut in mod-bit plies around a curb and over a cant strip and flash at top edge.

Trade Terms

Atactic polypropylene (APP)
Bill of lading
Bitumen
Bleed-out
Cant strip
Certified Roofing Torch Applicator (CERTA) Program
Fishmouth
Flash point (FP)
Mole run
Plies
Polymer-modified
Redundant
Scrim
Selvage edge
Styrene butadiene styrene (SBS)
Surfacing material
Tapered edge strip
Viscosity

Industry Recognized Credentials

If you are training through an NCCER-accredited sponsor, you may be eligible for credentials from NCCER's Registry. The ID number for this module is 16209. Note that this module may have been used in other NCCER curricula and may apply to other level completions. Contact NCCER's Registry at 1.888.622.3720 or go to **www.nccer.org** for more information.

You can also show off your industry-recognized credentials online with NCCER's digital badges. Transform your knowledge, skills, and achievements into badges that you can share across social media platforms, send to your network, and add to your resume. For more information, visit **www.nccer.org**.

VIDEOS AND DIGITAL RESOURCES

How to Access Resources

This craft has additional videos and resources to enhance your learning experience. To view these resources, scan the QR below. The videos and resources are separated by module.

You can scan this code using the camera on your phone or mobile device to view these videos and resources.

Contents

1.0.0 Modified Bitumen Roof Systems 1
1.1.0 Mod-Bit Components 2
1.1.1 Sheets and Plies 2
1.1.2 Bitumen 3
1.1.3 Surfacing 3
1.1.4 Flashings 4
1.1.5 Accessories 6
2.0.0 Installation and Repair 7
2.1.0 Tools and Equipment 7
2.1.1 Hand Tools 7
2.1.2 Power Tools 7
2.2.0 Installation Procedures 8
2.2.1 Safety Considerations 9
2.2.2 Installation Outline 10
2.2.3 Installing a Base Sheet 10
2.2.4 Installing Modified Base Sheets 10
2.2.5 Installing Cap Sheets 11
2.3.0 Basic Repair Methods 12
2.3.1 Cap Sheet Repair 12
2.3.2 Three-Course Repair 13
2.3.3 Fishmouths and Mole Runs 13

Figures

Figure 1 Mod-bit membrane installation 1
Figure 2 Components of a mod-bit system 3
Figure 3 Example of a mod-bit flashing detail drawing 5
Figure 4 Example of a membrane flashing in a mod-bit roof system 6
Figure 5 Mop and mop cart 8
Figure 6 Seven-torch modified bitumen applicator 8
Figure 7 Kettle 8
Figure 8 Roofing conveyor 9
Figure 9 Basic example of a mod-bit roof system 11
Figure 10 Hot-mopped and cold-applied bitumen 12
Figure 11 Embedding granules to prepare surface for adhesion 12

SECTION ONE

1.0.0 MODIFIED BITUMEN ROOF SYSTEMS

Objective

Identify types of modified bitumen (mod-bit) roof systems and their basic components.

a. List and describe components of a mod-bit membrane.

Trade Terms

Atactic polypropylene (APP): A plastic-like polymer used to modify bitumen.

Bill of lading: A detailed list of items shipped; it is delivered by the carrier with the shipment.

Bitumen: An asphalt-based product that provides waterproofing and adhesion.

Cant strip: A long, triangular object placed in 90-degree bends before applying the membrane to reduce the stress on the membrane.

Flash point (FP): The temperature at which a material can ignite.

Plies: Layers of reinforcement in a built-up or mod-bit roof system.

Polymer-modified: A term used to describe bitumen that has been modified with polymers (such as APP or SBS) to add plastic or rubber qualities desirable for roofing.

Redundant: Having multiple layers for extra protection.

Scrim: Reinforcement fibers in modified bitumen sheets.

Selvage edge: The edge of a mod-bit sheet with no granules, allowing it to be heat welded to the next sheet.

Styrene butadiene styrene (SBS): A rubber-like polymer used to modify bitumen.

Surfacing material: Protects the membrane from ultraviolet (UV) degradation and weather erosion. A variety of materials are used for surfacing, such as a mineral aggregate, a liquid-applied coating, or a sheet material.

Tapered edge strip: A rigid board shaped with built-in slope and used in penetration flashings to provide slope toward drains.

Viscosity: The consistency of a liquid. Liquids with high viscosity are thicker, while liquids with low viscosity are thinner.

Modified bitumen (mod-bit) roof systems (*Figure 1*), an evolution of built-up roof (BUR) systems, are asphalt based and designed primarily for low-slope or flat roof systems. Like BUR membranes, mod-bit membranes are made up of bitumen, reinforcement, and surfacing material. However, unlike BUR systems that are installed by rolling layers of reinforcement into hot bitumen, mod-bit systems are made of sheets that are pre-manufactured with reinforcement and bitumen. These sheets may be installed using a torch that heats up the bitumen in the sheet, causing it to stick to the roof, or they may be adhered with hot-mopped or cold-applied bitumen. On average, mod-bit systems contain fewer layers (or plies) than BUR systems.

The bitumen in mod-bit roof systems is modified with polymers to make it stronger. The sheets contain layers of the modified bitumen on both sides of a reinforcement fabric, typically either polyester, fiberglass, or a combination of the two.

Figure 1 Mod-bit membrane installation.

Did You Know?

The National Roofing Contractors Association (NRCA) Mission Statement

NRCA informs and assists the roofing industry, acting as its principal advocate, and helping members serve their customers. NRCA continually strives to enhance every aspect of the roofing industry and is also one of the construction industry's most respected trade associations and the voice of roofing professionals and leading authority in the roofing industry for information, education, technology, and advocacy. It represents all segments of the roofing industry, including contractors; manufacturers; distributors; architects; consultants; engineers; building owners; and city, state, and government agencies. NRCA's mission statement is as follows:

"Since 1886, the National Roofing Contractors Association has been the home for generations of entrepreneurial craftsmen and enterprises who shelter and protect America's families and businesses and each other. Our vision is the recognition of our members as professionals and to unite the industry to that purpose."

For more information on the NRCA, visit **www.nrca.net/about.**

Figure Credit: Courtesy of the National Roofing Contractors Assoc.

Originally introduced in Europe in the mid-1960s, mod-bit systems began being used in the United States in the mid-1970s. Mod-bit systems are known for being extremely durable, weather-resistant, and for having low maintenance requirements. A finished mod-bit roof system typically has two, three, or four plies. A two-ply system consists of a modified base sheet and a modified bitumen membrane, or cap sheet. Additional redundant plies of modified base sheets can be added for increased strength. Mod-bit can be installed using a variety of methods: hot-applied, cold applied, hot-air welded, self-adhered, or torch-applied.

1.1.0 Mod-Bit Components

Mod-bit systems include the following primary components (*Figure* 2):

- Modified base sheet
- Additional (if needed) modified bitumen membranes, or mid ply
- Cap sheet, or top ply
- Modified bitumen
- Surfacing
- Flashings (membrane or liquid-applied)
- Accessories

1.1.1 Sheets and Plies

A mod-bit roof system uses layers of modified bitumen sheets, or membranes, to create a strong and weatherproof roof. The polymer modification process adds tensile strength and elongation to the sheets. These sheets or plies are in the following categories:

- *Modified base sheet* — The first sheet, which is usually mechanically attached to a nailable substrate. Base sheets serve several purposes, including the following:
  - Separating a roof system from a substrate
  - Providing support over slightly rough or irregular substrates
  - Serving as an attachment or base layer for adhering additional layers of roofing on nailable roof decks

- Serving as a strengthening first layer in BUR and mod-bit roof systems, temporary roof systems, and built-up vapor retarders
- Serving as a single-layer vapor retarder

- *Modified bitumen membranes* — Modified bitumen membranes are factory-fabricated sheets consisting of a polymer-modified bitumen typically reinforced with either polyester, fiberglass, or a combination of the two. The reinforcement in modified bitumen sheets is called scrim.
- *Cap sheet* — The final and top sheet that covers and protects the other layers below it. Cap sheets can be smooth-surfaced or coated for added strength and durability. Coated cap sheets can be foil-faced or surfaced with different types of mineral granules. Granule-surfaced cap sheets have smooth edges, allowing them to be heat welded to the next sheet. This is called the selvage edge.

1.1.2 Bitumen

Bitumen is a versatile, asphalt-based material with both weatherproofing and adhesive properties. Asphalts are derived from petroleum. Hot-applied asphalts are thermoplastic in nature, which means they become fluid with heat and revert to a solid-state when cool. Different types of bitumen vary based on viscosity and melting temperature.

Hot asphalt is heated in a kettle or tanker and pumped to the roof through a supply pipe. Roofers then pour it into luggers, buckets, or mop carts and apply it to the substrate using application equipment such as mops. It must be applied to the roof at a specified application rate and temperature.

> **WARNING!**
>
> Operating a kettle requires special training and special attention to safety.
>
> Consult the product packaging, bill of lading, or Safety Data Sheet (SDS) to ensure that the flash point (FP) of bitumen is not reached. When overheated, kettles can explode. (As a general rule, the limit is the flash point minus 50°F [~28°C].)

Modified bitumen is created by adding polymers to the bitumen. This modification results in a stronger, longer-lasting product. There are two common types of modified bitumen: styrene butadiene styrene (SBS) and atactic polypropylene (APP).

SBS-modified bitumen is modified with synthetic rubber in the form of styrene butadiene styrene. This gives the bitumen rubber-like properties and improves its resistance to aging and weathering. Typically, SBS mod-bit sheets are installed by hot mopping asphalt, torching, or adhering with cold-process adhesives. They may also be self-adhered. If SBS mod-bit sheets do not have factory-applied mineral granule or foil-faced surfacing, they will need to have a field-applied ultraviolet protective surfacing applied.

APP-modified bitumen is modified with plastic in the form of atactic polypropylene. This improves the bitumen's performance by increasing its UV resistance, its flexibility at low temperatures, and its flow resistance at high temperatures. APP mod-bit sheets are typically applied by torching. Installers use heat to soften the modified bitumen on the bottom of the sheet, creating a molten adhesive, which flows on top of the substrate and subsequently cools to form a waterproof adhesive bond. Some APP sheets can also be applied with cold-process adhesives.

1.1.3 Surfacing

Surfacing is often required for mod-bit cap sheets to protect the roof system from UV rays, which can eventually wear down the strength and durability of the system. There are three main types of mod-bit cap sheet surfacing: mineral granules, foil-faced surfacing, and liquid-applied surfacing. Mineral granules and foil-faced surfacing are applied in the factory; coatings may be applied in the factory or in the field after the roof system has been installed:

- *Mineral granules* — Granule-surfaced mod-bit sheets are created in the factory by embedding ceramic-coated mineral granules into the top surface of the cap sheet. This is very similar to how asphalt shingles are created in the factory and a variety of colors are available as well. This is generally the most common type of surfacing for mod-bit roofs.

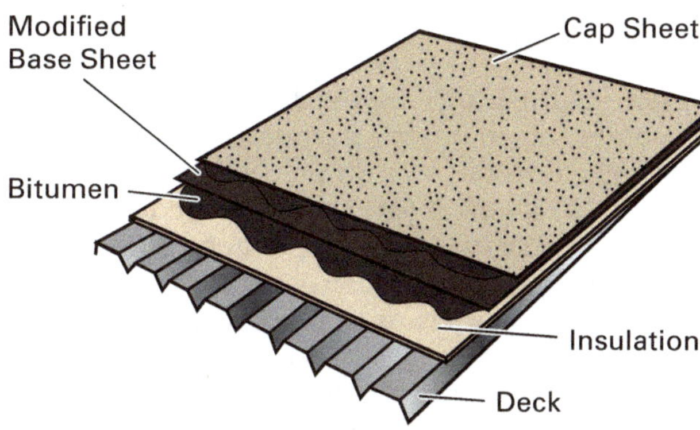

Figure 2 Components of a mod-bit system.

Did You Know?

Mod-Bit Origins

Evolving from one of the oldest roof systems, built-up roofing (BUR), modified bitumen (mod-bit) membranes were created as a soft polymer-asphalt mix known as *atactic polypropylene* (*APP*) by Romolo Gorgati in Italy in 1967. By 1969, mod-bit membranes were available in Germany, and then most of Europe soon after. Following the oil embargo in 1973, rising petroleum prices drove the development of mod-bit membranes in the United States. However, it took until the 1980s for these membranes to come into prominence as a roofing product in the United States. Mod-bit is typically easier to install and provides enhanced protection compared to BUR systems, even in very cold temperatures.

Figure Credit: Courtesy of the National Roofing Contractors Assoc.

- *Foil-faced surfacing* — Foil-faced mod-bit sheets are also created in the factory by laminating metallic foil onto the top surface of the cap sheet. Aluminum foil is the most common, but copper, stainless-steel, plastics, and other materials are also used. A benefit of foil-surfaced sheets is that they will provide UV protection for the life of the roof system, whereas granule-surfaced sheets tend to lose their granules over time, which exposes the underlying asphalt and shortens the life span of the roof. The drawbacks, however, are that foil-surfacing is relatively expensive and does not stand up to foot traffic very well.
- *Liquid-applied surfacing* — For liquid-applied surfacing, mod-bit sheets are coated in the field after the roof system is built. This method requires thorough preparation and installation as mod-bit sheets do not hold coatings as well as unmodified sheets.

1.1.4 Flashings

Membrane flashings are used to terminate a mod-bit membrane at a roof's perimeters, penetrations, and interruptions. Membrane flashings typically include a base sheet (or multiple base layers) and a cap sheet. Flashings may also include additional components, such as a **cant strip** to alleviate the stress on the membrane against walls, or **tapered edge strip** to provide slope toward a drain. Metal flashing components include drip edges, copings, and counterflashings. All these components work together to keep water out of perimeters, penetrations, and interruptions. An example of a flashing detail in a mod-bit system is shown in *Figure 3*.

> **NOTE**
> Metal roofing components and flashings are described in NCCER Module 16107, *Sheet Metal in Roofing*.

Membrane Flashings

The following types of sheets are used for membrane flashings (*Figure 4*) in mod-bit systems:

- *APP-modified bitumen sheets* — Smooth- or granule-surfaced sheets available with fiberglass and/or polyester reinforcements.
- *SBS-modified bitumen base sheets* — Among the most common bituminous roofing sheets used for mod-bit.
- *SBS-modified bitumen cap sheets* — Granule-surfaced or metal foil-surfaced sheets available in different weights with fiberglass or polyester reinforcements.

Liquid-Applied Flashings

Liquid-applied flashings typically consist of a one- or two-component resin coating. They generally consist of a base coat, polyester fabric reinforcement, and a top coat. In some systems, they may include surfacing, such as a coating or granules.

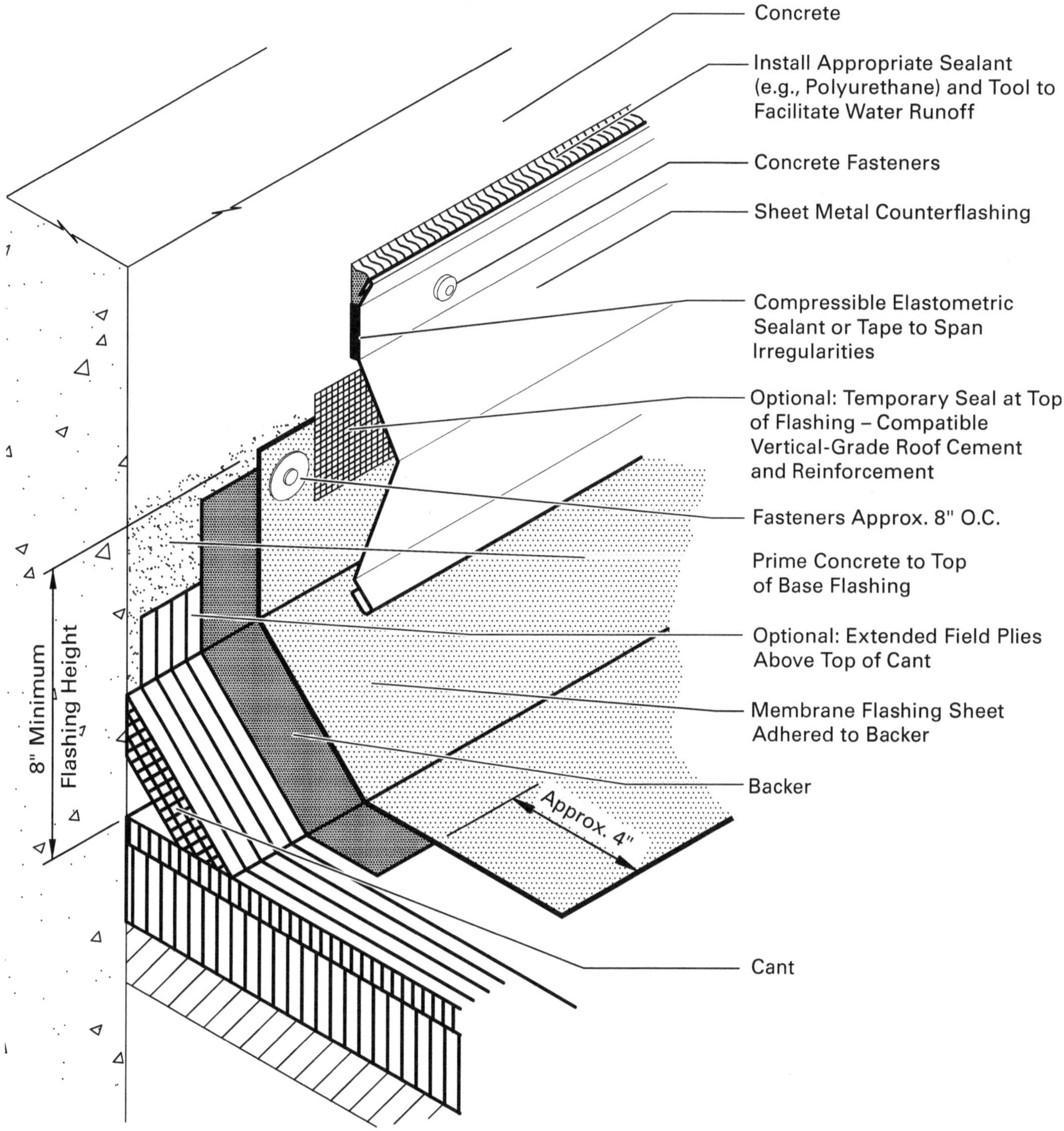

Notes:

1. This detail is applicable for hot-, cold- or torch-applied flashing systems.
2. This detail does not allow for differential movement between the deck and wall. Counterflashing detail may be a two-piece reglet and counterflashing.

Figure 3 Example of a mod-bit flashing detail drawing.

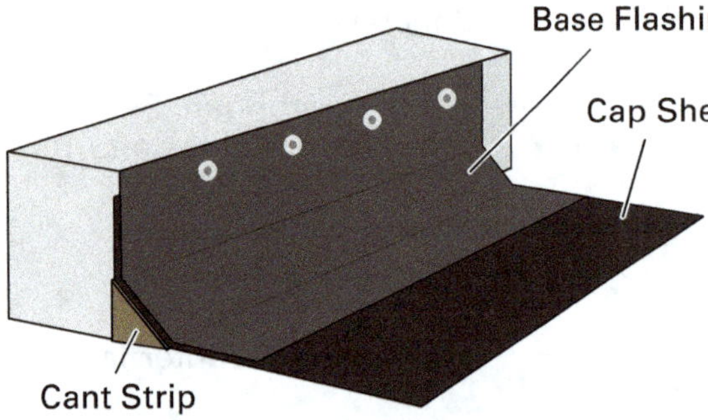

Figure 4 Example of a membrane flashing in a mod-bit roof system.

Liquid-applied flashings are commonly used in situations where penetrations or roof surfaces are irregularly shaped and difficult to flash using membrane flashing.

1.1.5 Accessories

Accessory products commonly used when constructing mod-bit roof membranes include roof cements, adhesives, and asphalt walkway pads.

1.0.0 Section Review

1. To make it stronger and more flexible, the bitumen in mod-bit roof systems is modified with _____.
   a. adhesive
   b. granules
   c. polymers
   d. sealant
2. Where in a multi-ply mod-bit roof system are cap sheets installed?
   a. As the top layer
   b. Onto the deck
   c. On the insulation
   d. As the bottom layer

SECTION TWO

2.0.0 INSTALLATION AND REPAIR

Objective

Explain how mod-bit roof systems are installed and identify common repair methods.

a. Identify tools and equipment used to install mod-bit roof systems.
b. Describe how mod-bit roof systems are installed.
c. Explain basic repair methods for damaged mod-bit roof systems.

Performance Tasks

1. Miter a cant strip to a vertical projection on an inside or outside corner.
2. Using cold or hot bitumen, install a mod-bit roof on a mock-up or jobsite.
3. Install a plumbing vent flashing.
4. Cut and install a drip edge corner.
5. Install and strip in a drip or metal edge.
6. Install a cap sheet surface using cold or hot bitumen.
7. Cut in mod-bit plies around a curb and over a cant strip and flash at top edge.

Trade Terms

Bleed-out: A continuous line of bitumen at the edge of a seam where it has flowed out of the lap and is visible on the outside, indicating that the proper amount of bitumen was used.

Certified Roofing Torch Applicator (CERTA) Program: A training program by the National Roofing Contractors Association (NRCA) that teaches critical safety hazards, considerations, and practices for all torching work.

Fishmouth: An opening in the edge of a membrane caused by wrinkling or shifting of sheets during installation.

Mole run: A long wrinkle in a roof membrane.

A wide variety of tools are required to install mod-bit roof systems. Each must be used carefully and safely. To install quality roof systems, roofers must be familiar with each of these tools and the steps involved in using them. They must also be able to fix mistakes that occur during installation, such as wrinkling and openings in membrane sheets.

2.1.0 Tools and Equipment

Roofers use many tools to install mod-bit systems. Some are used specifically for hot bitumen or cold-applied bitumen. It is important to be familiar with each of these tools and be able to use them safely.

> **NOTE**
> Basic tools used for construction work are covered in NCCER Module 00103, *Introduction to Hand Tools*, and NCCER Module 00104, *Introduction to Power Tools.*

2.1.1 Hand Tools

Roofers must be familiar with the following hand tools for mod-bit construction:

- *Mop* — Used to apply hot bitumen (*Figure 5*).
- *Bucket* — Used to transport bitumen (*Figure 5*).
- *Pouring can* — Used to transport bitumen.
- *Hook* — Used to roll out membrane sheets.
- *Felt slitter* — Used to cut sheets.
- *Utility knife* — Used to cut sheets.
- *Hot machine* — Used to heat bitumen (also called a *chain mop* or *johnny mop*).
- *Torch* — Used to heat saturated felts and make them pliable or to heat bitumen to a liquid state (*Figure 6*).
- *Hot-air gun* — Used to heat saturated felts and make them pliable or to heat bitumen to a liquid state.
- *Trowel* — Used to press or smooth down sheets around flashings or tight areas. Types of trowels include margin trowels (square-shaped) and linoleum trowels (used for cold-applied systems).
- *Notched squeegee* — Used for spreading cold-applied bitumen.
- *Broom* — Used to smooth down sheets and ensure tight adhesion to the bitumen underneath.
- *Roller* — Used to smooth down sheets and ensure tight adhesion to the bitumen underneath.
- *Paint roller and pan* — Used to apply primer.
- *Chalk line* — Used to mark lines on the substrate.
- *Wheelbarrow* — Used to transport materials, such as gravel.

2.1.2 Power Tools

Roofers must be familiar with the following power tools for mod-bit construction:

- *Kettle* — Used to heat asphalt or bitumen to a molten liquid and pump it to the roof through a pipe (*Figure 7*).

Figure 5 Mop and mop cart.

- *Sprayer* — Used for cold process installations, adhesives, and coatings.
- *Powered felt slitter* — Used to cut sheets.
- *Conveyor* — Used to transport roofing materials from the ground to the roof (*Figure 8*).
- *Roof hoist* — Used to lift roofing materials to the roof.

2.2.0 Installation Procedures

Steps for installing a mod-bit system may include mechanically attaching a base sheet, adhering any additional plies, adhering a cap sheet, and then surfacing to the top of the system. Installing mod-bit systems involves using dangerous equipment and materials, such as roofing torches and hot bitumen. It is critical for roofers to be aware of potential hazards and follow applicable safety procedures.

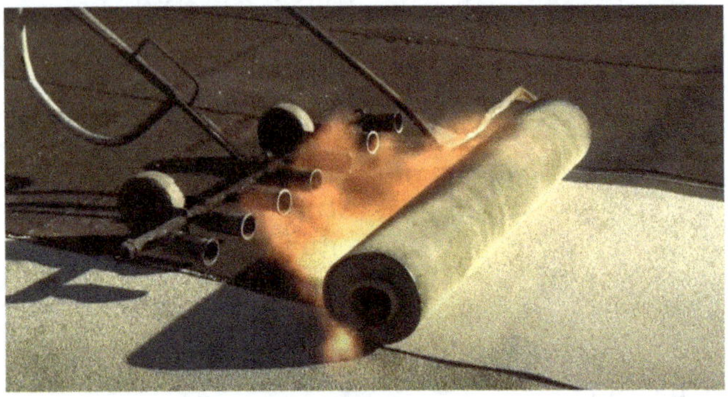

Figure 6 Seven-torch modified bitumen applicator.

Figure 7 Kettle.

Figure 8 Roofing conveyor.

Fire Triangle

Fire is a chemical reaction that requires the following three elements to be present: heat (or an ignition source), fuel, and oxygen. These three elements typically are referred to as the *fire triangle.*

Fire is the result of the reaction between the fuel and oxygen in the air. Scientists developed the concept of a fire triangle to aid in understanding of what causes fires and how they can be prevented and extinguished. Heat, fuel, and oxygen must combine in a precise way for a fire to start and continue to burn. If one element of the fire triangle is not present, fire will not start.

For example, some fire extinguishers smother a fire to remove oxygen, some cool fires to remove heat, and others can even change the composition of the fuel.

2.2.1 Safety Considerations

Installing mod-bit roof systems often involves the use of torching equipment, which comes with many possible fire hazards. Failure to recognize hazards, wear personal protective equipment (PPE), and follow safety practices can easily result in injury or death. However, with the right amount of care and training, torching work can be performed safely.

Always wear appropriate PPE when working with torching equipment, including work boots, long pants, long sleeves, heat-resistant gloves with tight cuffs, safety glasses and/or a face shield, and other protective clothing.

> **NOTE**
> NRCA's Certified Roofing Torch Applicator (CERTA) Program is a great resource for safety considerations associated with torching. Some insurance companies require CERTA training for contractors who use torches for roofing.

Be aware of the following torching-related safety hazards:

- Hot torch heads, trowels, or heated roofing membrane can all cause serious burns.
- Direct contact with liquid propane can cause a skin injury similar to frostbite.
- Fire is a serious hazard during torching operations. Observe the following safety considerations:
  - Heat and flame from a torch head can penetrate gaps and openings and ignite material far below the work surface.
  - If you see smoke, report it to your supervisor immediately. Smoke may contain poisonous gases that can incapacitate you quickly.
  - Any time you use torching equipment, there must be a *minimum* of two 4-A:60-B:C rated fire extinguishers within 10' of torching operations. Fire extinguishers must be readily accessible, with no obstacles blocking the path to reach one. To prevent them from exploding, do not expose fire extinguishers to open flame or heat.
  - Never use water to extinguish or cool hot bitumen. This will cause it to pop violently or explode.
- A fire watch should be assigned during lunch breaks and at the end of the shift for a minimum of two hours after torching to monitor the area for signs of fire.

Installing mod-bit roof systems often involves working with hot bitumen. Handling hot bitumen is extremely dangerous as it may reach temperatures over 500°F (260°C). Always be aware of the following safety considerations when working with hot bitumen:

- If bitumen is heated above the flash point, it may ignite and cause fires or explosions. Consult the bitumen packaging or SDS to ensure that the temperature of bitumen is always at least 50°F (~28°C) below the flash point.
- To protect yourself from burns, use extreme care and wear PPE when working with hot bitumen. Necessary PPE may include work boots, long pants, long sleeves, heat-resistant gloves with tight cuffs, safety glasses and/or a face shield, and other protective clothing. Clothing must be made of a material that doesn't melt, such as cotton. If hot bitumen gets on your skin, seek medical attention immediately.

NOTE

Gauntlet style gloves and types that do not have tight cuffs may fill up with molten liquids.

- Always stay hydrated by drinking enough water, especially when working with hot bitumen on a hot day.
- Mops are subject to spontaneous combustion if not periodically spun out and laid on a noncombustible surface. As the outer layer of bitumen on the mop cools, the heat of the bitumen underneath may cause the outer layer to crack and explode. To prevent this, spin out mops regularly.

2.2.2 Installation Outline

The following is an example of a very general procedure for installing a mod-bit system:

NOTE

Always refer to all applicable documentation when installing a roof system, including building codes, manufacturer instructions for equipment and materials, construction drawings, and project specifications.

Step 1 Make sure the substrate is clean, dry, and in good condition.

Step 2 Install insulation and/or cover boards as applicable or mechanically attach a base sheet. Modified base sheets can be installed using hot, cold, or torch-applied methods, as well as self-adhered.

Step 3 Seal gaps by applying adhesive and spreading it smooth with a trowel.

Step 4 Install the specified number of modified base sheets in shingle fashion (*Figure 9*).

- Mop the hot bitumen or use a notched rubber broom to spread cold bitumen.
- Roll the sheets into the bitumen.

Step 5 Apply the cap sheet.

Step 6 Flash any penetrations.

NOTE

Some flashings are installed prior to the cap sheet installation.

Step 7 If the cap sheet does not have factory mineral granule surfacing or foil-faced surfacing, then a surface coating must be installed in the field.

WARNING!

Handling hot bitumen is extremely dangerous. Always follow applicable safety precautions when working with hot bitumen.

2.2.3 Installing a Base Sheet

A base sheet/ply is used as the first layer over a nailable or concrete deck. It can be a modified base sheet or a conventional asphalt base sheet. Roof designers specify the type of base sheet and attachment method that should be used.

Base sheets are often mechanically attached to nailable substrates, although they can also be hot mopped, torch applied, hot-air welded, cold applied, or adhesive backed. A solid base sheet installation allows the subsequent sheets to adhere properly. Before installation, the roof deck should be checked to ensure it is dry, smooth, and free of debris. After the base sheets are rolled onto the roof surface, they should be given time to relax before fastening.

2.2.4 Installing Modified Base Sheets

While mod-bit roof systems are typically a two-ply system consisting of a base sheet and a surfaced cap sheet, additional modified base sheets

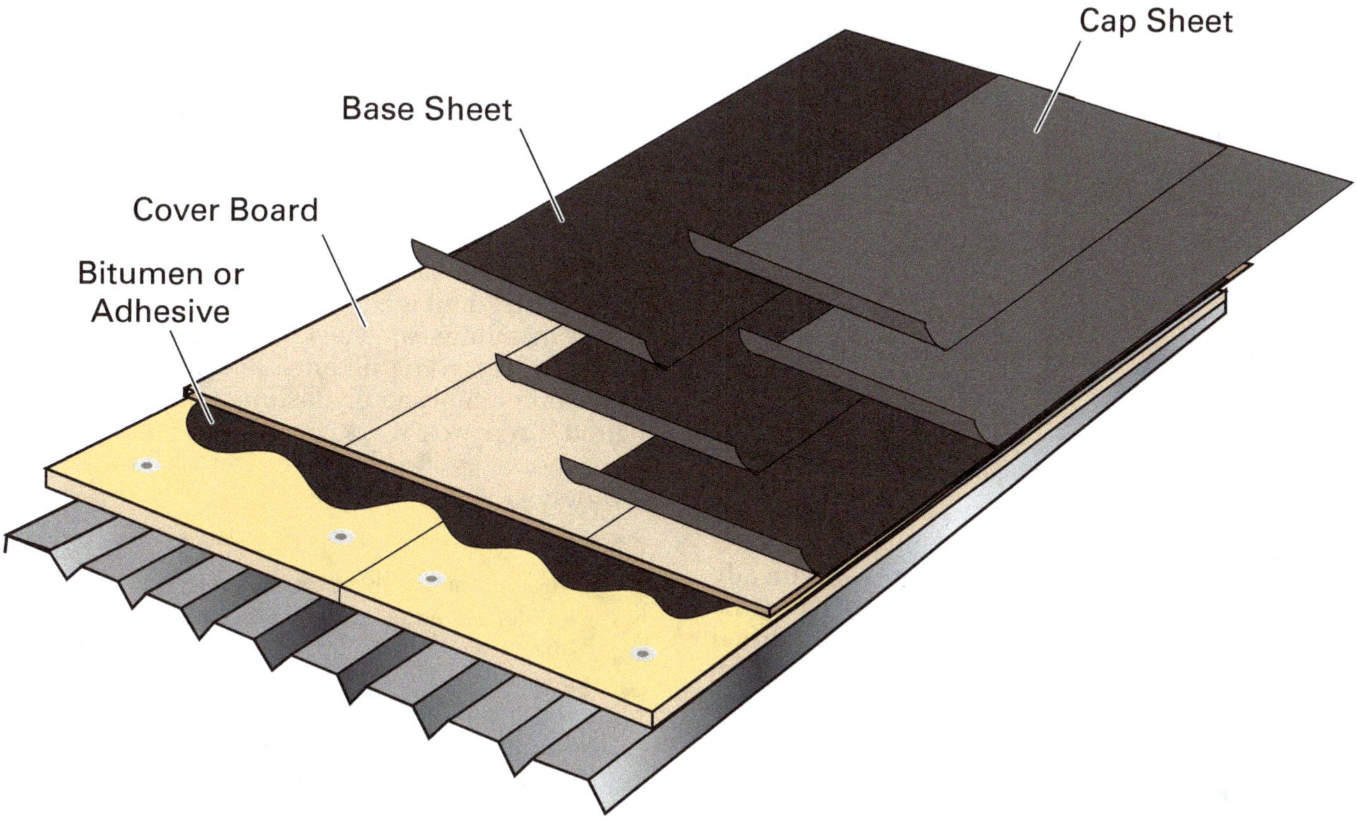

Figure 9 Basic example of a mod-bit roof system.

or mid-plies can be added to create the desired thickness and ensure watertightness. Modified base sheets can also be hot mopped, torch applied, hot-air welded, cold applied, or adhesive backed (*Figure 10*). These installation processes use different tools and come with unique safety considerations.

Hot-mopped bitumen must be just the right temperature and viscosity. If it is too hot, it will be too thin to provide enough coverage. If it is not hot enough, it may not spread or adhere effectively. Hot-mopped bitumen is transported on the roof in buckets. A roofer will dip a mop in the mop cart or bucket and spread it on the substrate while other crew members roll in the sheets and cut them to fit as needed. Roofers will snap chalk lines or use lines applied by manufacturer as guides to install sheets in a straight line.

> **WARNING!**
> Mops may seem heavier throughout a day's work. Be aware of dehydration and overexertion by drinking water, using good posture, and taking breaks when needed.

For cold-applied installations, roofers typically use a notched rubber squeegee or brush to evenly coat the substrate. The sheets are then rolled into the modified bitumen and pressed into it with a broom or roller to ensure even adhesion. Using cold-applied bitumen is less dangerous than using hot bitumen, but the process takes longer.

2.2.5 Installing Cap Sheets

Cap sheets are the top layer of the roof system and are used to create a weatherproof surface. Before applying modified cap sheets, the cap sheet should be unrolled, cut into workable lengths, and allowed to lie flat. This allows it to relax and reduces the possibility of it curling during installation. Then, they are applied to the roof with bitumen adhesive just like the ply sheets. They are usually installed with 4" side laps and 6" end laps.

Cap sheets are often be covered with surfacing to protect the roof system from UV rays. Factory surfaced cap sheets can contain mineral granule surfacing or foil-faced surfacing. Cap sheets that do not come surfaced from the factory may be coated with surfacing in the field after the roof system has been installed. Field surface coating can include a flood coat of hot modified bitumen and aggregate, as well as elastomeric, aluminum, or liquid-applied coatings.

Coatings are described in detail in NCCER Module 16210, *Liquid-Applied Roofing.*

2.3.0 Basic Repair Methods

Mistakes happen even under the best conditions for roof installation. To ensure a quality and efficient job, workers on mod-bit roof systems should know how to repair any damage created during the installation process.

It is important to note that SBS and APP polymer-modified products may have slightly different repair procedures. For instance, application using hot asphalt is only appropriate for SBS polymer-modified sheets and specifically SBS polymer-modified sheets surfaced with fine mineral matter (sanded). APP polymer-modified and SBS polymer-modified sheets with film surfacing are intended for heat welding using a roofing torch.

(A) Hot-Mopped Bitumen

(B) Cold-Applied Bitumen

Figure 10 Hot-mopped and cold-applied bitumen.

WARNING!

Use extreme care when working with torches. NRCA's CERTA Program is a great resource for safety considerations associated with torching.

2.3.1 Cap Sheet Repair

Cap sheet patches are commonly used to cover and repair damaged areas of a mod-bit roof. Before installing cap sheet patches, remove rough spots, debris, contaminants, and loose surfacing from the membrane or flashing surface to be repaired. Prepare a work area that extends at least 6" or more beyond the perimeter of the repaired location.

NOTE

Always check applicable manufacturer guidelines and company procedures for exact measurements.

To promote complete adhesion of a patch, it is necessary to prepare the surface where the patch will tie into the existing membrane. If the membrane has a granule surfacing, this includes embedding the granules with a heated trowel or scraping them off and priming the surface (*Figure 11*).

Next, it is necessary to lightly prime the surface of the membrane with asphalt primer to enhance the adhesion of the cap sheet patch. The final step is installing the patch over the repair area with cold-applied adhesive, hot asphalt (SBS only), or by heat welding—as appropriate for the materials used—and then reapply surfacing as required.

Figure 11 Embedding granules to prepare surface for adhesion.

2.3.2 Three-Course Repair

Three-coursing is commonly used to repair mod-bit roofs. This repair involves successive applications of a base course of roof cement, reinforcing fabric, and a second course of roof cement. Modified bitumen roof cement must be used on mod-bit membranes. Where a three-course repair will remain exposed, it should additionally include an application of a surfacing material.

2.3.3 Fishmouths and Mole Runs

A **fishmouth** is a raised opening in the lapped edge of a membrane sheet or flashing. A **mole run** is a long ridge or wrinkle in a roof membrane. Fishmouths and mole runs pose a risk to weatherproofing and should be cut and patched. The following is a general procedure for repairing a fishmouth or mole run:

Step 1 Remove rough spots, debris, contaminants, and loose surfacing from the membrane or flashing surface to be repaired. Granule-surfaced membranes should be washed, brushed, broomed, or blown clean.

Step 2 Cut the fishmouth or mole run down the center to lay the material flat. Remove any debris from under the fishmouth and ensure it is dry, and then embed the granules for a smooth surface.

Step 3 Use the appropriate adhering method for the type polymer-modified bitumen patch being used and ensure it extends at least 6" beyond any part of the cut and is well adhered.

NOTE

Always check applicable manufacturer guidelines and company procedures for exact measurements.

Step 4 To promote complete adhesion of a patch, prepare the surface where the patch will tie into the existing membrane. Different steps are required for different surfacing types, as follows:

- *Granule-surfaced cap sheets* — Apply heat and press granule surfacing into the bitumen. Where a patch will be installed in cold-applied adhesive, prime the surface of the membrane with asphalt primer and allow it to dry.
- *Aluminum coatings* — Use a stiff-bristle broom to release and sweep away loose coating. Use a hot-air gun or torch to heat the patch area only until the surface turns black.
- *Elastomeric coatings* — Use a hot-air gun or torch to heat the surface sufficiently to scrape the coating off.

Step 5 If the granules cannot be removed or embedded, install the patch in cold-applied adhesive over a lightly primed surface.

CAUTION

Primer contains solvents and is used to enhance adhesion; however, overuse of primer can harm the membrane.

WARNING!

Additional PPE may be needed to safely handle primer. Primer is also highly flammable, so be aware of any heat sources and do not place primer containers close to them. Fumes from some primers may collect in low-lying areas and be ignited by static sparks, causing a fire. Always consult the SDS for proper PPE and related hazards.

Step 6 Where plies are removed from the membrane, install the same number of plies of reinforcing material in cold-applied or hot bitumen. Extend the bottom ply at least 6" beyond the area to be repaired and each succeeding ply at least 3" beyond the previous ply.

Step 7 Check the edges of the patch for voids or delamination. If voids are present, correct the deficiency, ensuring the proper **bleed-out** is visible.

NOTE

If patching over a vertical substrate, fasten the top of the patch to the wall or curb approximately 8" (depending on manufacturer instructions) on center using fasteners suitable for the substrate.

Step 8 Reapply the surfacing as required.

2.0.0 Section Review

1. What tool is used to heat mod-bit sheets and make them pliable or to heat bitumen to a liquid state?
   a. Kettle
   b. Hot-air gun
   c. Hot machine
   d. Hot lugger

2. What is the top layer of the roof system that is used to create a weatherproof surface called?
   a. Modified base sheet
   b. Mid ply
   c. Base sheet
   d. Cap sheet

3. You are repairing a deficiency in a mod-bit membrane, but the granules cannot be removed or embedded. What should you do?
   a. Use a torch and a trowel to smooth out the surface.
   b. Install the patch without priming.
   c. Lightly prime the surface and install the patch in cold-applied adhesive.
   d. Use a hot-air welder to melt the membrane before installing the patch.

Review Questions

1. The top sheet in a mod-bit system is called a _____.
   a. flood coat
   b. cap
   c. ply
   d. mat

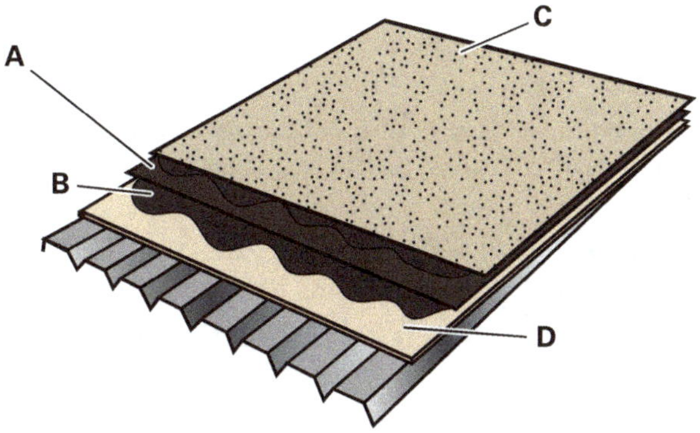

Figure RQ01

2. Refer to *Figure RQ01*. Which component is the bitumen?
   a. A
   b. B
   c. C
   d. D
3. Refer to *Figure RQ01*. Which component is the insulation?
   a. A
   b. B
   c. C
   d. D
4. Refer to *Figure RQ01*. Which component is the modified base sheet?
   a. A
   b. B
   c. C
   d. D
5. What is the *minimum* required amount of fire extinguishers that must be within reach while using torching equipment?
   a. None
   b. 1
   c. 2
   d. 3
6. Which of the following statements about bitumen is *true*?
   a. It is made of pine tar.
   b. It becomes fluid when heated and solid when cool.
   c. It is not used on mod-bit systems.
   d. It is a fastener.
7. Which of the following tools is used to press or smooth down sheets around flashings or tight areas?
   a. Trowel
   b. Hot-air gun
   c. Johnny mop
   d. Sprayer
8. Prior to installing them, mineral-surfaced cap sheets should be _____.
   a. coated with two-component adhesive
   b. washed with water
   c. unrolled and allowed to lie flat
   d. placed in a humid area
9. Which of the following is an opening in the edge of a membrane sheet?
   a. Mole run
   b. Fishmouth
   c. Puncture
   d. Ridge
10. How far should a patch extend beyond any part of the membrane damage?
    a. At least 1"
    b. At least 2"
    c. At least 6"
    d. At least 10"

Trade Terms Introduced in This Module

Atactic polypropylene (APP): A plastic-like polymer used to modify bitumen.

Bill of lading: A detailed list of items shipped; it is delivered by the carrier with the shipment.

Bitumen: An asphalt-based product that provides waterproofing and adhesion.

Bleed-out: A continuous line of bitumen at the edge of a seam where it has flowed out of the lap and is visible on the outside, indicating that the proper amount of bitumen was used.

Cant strip: A long, triangular object placed in 90-degree bends before applying the membrane to reduce the stress on the membrane.

Certified Roofing Torch Applicator (CERTA) Program: A training program by the National Roofing Contractors Association (NRCA) that teaches critical safety hazards, considerations, and practices for all torching work.

Fishmouth: An opening in the edge of a membrane caused by wrinkling or shifting of sheets during installation.

Flash point (FP): The temperature at which a material can ignite.

Mole run: A long wrinkle in a roof membrane.

Plies: Layers of reinforcement in a built-up or mod-bit roof system.

Polymer-modified: A term used to describe bitumen that has been modified with polymers (such as APP or SBS) to add plastic or rubber qualities desirable for roofing.

Redundant: Having multiple layers for extra protection.

Scrim: Reinforcement fibers in modified bitumen sheets.

Selvage edge: The edge of a mod-bit sheet with no granules, allowing it to be heat welded to the next sheet.

Styrene butadiene styrene (SBS): A rubber-like polymer used to modify bitumen.

Surfacing material: Protects the membrane from ultraviolet (UV) degradation and weather erosion. A variety of materials are used for surfacing, such as a mineral aggregate, a liquid-applied coating, or a sheet material.

Tapered edge strip: A rigid board shaped with built-in slope and used in penetration flashings to provide slope toward drains.

Viscosity: The consistency of a liquid. Liquids with high viscosity are thicker, while liquids with low viscosity are thinner.

Additional Resources

This module presents thorough resources for task training. The following reference material is suggested for further study.

National Roofing Contractors Association (NRCA), **www.nrca.net**.
NCCER Module 00103, *Introduction to Hand Tools*.
NCCER Module 00104, *Introduction to Power Tools*.
NCCER Module 16107, *Sheet Metal in Roofing*.
NCCER Module 16208, *Built-Up Roof Systems*.
NCCER Module 16210, *Liquid-Applied Roofing*.
NRCA Repair Manual for Low-Slope Membrane Roof Systems, NRCA.
NRCAnews. "CERTA Training Video." YouTube video. 26:03. April 12, 2018. **www.youtube.com**.

Figure Credits

iStock@peuceta, Module Opener
Courtesy of the National Roofing Contractors Assoc., Figures 1, 3, 5–7, 11
Silver Systems Incorporated, Figure 8
Courtesy of GAF, Figure 10 (A)
iStock@Niko_Cingaryuk, Figure 10 (B)

Section Review Answer Key

SECTION 1.0.0

Answer	Section Reference	Objective
1. c	1.0.0	1
2. a	1.1.1	1a

SECTION 2.0.0

Answer	Section Reference	Objective
1. b	2.1.1	2a
2. d	2.2.5	2b
3. c	2.3.3	2c

User Update

NCCER makes every effort to keep its textbooks up-to-date and free of technical errors. We appreciate your help in this process. If you find an error, a typographical mistake, or an inaccuracy in NCCER's curricula, please submit a User Update form by visiting **https://www.nccer.org/olf**. You can also scan the QR code using the camera on your phone or mobile device to access the form.

Liquid-Applied Roofing

Overview

Liquid-applied roofing is an approach that dates back centuries. Liquid roofing products have the distinct advantage of being free of seams. There are two primary applications for liquid-applied roofing products: coatings for restoration and liquid-applied membranes as a primary weather protection system. Liquid-applied roofing is a rapidly growing part of the roofing industry.

Module 16210

Trainees with successful module completions may be eligible for credentialing through the NCCER Registry. To learn more, go to **www.nccer.org** or contact us at 1.888.622.3720. Our website, **www.nccer.org**, has information on the latest product releases and training.

Your feedback is welcome. You may email your comments to **curriculum@nccer.org**, send general comments and inquiries to **info@nccer.org**, or fill in the User Update form at the back of this module.

This information is general in nature and intended for training purposes only. Actual performance of activities described in this manual requires compliance with all applicable operating, service, maintenance, and safety procedures under the direction of qualified personnel. References in this manual to patented or proprietary devices do not constitute a recommendation of their use.

16210 V1.0

16210
LIQUID-APPLIED ROOFING

Objective

Successful completion of this module prepares you to do the following:

1. Describe types of liquid-applied roof systems and how they are properly applied.
   a. Differentiate between roof coatings and liquid-applied roof systems.
   b. Identify basic tools used to apply and measure the thickness of roof coatings and liquid-applied roof systems.
   c. Describe various coatings and their application considerations.
   d. Outline the general steps in liquid-membrane roof system application.

Performance Tasks

Under supervision, you should be able to do the following:

1. Prepare the seams and penetrations of a single-ply membrane roof for a coating per the product manufacturer's instructions.
2. Complete the steps required for an adhesion test.
3. Apply the base coat of a roof coating to the proper wet film thickness using the manufacturer-specified application rate.

Trade Terms

Adhesion test
Application rate
Dry film thickness
Elastomeric
Flashing-grade coatings
Liquid-applied roof membranes
Mils
Monolithic
Overspray
Roof coatings
Solvent-based
Thermography
Volatile organic compounds (VOCs)
Water-based
Wet film thickness

Industry Recognized Credentials

If you are training through an NCCER-accredited sponsor, you may be eligible for credentials from NCCER's Registry. The ID number for this module is 16210. Note that this module may have been used in other NCCER curricula and may apply to other level completions. Contact NCCER's Registry at 1.888.622.3720 or go to **www.nccer.org** for more information.

You can also show off your industry-recognized credentials online with NCCER's digital badges. Transform your knowledge, skills, and achievements into badges that you can share across social media platforms, send to your network, and add to your resume. For more information, visit **www.nccer.org**.

Videos and Digital Resources

How to Access Resources

This craft has additional videos and resources to enhance your learning experience. To view these resources, scan the QR below. The videos and resources are separated by module.

You can scan this code using the camera on your phone or mobile device to view these videos and resources.

Contents

1.0.0 Liquid-Applied Roofing 1
  1.1.0 Roof Coatings and Liquid-Applied Roof Membranes 2
    1.1.1 Elastomeric and Reflective Properties 2
  1.2.0 Tools and Testing 2
    1.2.1 Basic Hand Tools 3
    1.2.2 Power Tools 4
    1.2.3 Coating Measurement 5
  1.3.0 Roof Coating Products 6
    1.3.1 Coating Types 7
    1.3.2 Substrates 9
    1.3.3 Coating Application 12
  1.4.0 Liquid-Applied Roof Membranes 14

Figures

Figure 1 Liquid-applied roof membrane 2
Figure 2 Roof coating project 2
Figure 3 Roller 4
Figure 4 Applying a coating with a nap roller 4
Figure 5 Applying a coating with a brush 4
Figure 6 Paddle mixer 4
Figure 7 Roofing sprayer 5
Figure 8 Coating application by sprayer 5
Figure 9 Wet film gauge 6
Figure 10 Reading a wet film gauge 6
Figure 11 Metal roof with urethane coating 9
Figure 12 PMMA-coated concrete roof 9
Figure 13 IR thermography scan 10
Figure 14 Adhesion test 12
Figure 15 Small roll of polyester roofing membrane used for detailing 14
Figure 16 Feather the fabric around protrusions 15
Figure 17 Example of a liquid-applied pipe flashing detail 15
Figure 18 Polyester roofing membrane being placed in the base coat 16

SECTION ONE

1.0.0 LIQUID-APPLIED ROOFING

Objective

Describe types of liquid-applied roof systems and how they are properly applied.

a. Differentiate between roof coatings and liquid-applied roof systems.
b. Identify basic tools used to apply and measure the thickness of roof coatings and liquid-applied roof systems.
c. Describe various coatings and their application considerations.
d. Outline the general steps in liquid-membrane roof system application.

Performance Tasks

1. Prepare the seams and penetrations of a single-ply membrane roof for a coating per the product manufacturer's instructions.
2. Complete the steps required for an adhesion test.
3. Apply the base coat of a roof coating to the proper wet film thickness using the manufacturer-specified application rate.

Trade Terms

Adhesion test: A test performed on an existing roof to verify that a substrate can successfully bond to a liquid coating. Also called a *peel test*.

Application rate: The average quantity (mass, volume, or thickness) of material applied per unit area, typically specified in gallons per square or liters per square meter.

Dry film thickness: The thickness, expressed in mils, of an applied and fully cured liquid coating.

Elastomeric: Able to be stretched at least twice its original size and return to its original shape.

Flashing-grade coatings: Liquid coatings specifically designed for areas where the primary roof covering is interrupted or terminated, such as protrusions and damaged areas. These are generally applied before base and finishing coats and can be applied to vertical surfaces since they have a thicker consistency.

Liquid-applied roof membranes: Continuous roof membranes constructed in place with a liquid coating and a reinforcing material. The liquid coating is available as a one- or two-component product and is typically applied in two coats. Depending on the chemistry, a catalyst or hardener may be added to induce the curing process. The reinforcement is typically polyester or fiberglass mat set into a still-wet base coat.

Mils: A unit of measurement equal to one-thousandth of an inch, or 0.001" (0.0254 mm).

Monolithic: Formed or composed of a single material; seamless.

Overspray: Deposits of sprayed material that fall outside the intended area.

Roof coatings: Liquids containing essential weatherproofing properties that are applied to roofs to protect, restore, prolong, or enhance performance.

Solvent-based: Containing a petroleum-based solvent as a liquifying agent, leading to a higher content of organic compounds (see *volatile organic compounds*).

Thermography: The creation of digital images using infrared cameras to detect the amount of heat energy emitted.

Volatile organic compounds (VOCs): Compounds and chemicals that easily evaporate at room temperature. They are found in products such as solvents, paints, and adhesives.

Water-based: Having water as a liquifying agent. Although water can also be considered a solvent, the term *solvent* typically refers to petroleum-based products.

Wet film thickness: The thickness, expressed in mils, of a liquid coating as applied (wet).

Although the modern liquid-applied roofing market rapidly expanded in the twentieth century, the concept of liquid weatherproofing was born long ago. Natural liquid bitumen was used for roof waterproofing in the 1800s. Its use as a boat sealant dates back to ancient times, according to the Liquid Roofing and Waterproofing Association.

Liquid-applied roofing is one of the fastest-growing segments in the roofing industry. Innovations in chemistry have led to a diverse market with products that suit a variety of roofing applications and scenarios.

1.1.0 Roof Coatings and Liquid-Applied Roof Membranes

There are two branches of the liquid-applied roofing market: liquid-applied roof membranes and roof coatings. A liquid-applied roof membrane combines liquid materials with reinforcement to form a monolithic roof protection system. This module uses the following terminology:

- *Liquid-applied roofing* — General term for all liquid roofing products and systems (includes both coatings and liquid-applied roof membranes).
- *Liquid-applied roof membranes* — Refers to a complete roof system in which liquid-applied roofing products are reinforced to provide the primary weatherproofing membrane.
- *Roof coatings* — Refers to liquid-applied coatings that are applied over roof systems to protect, restore, prolong, or enhance performance.

Liquid-applied roof membranes (*Figure 1*) are typically used on commercial low-slope roofs. An advantage of these systems is that there are no seams, since most roof failures begin at seams and joints. They are also lightweight, handle traffic well, and can be recoated at a reasonable cost to extend their lifetime. They may be installed over many existing roof types, eliminating the need for expensive and time-consuming roof demolition.

In contrast, a roof coating is applied over a roof system to refresh its protection from the elements (*Figure 2*). Roof coatings are used to restore aged roof systems and reflect heat from the sun. They are a cost-effective option to extend the life of an existing roof. Coatings can literally add decades to the life of a roof system.

However, not all roofs are good candidates for a restorative coating. For example, if a roof system is saturated with water from leaks, a coating will not restore it. The only practical option is complete removal and replacement. For a roof to benefit from a coating, it must be reasonably sound.

1.1.1 Elastomeric and Reflective Properties

Elasticity and reflectivity are important properties for liquid-applied roofing products. Roof coatings are elastomeric, which means they can stretch. Because all buildings are subject to movement from thermal expansion and contraction, a roof system with elastic properties can reduce the potential for cracks and damage. An elastomeric coating can stretch to accommodate thermal growth and contract as the surface beneath contracts.

Reflectivity is a roof's ability to reflect light and heat. Reflectivity can help reduce the heat transferred to the space below. This saves energy costs and prevents wear and tear on HVAC equipment. Light colors, like white, are more reflective than dark colors and are often chosen for liquid-applied roofing projects.

Both elasticity and reflectivity are affected by time. The roof surface can become soiled and stained, reducing reflectivity. Ultraviolet (UV) light from the sun damages virtually all materials over time, affecting a coating's elasticity. Both properties, however, can be restored by reapplying the coating before extensive damage occurs.

1.2.0 Tools and Testing

When working with liquid-applied products, you should be familiar with the basic tools used to apply them. You must also know how coating thickness is measured and how it relates to coating application.

Figure 1 Liquid-applied roof membrane.

Figure 2 Roof coating project.

Did You Know?

Central Park Avenue South

Central Park Avenue South in New York City was constructed as the Barbizon Plaza Hotel in 1930, but it was renovated as a luxury condominium residence in 1988. Its unique roof was rebuilt to accommodate HVAC systems.

By 2014, the roof had developed severe leaks. A liquid-applied membrane roof was chosen for its renovation. This roof used a fleece-reinforced membrane and a custom-color polyurethane finish coating. The work was conducted from swinging scaffolding.

Figure Credit: Courtesy of Sika USA

Figure Credit: Courtesy of Sika USA

1.2.1 Basic Hand Tools

Hand tools for applying liquid material are typically designed for one-time use. They are often easier to replace than they are to clean.

Rollers (*Figure 3*) and brushes are often used to apply the coatings. Although power sprayers work faster, overspray from them can be a serious problem. If overspray is likely to occur, use basic hand tools for application instead.

Rollers and brushes are also more effective when a small area is involved, eliminating the time and effort required to set up and clean spraying equipment. A roller installation is shown in *Figure 4*, and a brush installation is shown in *Figure 5*.

Brushes and rollers are dipped into buckets or large roller pans of the product. Five-gallon buckets of the coating are generally prepared as needed during the installation.

A more efficient roller application can be done with a tank spreader. Coating is poured into the tank above the roller, and a controlled amount of product drains down to the roller. Also note the profiled roller used on the metal roof. Rollers are available in a variety of shapes to improve coverage over unique surfaces. For example, a flat roller is not as effective along the profile of a metal roof.

The tank of the spreader can be replaced with just a few simple pieces of hardware. The flow is controlled by the size of the holes drilled into the tank. Once the job is done, the tanks must be properly cleaned to ensure all the holes remain open. Contractors may keep several tanks on hand with different hole sizes for different products. The roller cover beneath the tank is also easy to change.

Another roller tool is designed as an accessory for use with a coating pump, like those used with sprayers. A controlled-flow roller frame screws directly onto the end of a coating pump hose. The pump keeps the roller wetted with coating as it is rolled onto the surface.

Figure 3 Roller.

Figure 4 Applying a coating with a nap roller.

Figure 5 Applying a coating with a brush.

1.2.2 Power Tools

Power tools used in liquid-applied roofing need to be cleaned thoroughly after each use and maintained per the manufacturer's instructions. The simplest power tool to mention is the paddle mixer (*Figure 6*). The paddle mixer isn't a power tool itself. It is an accessory used *with* a power drill to thoroughly mix coatings. For example, epoxy coatings come in two parts that must be thoroughly mixed before application. Solvent-based coatings may also be mixed on the jobsite.

To avoid slinging coating around, never withdraw a paddle mixer from the coating while it is still spinning. Always wear all necessary PPE when mixing and applying coating products.

The most common power tool associated with liquid-applied roofing is a power sprayer (*Figure 7*). An airless sprayer pressurizes the coating to several thousand pounds per square inch (psi) and routes it through a hose and out through the spray nozzle. The tip is designed to break up the coating evenly into a fan-shaped spray pattern of tiny droplets. The reduction in pressure as the coating leaves the nozzle causes it to break into droplets.

Project planners consider the application method long before the job starts since it affects the estimate and bidding process. Using a sprayer is fast and efficient (*Figure 8*). However, overspray is a very important consideration. Overspray occurs when sprayed material lands outside of the intended area. Since the coating is broken into tiny droplets, the spray can be carried a significant distance by even a light wind.

Figure 6 Paddle mixer.

Figure 7 Roofing sprayer.

Figure 8 Coating application by sprayer.

While overspray may not seem like a problem on a large roof, it can cause property damage to cars, windows, and building finishes. This damage can occur even in places you can't see. The cost to remove overspray from a parking lot full of cars can be devastating and take months. Using a sprayer to apply coatings is a risk that must be considered before use. The risk should also be assessed as the work progresses. Project specifications, company policy, or the work order usually dictate the maximum allowable wind speed.

CAUTION

Before using a pressurized spray to apply a coating, evaluate the wind speed and direction carefully. Do not use a sprayer when the conditions can lead to property damage. If the conditions suggest that overspray could result in property damage, notify your supervisor immediately and ask for guidance.

WARNING!

Before maintaining or cleaning any sprayer, ensure that all sources of energy are disconnected. Ensure that pneumatic hose connections are secured and cannot whip about if the connection is broken. Electric sprayers must be properly grounded and connected to ground-fault circuit interruption (GFCI) protected circuits. Always follow applicable manufacturer instructions for cleaning and maintenance of spray equipment.

To reduce overspray, protective boards can be placed around the roof perimeter. Overspray monitors can also assist by following along with handheld screens to help capture airborne coating.

High-quality and high-volume spray equipment is expensive and must be handled with care. This includes regular maintenance and proper cleaning. When cleaning and servicing spray equipment, follow the manufacturer's guidance carefully and give it the attention it deserves. No craftworker is any better than the tools that serve them.

1.2.3 Coating Measurement

Manufacturers specify the proper thickness of their coatings in **mils**. A mil is equal to 0.001" (one-thousandth of an inch, or 0.0254 mm). For example, if a coating is 30 mils thick, it is 0.030" (0.762 mm) thick.

The final thickness of a coating is called its **dry film thickness** (that is, its thickness after it cures). The thickness of a coating after it cures can be substantially less than when it is still wet. This is because the liquid in the coating evaporates after application, leaving only the solids behind.

The change in coating thickness from when it is applied to when it is cured can be calculated ahead of time. For that reason, a measure of **wet film thickness** is used to determine what the dry film thickness will be. On the job, you are often required to measure the wet film thickness as you apply a coating, to ensure the right amount of product is being consistently applied.

The wet film thickness of the coating is checked using a wet film gauge (*Figure 9*). Each tooth of the gauge is a bit shorter than the adjacent tooth and is marked with its distance from tip to base. The gauge is carefully placed down in the coating at a 90-degree angle to the roof surface, then lifted straight up. The measurement is found by looking for the last tooth with coating on it and the adjacent tooth that is clean. The coating thickness lies between these two measurements, as shown in *Figure 10*. After taking a measurement, remember to wipe the gauge clean before the product dries.

In this example, the wet film thickness would need to reach 102 mils to result in a dry film thickness of 46 mils.

> **NOTE**
>
> The figures used in these examples should not be considered average or recommended. You must use the information provided for each individual product, as dry film thickness and the percentage of solids vary significantly. Also note that a specified dry film thickness for a coating may be reported as the result of multiple coats rather than a single application.

Dry film thickness can also be tested with a simple electronic tool called a *digital coating thickness gauge*. However, they can only be used with a metal substrate.

1.3.0 Roof Coating Products

Some of the earliest roof coatings included naturally occurring substances such as clay, gelatin, and beeswax. Today, advances in chemistry and synthetic materials have led to products that offer unique benefits and greater longevity. Common types of coatings include acrylic, urethane, silicone, and polymethyl methacrylate (PMMA).

Figure 9 Wet film gauge.

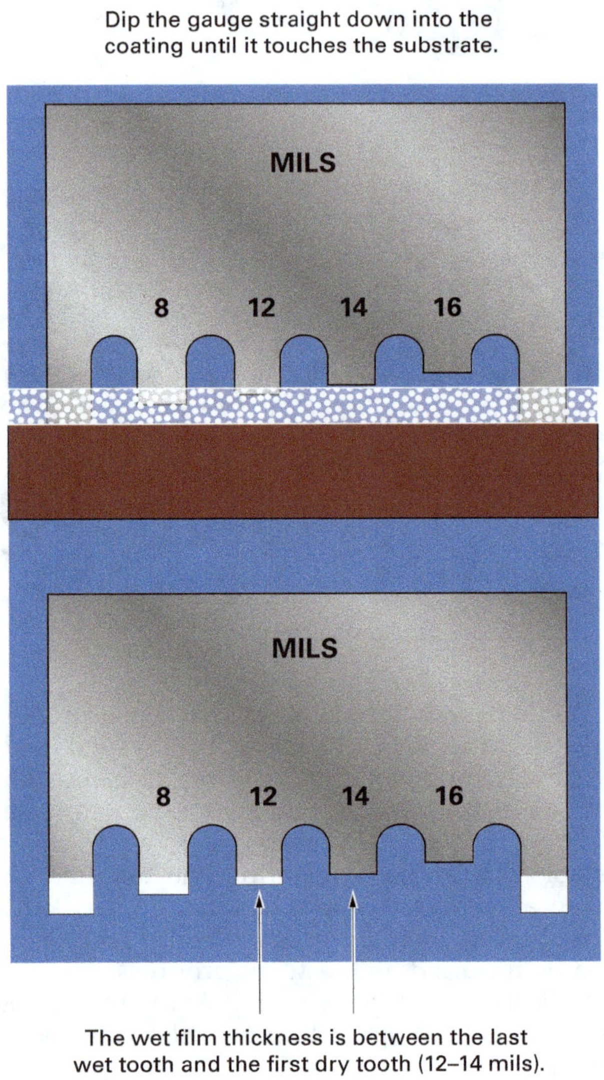

Figure 10 Reading a wet film gauge.

Calculating Dry Film Thickness

You can use the wet film thickness to calculate the dry film thickness with some simple math. Product manufacturers indicate the percent of solids found in each coating in their specifications and catalog data. The dry film thickness can be determined by multiplying the wet film thickness by the percent of solids.

Example:

You have measured a wet film thickness of 40 mils. The product label indicates the percent of solids in the coating is 54%. Begin by changing the percent of solids to a decimal value. A percentage is changed to a decimal value by dividing by 100, as shown here:

$$\text{Dry film thickness} = \text{Wet film thickness} \times (\text{percent of solids} \div 100)$$
$$\text{Dry film thickness} = 40 \text{ mils} \times (54 \div 100)$$
$$\text{Dry film thickness} = 40 \text{ mils} \times 0.54$$
$$\text{Dry film thickness} = 21.6 \text{ mils}$$

The dry film thickness for this coating after curing should be **21.6 mils** if it is applied at a wet film thickness of 40 mils.

You can also determine the wet film thickness needed to produce the dry film thickness required. This requires two pieces of information—the specified dry film thickness for an effective treatment and, again, the percent of solids. Then divide the wet film thickness by the percent of solids as a decimal.

Example:

The required dry film thickness for a roof is 46 mils, and the percent of solids in the coating is 45%. The wet film thickness at which the coating should be applied is calculated as follows:

$$\text{Wet film thickness} = \text{Dry film thickness} \div (\text{percent of solids} \div 100)$$
$$\text{Wet film thickness} = 46 \text{ mils} \div (45 \div 100)$$
$$\text{Wet film thickness} = 46 \text{ mils} \div 0.45$$
$$\text{Wet film thickness} = 102.2 \text{ mils}$$

To achieve a dry film thickness of 46 mils, the coating should be applied with a wet film thickness of **102.2 mils**.

1.3.1 Coating Types

Coatings may be water-based or solvent-based. Solvents are substances that dissolve or disperse other substances. In general, water-based products are safer to work with and apply. Some products must be prepared on-site by mixing two or more components. Mixing the components initiates a chemical reaction, and curing begins. As it cures, the water or solvent evaporates and leaves the solids behind.

Coatings come in a variety of colors. White is a popular option because it reflects UV rays, which can save energy spent on cooling the building. Building owners, however, may have additional priorities to consider.

Coatings contain volatile organic compounds (VOCs). VOCs are gases that are released into the air from products or processes. Different products contain different levels of VOCs. In general, water-based products contain lower VOCs than solvent-based products. In some cases, it may be necessary to temporarily shut down ventilation units while applying coatings until the fumes settle.

WARNING!

Before applying a coating, review the Safety Data Sheet (SDS) and take the necessary precautions. Coatings may be highly flammable and/or an inhalation hazard. Ensure that any electrically powered equipment is properly grounded and keep the equipment as far from the product as possible.

CAUTION

Always check manufacturer instructions for the coating product you are using for exact application temperatures and other specific considerations.

Acrylic

Acrylic products are water-based coatings. They are cost effective, versatile, and easy to work with. Acrylics are currently the most widely used coating type. Other products may be more durable, but acrylic products make up for that in cost and ease of handling.

Acrylic coatings should be applied at temperatures of 50°F (10°C) or above. This is common for water-based products. The time required for curing depends on the temperature and humidity levels. The ideal curing conditions are warm weather with low humidity since the water in the coating will evaporate quicker if the air is relatively dry.

Acrylic coatings are applied to flat and low-slope roofs and over many existing substrates, such as built-up, polymer-modified bitumen, single-ply, metal, and concrete.

Compared to products such as silicone and urethane, acrylic products contain a low percent of solids. That means they experience greater shrinkage as they cure. They are relatively thin liquids that spread easily and typically require multiple coats. Manufacturers generally require at least 24 hours between coats. A base coat with a compatible primer is usually applied first.

Urethane

Urethanes are two-part products, which means they are packaged in separate cartridges (often labeled *Part A* and *Part B*) that must be properly mixed when applied.

Urethanes perform well where foot traffic and hail are concerns. They also provide good service in wet conditions. Urethane products typically contain a high volume of solids, so they offer more flexibility to withstand thermal expansion and contraction.

Urethane can even be used in applications involving potable water. It is used on metal roofs (*Figure 11*) and over spray polyurethane foam (SPF) roofs. Urethane is also used over concrete, steel, wood, and fiberglass substrates. The material is reflective, resists UV damage, and adheres to properly prepared substrates very well.

Urethane coatings should be applied at temperatures of 50°F (10°C) or above and at low humidity levels. As with acrylics, additional coats should be applied at intervals of at least 24 hours.

Urethane's disadvantages include a potentially higher price, tendency to chalk over time, and a very strong smell (that can be unpleasant but not harmful). Commercial facilities can draw the odors indoors through outside air intakes of HVAC systems. Before applying the product, ensure that the odor can be kept away from intakes, or ask that the intakes be shut or blocked during the application process.

CAUTION

Do not block the outside air intakes for HVAC systems without informing the building owner. This may cause functional problems as well as alerts and error messages in building management systems.

Silicone

Silicone is a unique material with desirable characteristics as a coating. For example, it is less susceptible to peeling than other products, giving it good ponding resistance. It offers longevity and doesn't get hard and brittle over time.

Silicone coatings cure quickly and can defend against rain faster than most coatings. Silicone naturally resists algae, mildew, and mold.

One disadvantage of silicone is its incompatibility with other coatings and with previously applied layers of itself. It does not generally adhere well to other coatings or to cured layers of silicone. This can make spot repairs challenging. To ensure proper adhesion, apply subsequent coats as soon as the previous coat has cured.

As with acrylic and urethane, silicone should generally be applied at temperatures of 50°F (10°C) and above.

PMMA

PMMA stands for polymethyl methacrylate—a synthetic resin. This product is often used to seal around flashing. While many PMMA products are flashing-grade coatings, others are designed for broad application as a roof-wide coating. As a relatively new introduction to the industry, it serves a niche market as a unique solution to difficult roofing conditions. Roofing design professionals look to PMMA for the some of the most difficult and harshest roofing challenges. Green rooftops are an example of application where PMMA will excel.

Because PMMA is not water-based, it can be applied at temperatures below freezing. There are specific formulas available that are designed for low or high temperatures. PMMA can cure more quickly than other coatings. Once it is applied, its ability to cure is not affected by humid or rainy conditions, making it a good choice if precipitation is possible during the curing process. Like urethanes, PMMA has good hail and abrasion resistance. However, it can also have a strong and unpleasant smell. *Figure 12* shows a concrete roof coated with a PMMA product.

Figure 11 Metal roof with urethane coating.

Figure 12 PMMA-coated concrete roof.

PMMA cures very quickly after it is mixed. For this reason, PMMA products have specific guidelines for when the next layer can be applied (for example, between 24 and 48 hours). PMMA can also cure quickly in hot climates. It is best to choose and apply a product formula best suited to the current climate. When working with PMMA, work quickly and in small batches.

PMMA products are often more expensive than others, and suitable air-purifying or supplied-air respirators may be required where airborne concentrations could exceed exposure limits.

1.3.2 Substrates

Coatings are often applied over existing roof systems. The substrate material and condition, however, are crucial factors in a successful application. It is important to be familiar with the acceptable substrates and the conditions required. Any deficiencies in the existing roof need to be repaired before a coating is applied.

The first consideration is to ensure that the substrate is compatible with the chosen coating. This is typically resolved well before the job begins. Manufacturers specify what substrates are compatible for each product. Failure to follow these guidelines can cause warranty claims to be rejected. Never apply a coating to a substrate that is not clearly approved by the product manufacturer.

Every type of substrate must also be clean and completely dry before a coating is applied. Coatings cannot adhere to a dirty roof. The reliability and final appearance of every coating relies on the material beneath being clean and completely dry. Substrates may be cleaned by broom cleaning or low-pressure power washing.

WARNING!

Substrates may be slippery when wet.

One common surface condition to check for prior to application is ponding water. If any water remains on the roof 48 hours or more after rainfall, it is considered ponding. These areas should be repaired to prevent future ponding. Although some coatings handle ponding rather well once they are cured, that's not a valid reason to ignore it. Evidence of areas where water tends to pond includes algae as well as black and red stains.

The drainage of condensate water from HVAC equipment must also be addressed. A winter survey will not reveal problems with condensate drainage, but craft professionals conducting surveys are experienced enough to recognize a potential problem from roof-mounted HVAC systems.

Ensuring that the substrate surface is dry is only the beginning. When required, a *moisture survey* may be conducted to determine the amount of water trapped beneath the surface. Moisture held beneath the surface may be indicated by soft or collapsed insulation and protruding fasteners under the membrane. The moisture survey is often a factor in the product warranty provided by the manufacturer. Applying a coating over a saturated roof traps the water in place, creating a new problem rather than solving an old one. Water beneath the substrate surface also degrades the roof deck and can destroy its structural integrity. Further, wet insulation loses virtually all of its thermal value. Water in the insulation is a serious liability for the owner, and not just a factor in coating application.

Moisture surveys are conducted using one of these common methods:

- *Nuclear scan* — Nuclear moisture-density meters offer accuracy and work well with all types of roofs. They identify wet areas through the detection of hydrogen atoms, which comprises about two-thirds of water's structure. A grid pattern with points established 5' to 10' apart determines the measuring points. Once the data is analyzed and plotted, wet locations are shown as well as the extent of the problem.
- *Infrared (IR) scan* — IR scans (*Figure 13*) look for heat loss from the building using a process called **thermography**. Water readily transfers heat, while dry insulation does not. Scans can be done while walking the roof or from the air.
- *Core cuts* — While physically cutting into the roof may seem primitive in comparison to the other methods, core cuts provide hard evidence of how much water is trapped beneath the surface. Some manufacturers may require core cuts in addition to non-destructive tests. Core cuts are taken from various areas of the roof to verify the accuracy of machine readings. They also reveal the characteristics of the substrate and insulation, providing key information to estimators and project managers.

NOTE

Manufacturers typically reserve the right to determine the moisture survey method used for various substrates and coatings as a basis for warranty length.

The results of the moisture survey generally determine what happens next. Although project specifications and manufacturer requirements vary, it is generally considered best to remove the entire roof if 25 percent or more of it is found to be wet beneath. Otherwise, only the wet areas are repaired before a coating is applied.

There are unique coating-related considerations for the following different substrates:

- Metal
- Single-Ply
- Asphaltic
- Spray Polyurethane Foam (SPF)
- Structural Concrete

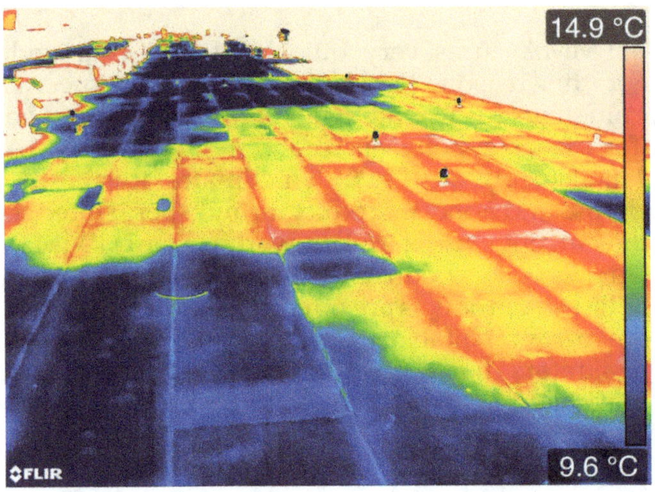

Figure 13 IR thermography scan.

Did You Know?

Thermography

Thermography is used in a wide variety of applications today. Electricians use it to evaluate switchgear and power transfer equipment, looking for hot spots in high-current contacts and wiring connections that would indicate a problem. Industrial maintenance specialists apply thermography to operating equipment to evaluate their condition. Points of substantial heat gain and loss from homes and businesses can be identified and resolved. Thermography is also used in airborne search and rescue to detect the heat from people and animals in the wilderness.

In this image, note the bright yellow in the upper right corner of the thermal image. It shows that one wiring connection is significantly hotter than the others. Thermography technology may also be used with smartphones and drones, making this technology more portable and easy to bring to the field.

Figure Credit: iStock@undefined undefined

Metal

Acrylic coatings are traditionally used on metal substrates. Silicone, urethane, or PMMA can also be used.

Craft professionals must consider the following before applying a coating directly to a metal substrate:

- Fasteners should be tightened or replaced as necessary.
- Some metal roofs have an existing finish to consider; a primer is often required.
- Metal substrates are susceptible to rust. Rusty or corroded areas must be primed or replaced before a coating is applied.
- The surface temperature should be under 105°F (~40°C) for the coating to successfully adhere to the metal. (Check manufacturer instructions for specific temperature restrictions.)

Single-Ply

Single-ply roof systems have one layer of waterproofing membrane material. Craft professionals should be familiar with the common single-ply membranes such as TPO, PVC, and EPDM. Always follow the coating manufacturer's guidelines regarding compatibility with these materials. Acrylic, urethane, and silicone coatings can generally be used on single-ply substrates.

Seams in single-ply membranes need careful scrutiny. The coating should not be considered a reliable method of sealing deteriorating seams or damaged areas. Open seams, tears, cuts, and buckled membranes must be repaired with compatible materials before a coating is applied. Specific primers may be recommended for the various single-ply materials.

Asphaltic

As the name implies, asphaltic roof systems rely on asphalt for weatherproofing the membrane below. Common asphaltic systems include BUR (built-up roofing) and polymer-modified bitumen.

Asphaltic roof materials are typically offered with either a smooth or granulated coating. These surfaces must be swept vigorously, and the loose material collected during the cleaning process.

Check for blistered, buckled, or otherwise damaged areas and repair them with compatible materials. The new material should generally weather for at least 30 days before a coating is applied. Repair cracks larger than $^1/_{16}$" (1.6 mm) with a flashing-grade coating to create a smooth surface before applying the chosen monolithic coating.

Spray Polyurethane Foam

Silicone is often chosen for SPF substrates. Acrylic and urethane coatings are also common. If the foam has degraded in some areas, those areas should be re-foamed. If the foam has sustained water damage or had water trapped beneath it, the problem areas should be removed and re-foamed before a coating is applied.

Structural Concrete

Acrylic, urethane, silicone, and PMMA can all be applied directly to concrete substrates. Adhesion to concrete is generally sound. Any gaps larger than ¼" between panels or placements must be repaired with grout. Allow the grout to cure before applying the coating.

1.3.3 Coating Application

The most important consideration in coating application is to review and follow the manufacturer's guidelines. No one knows a coating product better than its manufacturer. Ignoring the guidelines can not only cause a loss of warranty protection, but also result in an otherwise good project to be ruined.

With the moisture surveys done and all necessary repairs complete, you're much closer to the real task of applying the coating. But another test is usually required, along with a lot of cleaning.

Adhesion Testing

An adhesion test is conducted to ensure the coating will stick to the substrate reliably. Engineers and designers generally conduct this test to ensure that the chosen product will adhere to the current surface. The number of adhesion tests required, if any, will be specified by the manufacturer of the coating. The basic steps in an adhesion test are as follows:

Step 1 Clean the substrate in the test area thoroughly, ideally using the same method you will use for the entire roof.

Step 2 If a primer is to be applied, followed by a base coat in which the fabric will be embedded, prime the area first. Use the same sequence of application steps planned for the rest of the roof.

Step 3 Cut a fabric strip about 12" long and 1" wide (~30 cm × ~2.5 cm).

Step 4 Brush the coating material over the substrate to create a patch that is roughly 6" (~15 cm) long and wide.

Step 5 Embed the first 6" of the fabric into the coating and then brush additional coating over it. Leave the remaining 6" of the fabric strip outside of the coated area. Don't let it blow around and get stuck in the coating while it is still wet.

Step 6 After the coating has cured as directed by the manufacturer (typically a number of days, but dependent on the product and the significance of the results), tie the free end of the fabric into a knot.

Step 7 Insert the hook of a fisherman's or digital pocket scale into the knot without puncturing the fabric, and slowly pull straight up (*Figure 14*). Add pressure slowly until the fabric begins to separate from the roof. Note the pounds of pull registered on the scale as the fabric begins to pull away. If the pull is greater than the minimum specified by the manufacturer, the adhesion is reliable enough to apply the product with confidence.

The fabric will separate in one of two ways. If the fabric and coating peel away together from the substrate below, the adhesion has been accurately measured. However, the fabric may separate from the coating while leaving a layer of coating beneath it, still attached to the substrate. In this case, you know that the adhesion strength is something higher than the pull you recorded, but not how much higher. If the pull was already above the minimum specified, it's considered a success.

Figure 14 Adhesion test.

Application Rate

Before beginning, you will also need to determine the application rate. Using wet film measurements is one way to keep an eye on it, as outlined earlier in this module. You can determine the wet film thickness based on the dry film thickness required and the percentage of solids in the coating. Then check the wet film thickness every now and then as you apply the coating.

Manufacturers typically report the application rate as gallons per square of roof area, or in liters per square meter. The best way to get started, then, is to compare your technique and application rate to the manufacturer's rate of coverage, as follows:

- Measure off a 10' × 10' area (equal in area to a square of roofing, or 100 square feet). You may use traffic cones to mark the corners or use any other suitable method to mark the perimeter.
- Prepare the manufacturer-specified volume per square in a separate bucket.
- Apply the coating while monitoring the consumption of product from the bucket. If you run out of product before the area is fully covered, you are moving the sprayer too slowly or not spreading the product enough with the roller. If you have product left over, not enough product is being applied. Adjust your technique accordingly. When you have it right, measure the wet film thickness in a few areas so you are aware of it as you work.

NOTE

If the coverage rate is reported in liters per square meter, consider marking off an area equal to four square meters and using four times the coating specified in the test. A single square meter is a small area for a test.

Application Procedure

Even during the planning stages, the weather must be considered. The roof surface must be dry, and it must remain dry for a manufacturer-specified period. It is rarely advisable to bet against the forecasts since a significant amount of labor and materials are at stake. Imagine the rain from a thunderstorm pounding a freshly applied coating across an entire roof, washing it down roof drains and pooling in low areas. Throughout the application process, be aware of the weather and monitor forecasts to avoid surprises.

Consider what you have learned thus far as you review this basic application procedure:

Step 1 Prepare the roof system according to the substrate material and surface conditions. Note that manufacturers provide step-by-step guidance for each type of substrate. Common preparations include actions previously discussed such as the following:
- Correct ponding water conditions.
- Repair seams, penetrations, and damaged areas.
- Replace any wet insulation.
- Secure and seal fasteners.

Step 2 Power wash the substrate and allow it to dry thoroughly. Note that manufacturers offer products specifically designed for this purpose that work best for the chosen coating. Use a manufacturer-approved product such as bleach to kill biological growth.

CAUTION

Excessive washing in one area and excessive pressure can damage the substrate. The maximum pressure should generally be between 2,000 and 3,000 psi. Concrete substrates can better withstand pressures at the upper end of this range.

Step 3 Treat any areas of contamination, such as rust, old adhesives, or asphalt, per the manufacturer's instructions. All offer treatment products designed to work best with their coatings.

Step 4 Determine the target wet film thickness. This information will generally be provided in the project specifications and communicated to you by the project manager.

Step 5 Prepare an area for a test of coverage per gallon.

Step 6 For two-part coatings, mix the components per the manufacturer's instructions.

Step 7 Apply the coating to the substrate using brushes, rollers, or spray equipment. If a test patch for coverage is being applied, stop when the bucket is empty and evaluate the results.

Step 8 Use a wet film gauge periodically as you work to determine whether the application rate is correct. Don't wait too long between tests. Adjust as necessary for a thicker or thinner coat.

Step 9 Apply subsequent coats as directed, allowing the specified time between coats. Two to three coats are typically required.

1.4.0 Liquid-Applied Roof Membranes

Liquid-membrane roof systems provide a fundamental layer of weatherproofing. Coatings alone help to restore the surface of an existing roof system, but they lack the strength and integrity to be considered primary weatherproofing.

The added strength comes from the addition of a reinforcing membrane to the coating. Common materials include spun and woven polyester fabrics (*Figure 15*) and fiberglass mats. Only a single layer of reinforcement is generally applied in the base coat (also referred to as a *foundation coat*). It is then covered with additional layers of the coating. The liquid materials are similar to those used for restorative coatings, although many products are designed specifically for liquid-membrane roof system construction.

Liquid-applied roof membrane systems are typically applied on top of a solid substrate. But it can also be the first and primary roof covering. They can be applied to existing roof systems to add integrity and protection to an older roof. It is important to note that the same preparations are required—the addition of fiber reinforcement does not reduce the need to eliminate ponding, repair existing leaks, and clean the surface thoroughly before application.

Figure 15 Small roll of polyester roofing membrane used for detailing.

The following steps summarize the installation of a liquid-membrane roof system. Again, all the same preparations required for a coating are required here, including moisture surveys, adhesion testing, and power washing for existing roof substrates being covered:

Step 1 Secure all existing flashing components and seams. Note that this is contrary to most roof systems, where flashings are applied after the field of roofing. Make incisions (referred to as *fingers*) along the edge of reinforcement material to create fanned strips, then form them around protrusions such as plumbing vents, as shown in *Figure 16*. Applying the membrane 4" (~10 cm) up the side of the protrusion is common. These locations may require flashing-grade coatings over the membrane before the base coat is applied. An example of a liquid-applied flashing detail is shown in *Figure 17*.

- Pipes may need to be sanded or primed.
- Gaps greater than ¼" should be filled.
- Base caps should be installed where needed.

Step 2 For two-part liquid membrane systems, mix the components per the manufacturer's instructions.

Step 3 Apply the base coat to the substrate and up to the penetration, following the manufacturer's guidelines for environmental conditions and application rate.

> **NOTE**
> Do not confuse the base coat with a primer. A primer may be required on some substrates before the base coat is applied. The fabric is not embedded in the primer.

Step 4 Lay down the reinforcement material into the wet base coat (*Figure 18*). Ensure the fabric is laid flat with no wrinkles or voids, and within the window of time allowed by the manufacturer. Follow the manufacturer's guidelines for how much overlap there should be between sheets.

Step 5 Add the number of subsequent coats as required by the manufacturer's guidelines, allowing the appropriate amount of time between coats.

The Flexible Advantage

One of the advantages to liquid-membrane roof systems is the ability to seamlessly cover roofs with a unique profile. The liquid-membrane approach is also used on a variety of other structures, both horizontal and vertical.

Figure Credit: Courtesy of Sika USA

Figure 16 Feather the fabric around protrusions.

Note: Refer to specific coating system for base coat and top coat products.

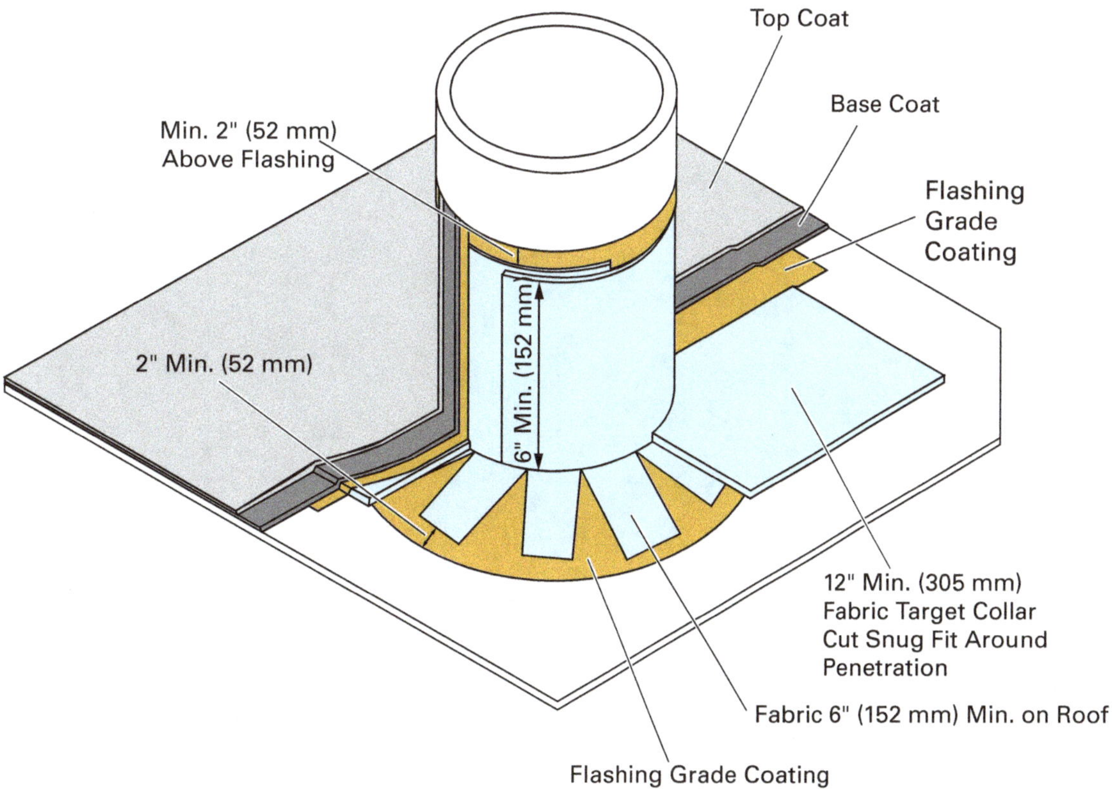

Figure 17 Example of a liquid-applied pipe flashing detail.

Figure 18 Polyester roofing membrane being placed in the base coat.

1.0.0 Section Review

1. Which of the following statements about liquid-applied roof systems is *true*?
   a. Liquid-membrane roof systems make up the largest percentage of the roofing market.
   b. A liquid-membrane roof system can only be applied to metal substrates.
   c. Liquid-membrane roof systems incorporate reinforcement into a roof coating.
   d. Roof coatings alone offer a greater level of protection and integrity than liquid-membrane roof systems.

2. Which of the following measures can be taken to reduce overspray?
   a. Place protective boards around the perimeter of the roof.
   b. Hold the sprayer closer to the substrate.
   c. Hold the sprayer farther from the substrate.
   d. Mix water with the coating before using it.

3. Generally, water-based products _____.
   a. are not very durable
   b. cannot be used where VOCs are restricted
   c. cannot be used on concrete substrates
   d. have lower VOCs

4. How many layers of coating can be applied to a liquid-membrane roof installation in an eight-hour workday?
   a. Only one.
   b. At least two.
   c. It depends on the substrate.
   d. It depends on the coating manufacturer's guidelines.

Review Questions

1. Liquid-applied roofing products are elastomeric, which means they _____.
   a. can stretch slightly
   b. are not affected by UV rays
   c. cannot be recoated
   d. reflect both heat and light

2. The characteristic of a coating that helps prevent cracks and tears due to expansion and contraction of the structure is called *reflectivity*.
   a. True
   b. False

3. A controlled-flow roller frame _____.
   a. has a built-in pump to keep the roller wet
   b. should never be used on a windy day
   c. is attached to a coating pump
   d. is only used to apply primers

4. The significant risk associated with using sprayers to apply coatings is _____.
   a. overspray on nearby surfaces
   b. product flammability
   c. hazardous fumes
   d. poor coverage

5. Which of the following statements about silicone coating is *true*?
   a. Silicone coatings should only be applied at temperatures of 80°F (27°C) and above.
   b. Silicone can be applied over any other type of coating without cleaning or priming.
   c. Silicone coatings cure quickly and are able to defend against rain faster than most coatings.
   d. A roof is usually primed with an acrylic coating before a silicone coating is applied.

6. One of the unique traits of PMMA coatings is that they _____.
   a. can be applied to any surface without cleaning
   b. have a very fast cure time
   c. require a week or more to cure
   d. can be mixed with either silicone or acrylic coatings for application

7. Which type of moisture survey provides hard evidence of how much water is trapped and reveals the composition of the roof below the surface?
   a. UV scan
   b. Infrared scan
   c. Nuclear scan
   d. Core cuts

8. An adhesion test is performed _____.
   a. to ensure the coating will stick to the substrate
   b. to keep overspray from sticking to surrounding buildings
   c. to ensure the substrate is fastened to the deck
   d. to ensure seams are weatherproof

9. To test the required application rate of a coating against your technique when the rate is specified in gallons per square, mark off an area that is _____.
   a. 1' × 1'
   b. 12' × 12'
   c. 4 square meters in size
   d. 100 square feet in size

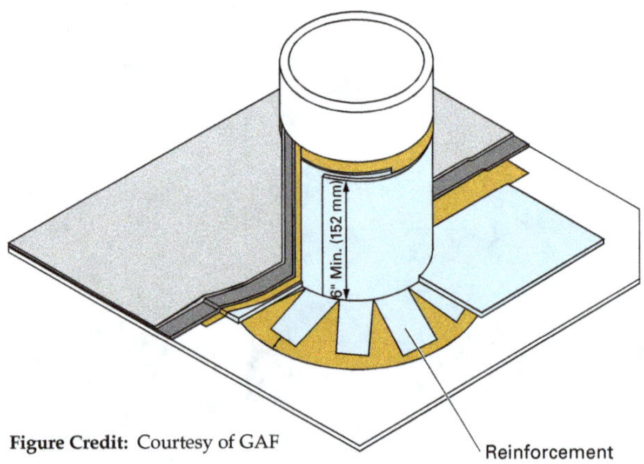

Figure Credit: Courtesy of GAF

Figure RQ01

10. The incisions in the reinforcement material shown in *Figure RQ01* are referred to as _____.
    a. slivers
    b. fingers
    c. pointers
    d. protrusions

Trade Terms Introduced in This Module

Adhesion test: A test performed on an existing roof to verify that a substrate can successfully bond to a liquid coating. Also called a *peel test*.

Application rate: The average quantity (mass, volume, or thickness) of material applied per unit area, typically specified in gallons per square or liters per square meter.

Dry film thickness: The thickness, expressed in mils, of an applied and fully cured liquid coating.

Elastomeric: Able to be stretched at least twice its original size and return to its original shape.

Flashing-grade coatings: Liquid coatings specifically designed for areas where the primary roof covering is interrupted or terminated, such as protrusions and damaged areas. These are generally applied before base and finishing coats and can be applied to vertical surfaces since they have a thicker consistency.

Liquid-applied roof membranes: Continuous roof membranes constructed in place with a liquid coating and a reinforcing material. The liquid coating is available as a one- or two-component product and is typically applied in two coats. Depending on the chemistry, a catalyst or hardener may be added to induce the curing process. The reinforcement is typically polyester or fiberglass mat set into a still-wet base coat.

Mils: A unit of measurement equal to one-thousandth of an inch, or 0.001" (0.0254 mm).

Monolithic: Formed or composed of a single material; seamless.

Overspray: Deposits of sprayed material that fall outside the intended area.

Roof coatings: Liquids containing essential weatherproofing properties that are applied to roofs to protect, restore, prolong, or enhance performance.

Solvent-based: Containing a petroleum-based solvent as a liquifying agent, leading to a higher content of organic compounds (see *volatile organic compounds*).

Thermography: The creation of digital images using infrared cameras to detect the amount of heat energy emitted.

Volatile organic compounds (VOCs): Compounds and chemicals that easily evaporate at room temperature. They are found in products such as solvents, paints, and adhesives.

Water-based: Having water as a liquifying agent. Although water can also be considered a solvent, the term *solvent* typically refers to petroleum-based products.

Wet film thickness: The thickness, expressed in mils, of a liquid coating as applied (wet).

Additional Resources

This module presents thorough resources for task training. The following reference material is suggested for further study.

National Roofing Contractors Association (NRCA), **www.nrca.net**.

NCCER Module 00103, *Introduction to Hand Tools*.

NCCER Module 00104, *Introduction to Power Tools*.

Figure Credits

Courtesy of Sika USA, Module Opener

Courtesy of GAF, Figures 1–5, 8, 9, 11, 15, 17, 18

iStock@DERO2084, Figure 6

Graco Inc, Figure 7

SOPREMA US, Figures 12, 16

Infrared Imaging Services LLC, Figure 13

iStock@Jiggo_thekop, Figure 14

Section Review Answer Key

Section 1.0.0

Answer	Section Reference	Objective
1. c	1.1.0	1a
2. a	1.2.2	1b
3. d	1.3.1	1c
4. d	1.4.0	1d

User Update

NCCER makes every effort to keep its textbooks up-to-date and free of technical errors. We appreciate your help in this process. If you find an error, a typographical mistake, or an inaccuracy in NCCER's curricula, please submit a User Update form by visiting **https://www.nccer.org/olf**. You can also scan the QR code using the camera on your phone or mobile device to access the form.

Roofing Service and Repair

Overview

Successful roof repairs do more than stop a leak—they restore peace of mind to customers. As the roofing industry has advanced, so have repair and maintenance technologies. There are many modern tools that help support and document repair work. However, old-fashioned problem-solving and observation skills remain the foundation of a successful roofing repair job. When repairing a roof, you must be able to recognize causes of leaks on common roof systems. You must also know the basic steps to fix them.

Module 16212

Trainees with successful module completions may be eligible for credentialing through the NCCER Registry. To learn more, go to **www.nccer.org** or contact us at 1.888.622.3720. Our website, **www.nccer.org**, has information on the latest product releases and training.

Your feedback is welcome. You may email your comments to **curriculum@nccer.org**, send general comments and inquiries to **info@nccer.org**, or fill in the User Update form at the back of this module.

16212 V1.0

16212

Roofing Service and Repair

Objectives

Successful completion of this module prepares you to do the following:

1. Describe roofing service and repair work.
   a. Explain the general process of investigating and completing repair work.
2. List and describe common steep-slope repair procedures.
   a. Describe common repairs for asphalt shingle roof systems.
   b. Describe common repairs for clay and concrete tile roof systems.
   c. Describe common repairs for architectural metal panel roof systems.
3. List and describe common low-slope repair procedures.
   a. Describe common repairs for thermoplastic roof systems.
   b. Describe common repairs for EPDM roof systems.
   c. Describe common repairs for built-up roof (BUR) systems.
   d. Describe common repairs for modified bitumen (mod-bit) roof systems.

Performance Tasks

Under supervision, you should be able to do the following:

1. Repair a flashing in an asphalt shingle roof system.
2. Repair a flashing in a clay or concrete tile roof system.
3. Repair a flashing in a metal panel roof system.
4. Repair a blister in a BUR or mod-bit roof membrane.
5. Patch a damaged thermoplastic or EPDM roof membrane.
6. Repair a flashing in a low-slope roof system.

Trade Terms

Blindside patch
Corrugated panels
Degranulate
Fishmouth
Flat bar
Hand sealing
Lap
Nailer
Perimeter metals
Roof repair
Roof service
Seam
Seam probe
Silicosis
Spudding
Standing seam panels
Strip flashing
Strip in
Three-course patch
Trim tile
Valley
Work order

Industry Recognized Credentials

If you are training through an NCCER-accredited sponsor, you may be eligible for credentials from NCCER's Registry. The ID number for this module is 16212. Note that this module may have been used in other NCCER curricula and may apply to other level completions. Contact NCCER's Registry at 1.888.622.3720 or go to **www.nccer.org** for more information.

You can also show off your industry-recognized credentials online with NCCER's digital badges. Transform your knowledge, skills, and achievements into badges that you can share across social media platforms, send to your network, and add to your resume. For more information, visit **www.nccer.org**.

Videos and Digital Resources

How to Access Resources

This craft has additional videos and resources to enhance your learning experience. To view these resources, scan the QR below. The videos and resources are separated by module.

You can scan this code using the camera on your phone or mobile device to view these videos and resources.

Contents

1.0.0 Introduction to Service and Repair 1
1.1.0 Repair Process 1
1.1.1 Leak Investigation 1
1.1.2 Jobsite Documentation 3
1.1.3 Career Opportunities 4
1.1.4 Repair Safety 7
2.0.0 Steep-Slope Repair 8
2.1.0 Asphalt Shingle Roof Systems 8
2.1.1 Damaged Shingle 9
2.1.2 Damaged Penetration Flashing 10
2.1.3 Missing Vent 10
2.1.4 Drip Edge Replacement 10
2.1.5 Valley Flashing Replacement 11
2.2.0 Clay and Concrete Tile Roof Systems 11
2.2.1 Damaged Tile 12
2.2.2 Damaged Penetration Flashing 12
2.2.3 Trim Tile (Hip, Ridge, or Rake) Replacement 13
2.2.4 Valley Flashing Replacement 13
2.3.0 Architectural Metal Panel Roof Systems 14
2.3.1 Damaged Penetration Flashing 14
2.3.2 Damaged or Missing Hip/Ridge Cap 14
3.0.0 Low-Slope Repair 16
3.1.0 Thermoplastic Roof Systems 16
3.1.1 Damaged Flashings 18
3.1.2 Damaged Perimeters 18
3.1.3 Seam Repair 19
3.1.4 Field Repair 20
3.2.0 EPDM Roof Systems 20
3.2.1 Damaged Flashings 20
3.2.2 Damaged Perimeters 21
3.2.3 Seam Repair 21
3.2.4 Field Repair 22
3.3.0 Built-Up Roof (BUR) Systems 22
3.3.1 Damaged Flashings 22
3.3.2 Damaged Perimeters 24
3.3.3 Liquid-Applied Penetration Flashing Repair 24
3.3.4 Field Repair 24
3.4.0 Modified Bitumen Roof Systems 25
3.4.1 Damaged Flashings 26
3.4.2 Damaged Perimeters 26
3.4.3 Penetration Repair 26
3.4.4 Field Repair 27

Figures

Figure 1 Roof service/repair process 2
Figure 2 Example of a paper work order 3
Figure 3 Example of an electronic work order 5
Figure 4 Example of job details in an electronic work ticket 5
Figure 5 Roofing repair careers include a variety of opportunities and settings 6
Figure 6 Steep-slope roofs 9
Figure 7 Components of a roof system 9
Figure 8 Basic steep-slope roof construction 9
Figure 9 Asphalt shingle roof systems 9
Figure 10 Torn shingle in need of replacement 10
Figure 11 Example of a metal drip edge 11
Figure 12 Open, no-cut metal valley flashing 11
Figure 13 Clay tile roof systems 12
Figure 14 Microscopic view of crystalline silica particles on a filter 12
Figure 15 Example of a broken roof tile 13
Figure 16 Metal roof installation 14
Figure 17 Example of a corrugated metal panel roof system 15
Figure 18 Example of low-slope roof 17
Figure 19 Thermoplastic roof system 17
Figure 20 Thermoplastic patch 18
Figure 21 Thermoplastic pipe flashing 19
Figure 22 Hot-air welding 19
Figure 23 Example of perimeter metal 19
Figure 24 Hot-air welding a blindside patch 20
Figure 25 EPDM roof system 21
Figure 26 Sealing a patch on an EPDM flashing 21
Figure 27 Open EPDM lap 21
Figure 28 BUR membrane 23
Figure 29 Prepare the surface by spudding or degranulating 23
Figure 30 Gravel stop 24
Figure 31 Flashing reinforcement cut with fingers at base of penetration 25
Figure 32 Damaged BUR field membrane 25
Figure 33 Mod-bit roof 25

Section One

1.0.0 Introduction to Service and Repair

Objective

Describe roofing service and repair work.

a. Explain the general process of investigating and completing repair work.

Trade Terms

Roof repair: Roofing work that involves fixing a deficiency, leak, or other damage to a roof system.

Roof service: Roofing work that involves routine maintenance of a roof system to prevent damage in the future and extend the life of the roof system.

Work order: A document that outlines the necessary tasks for a contracted job by showing the roofer where to look and what to do. On a roof repair job, the work order will outline the specific repair tasks the roofing company is expected to complete.

Roofing service and repair encompass two important segments of roofing work. Roof service is the routine preventative maintenance of a roof system. Roof repair (the main focus of this module) includes the repair and restoration of a deficiency, leak, or other damage to a roof system.

> **NOTE**
> This module is a brief introduction to the vast topic of roof repair. It describes basic concepts and examples of repairs. Repair careers require experience and practice beyond the scope of this module. However, with dedication and practice, a variety of opportunities and careers in the repair industry can be at your fingertips.

Common repairs on steep-slope and low-slope roofs involve removing damaged material and replacing it or patching over the damaged area. Roof repairs are smaller in scope than reroofing. Reroofing encompasses roof replacement or installing a new roof system over an aging one.

As always, keep safety at the forefront of your mind when stepping onto a roof. Remember that there may be damage underfoot that has not yet been discovered. The roofing company should conduct a safety inspection before work begins. Before repairs, visually inspect the underside of the deck for damage and moisture. It is also important to look and feel for excessive moisture during repairs. Document and report water damage and tell your supervisor if you notice damage to the deck.

1.1.0 Repair Process

The repair process typically begins when a customer finds a leak or damage and calls a roofing company to investigate. Once the damage has been identified, the roofing company can create a work order and dispatch workers to complete the repair (*Figure 1*).

The first step for roofing professionals completing a repair is to consult the work order. A work order gives important details about the job, such as a description of the leak or reported damage, the location of the building, and the type of roof.

Some work orders are paper forms, like the example in *Figure 2*. However, it is becoming more common to keep track of work orders and job details in online ticketing software that can be accessed on a mobile device (*Figure 3*). This allows roofers to reference and document information quickly during a job.

Work orders ensure that the customer and roofing company have the same expectations for the scope and cost of a repair job. Always talk to your supervisor if you encounter any issues outside of what's in the work order.

1.1.1 Leak Investigation

A leak investigation takes place prior to making the repair. In a leak investigation, a two-year roofing worker will sometimes accompany a more experienced roofer who diagnoses the problem. An eagerness to learn can help new roofers develop critical troubleshooting skills and advance their careers.

As with any detective work, leak investigation begins with asking good questions. You may ask the following questions when performing a leak investigation:

- *Where is the leak?*
- *How long has the roof been leaking?*
- *When does the roof leak?*
- *How old is the roof?*
- *Does the leak only happen during certain types of weather (rain, snow, wind, etc.)?*

The answers to these questions can help you determine and rule out potential causes or locations of the leak. Some leaks happen due to

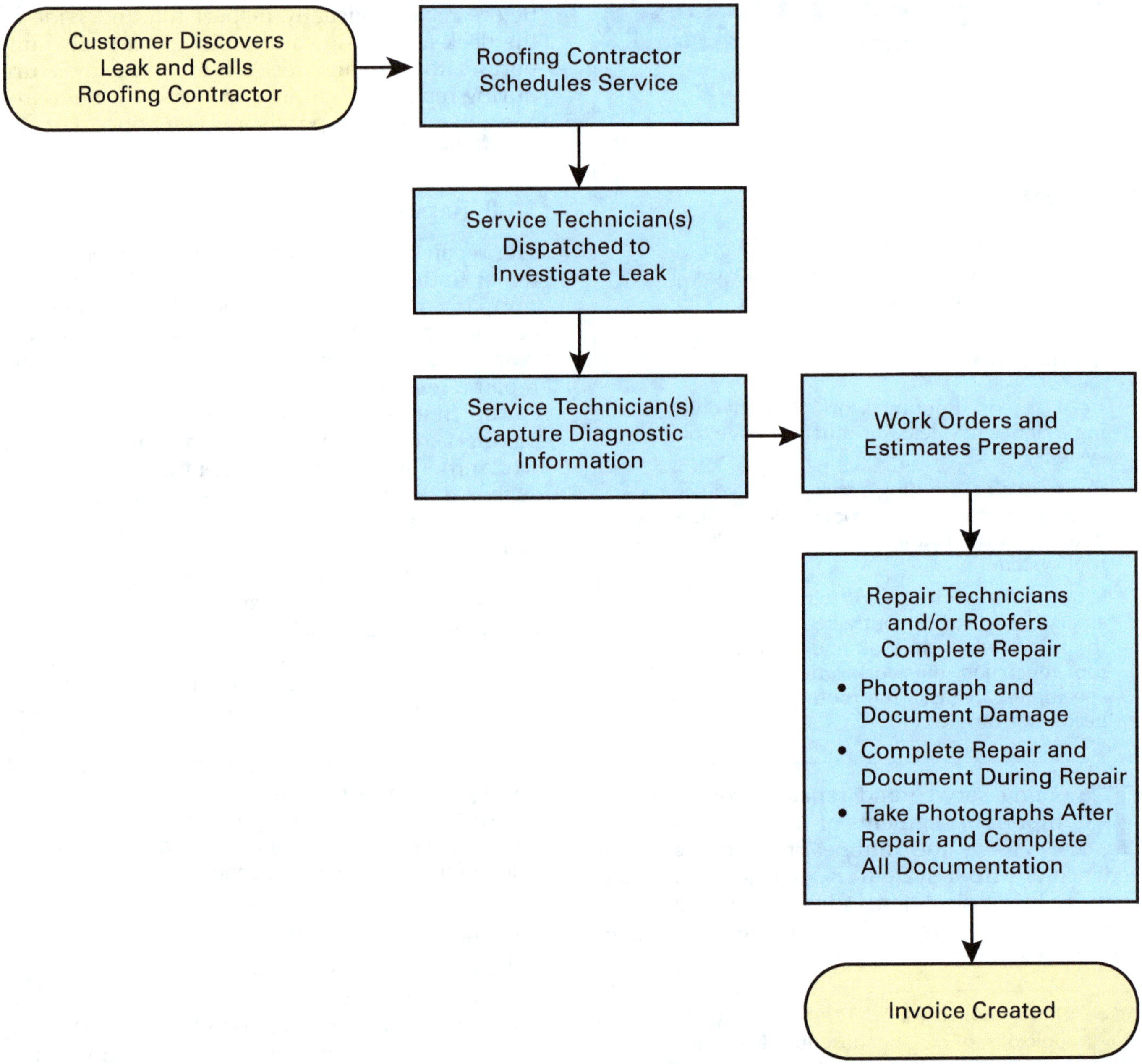

Figure 1 Roof service/repair process.

factors unrelated to the roof system. For example, if the leak only happens in the summer, it may be due to hot air coming in and forming condensation on the ceiling when it cools.

Locate the leak from the interior of the building first. Note the direction of any deck slope. For metal decks, note the direction of the flutes. These may provide an indication of the path water may travel.

After assessing from the inside, go outside the building and run a process of elimination to identify the source of the leak. Focus on penetrations and details first as leaks are most likely to occur there. Look for openings in the field, incorrectly installed components, and other potential deficiencies.

Water often travels across the roof before it leaks through, so the leak could appear several yards away from the damage. In some cases, a water test can help locate the leak. A water test is performed by running water over the roof with a hose. Typically, one worker will spray the water on top of the roof (starting downslope and slowly working upslope), while another stays inside the building to observe when the roof begins leaking. When the leak begins, the worker inside the building will notify the other via radio. This can help narrow down areas where the damage is located.

W.O.: 11712
Customer P.O.: M QCP 6101
Alt. #:

Deer Park Roofing, Inc.
FIELD TICKET SUMMARY
Commercial Service; Roof Repair Work

Reported: 08/18/2021
Scheduled 08/27/2021
Completed

SITE INFORMATION
SITE NAME: ABC Company
ADDRESS: 123 Main Street
Anywhere, USA 55555
ROOF AREA Main Roof Area

CUSTOMER INFORMATION
NAME: ABC Company
ADDRESS: 123 Main Street
Anywhere, USA 55555

ORIGINAL DESCRIPTION: Check In & Check Out with either Barbie or Chuck while on-site/leaving.

Apply Gaco tape/patch to repair area(s) previously coated in June - Geoff has details.

Install 3 square feet of flat stock to be fabricated on-site - Geoff has details on locations metal is to be installed.

Scope of Work:

1. Fabricate and install metal flashing detail to prevent birds from entering the building on the northeast corner of the upper roof area.

2. Seal the first three raised panels at the gutter edge to prevent bird entry.

3. Prepare the surface of the lap detail to accept new high-performance sealant extending approximately 4 feet further north and 4 feet further south than the existing repair in an attempt to minimize the leak.

****Also investigate leak above restroom in office portion of building. Barbie reported 8/25 - Check for defect in scope of work Proj# 39504?****

NTE $850
WORK NOTES: Couple voids in Gaco patch we put on a second coat and went 3 seams past

Materials 1 gallon Gaco patch and 2'x2' piece of metal

Dustin and Trent 11:30am to 3pm

TOTAL CHARGES	**$0.00**
GROSS PROFIT TOTAL	**$0.00**

Figure 2 Example of a paper work order.

It is important to be on the lookout for anything unexpected during repair work (for example, the use of incompatible fasteners leading to corrosion). Always notify your supervisor if you encounter unexpected issues or changes in the scope of the work.

1.1.2 Jobsite Documentation

Documentation builds trust between a roofing company and its customers and promotes efficient and effective service. Thorough documentation demonstrates that the work order was correctly completed. It may also provide support for amending the work order if unexpected issues are discovered on the jobsite.

Did You Know?

History of Roofing in the United States

Roofing in the United States dates back to the very first buildings constructed by settlers. The roofing industry has evolved to include a wide variety of materials and technologies.

1607	Development of eastern US begins when 105 people land at Jamestown, VA.
1608	First skilled builders arrive from Poland and install the first wood roofs.
1625	First use of slate roofs in Jamestown.
1680	Boston and New York restrict thatch use because of fire risk.
1763	New York City Hall roofed with copper.
1840	Composition roofing (BUR) introduced into the US.
1850	Roofing becomes specialized trade; breaks from carpentry industry.
1860	Prefabricated buildings with corrugated, galvanized roofs are introduced during California Gold Rush.
1887	First annual meeting at the National Roofing Contractors Association (NRCA).
1941	During WWII, labor shortage is critical. Women are hired as roofing material applicators for first time.
1958	Ingalls Rink, the first building to use a single-ply roof, constructed at Yale in New Haven, CT.
1970	Nixon signs Occupational Safety and Health Act. OSHA and the National Roofing Foundation are formed.
2001	Following 9/11, roofing industry volunteers reroof the Pentagon.
2005	Recycling of EPDM and PVC membranes gains traction in the roofing industry.
2008	The Center for Environmental Innovation in Roofing is formed.
2010	Roof Integrated Solar Energy (RISE) formed to evaluate and certify solar energy installers.

It is important to photograph the jobsite before, during, and after the repair work has been completed. If you encounter any unexpected issues or damage not covered in the work order, take photos and contact your supervisor. Written notes are also important documentation. You may be provided with a checklist of data, such as roof material conditions, to collect on the jobsite.

Many roofing companies use special apps or software to collect documentation and seamlessly share it internally or with the customer. Some even use drones to collect aerial photographs, take measurements, and more. Aerial views of the building (*Figure 4*) can help ensure you are in the right place and document where the damage is located on the roof.

1.1.3 Career Opportunities

Roofing repair is a respected and desirable career opportunity for experienced roofers (*Figure 5*). Repair professionals tend to have flexible schedules and varied work experiences. These experiences include traveling to different jobsites, learning about different roof systems, and representing their companies.

Repair professionals should have strong investigative and problem-solving skills. Successful leak investigation requires the knowledge and instinct to ask the right questions about what could be damaged in the roof system.

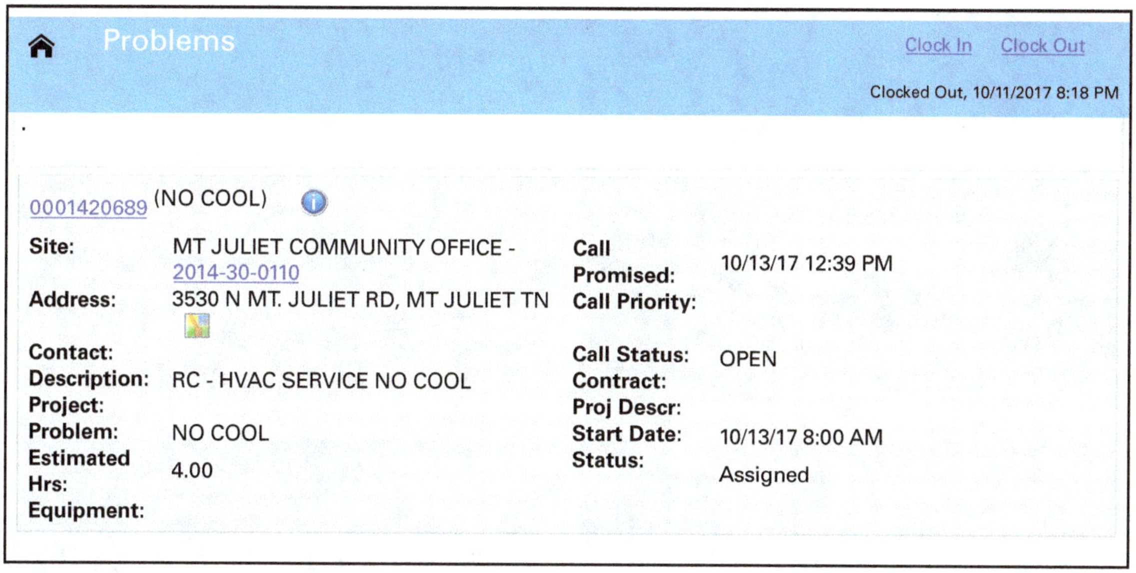

Figure 3 Example of an electronic work order.

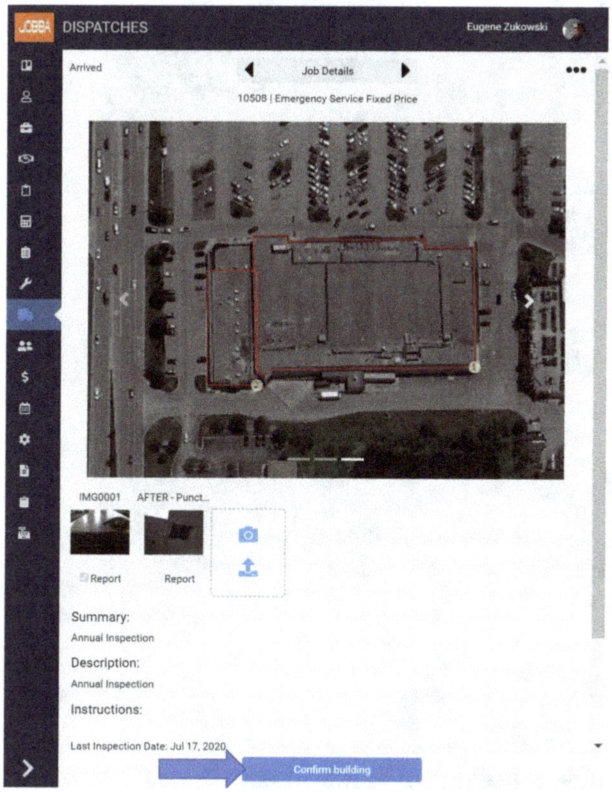

Figure 4 Example of job details in an electronic work ticket.

Because leak investigation requires direct consultation with the customer, communication and customer service skills are also key.

Repair careers typically require several years of on-site job experience as a roofing professional or apprentice. The following are some examples of roofing service/repair careers:

- *Service Technician* — Troubleshoots and repairs problems on roofs. Works independently, traveling to different worksites and communicating with customers.
- *Crew Leader / Foreman* — Leads and supervises repair crews and communicates with customers.
- *Roof Assessor* — Performs roof inspections and take-offs, takes core samples, and records important data. Sometimes serves as foreman if needed.
- *Superintendent* — Ensures jobs are completed according to company guidelines. May work with sales teams to recruit clients.
- *Estimator* — Performs inspections and project management duties, such as overseeing the bidding process and determining scope of work.

Figure 5 Roofing repair careers include a variety of opportunities and settings.

1.1.4 Repair Safety

Working on a damaged roof involves many related hazards. It is important for a repair technician to never go to a worksite alone—repairs should always be done by at least two people.

Repair workers are not exempt from safety guidelines. It is just as important to be safety conscious while doing repair work as for any other roofing job. No matter how much or how little time you spend on the roof, there are dangers that should not be ignored.

NOTE

General safety considerations for roofing work are described in detail in NCCER Module 16102, *Roofing Safety*, and NCCER Module 75901, *Fall Protection Orientation*.

Specific safety considerations for each roof system are covered in other *Roofing Level Two* modules. These modules are pointed out in each related section of this module.

1.0.0 Section Review

1. Which of the following is *not* a step involved in leak investigation?
   a. Asking questions about the leak and building.
   b. Photographing the roof after repairs have been made.
   c. Assessing the inside of the building.
   d. Assessing potential causes of the leak with a process of elimination.

Section Two

2.0.0 Steep-Slope Repair

Objective

List and describe common steep-slope repair procedures.

a. Describe common repairs for asphalt shingle roof systems.
b. Describe common repairs for clay and concrete tile roof systems.
c. Describe common repairs for architectural metal panel roof systems.

Performance Tasks

1. Repair a flashing in an asphalt shingle roof system.
2. Repair a flashing in a clay or concrete tile roof system.
3. Repair a flashing in a metal panel roof system.

Trade Terms

Corrugated panels: Architectural metal panels with alternating ridges and grooves that overlap at seams.

Flat bar: A long and narrow hand-held tool with an L-shaped curve at the head. In roofing repair, flat bars can be used for wedging between and separating attached components such as asphalt shingles, or for removing nails.

Hand sealing: The process of adding additional adhesive to ensure new shingles adhere to old shingles, especially in cold weather. Also called *hand tabbing*.

Nailer: A piece or pieces of dimensional lumber and/or plywood secured to a structural deck or walls that provide(s) a receiving medium for the fasteners used to attach tiles or panels.

Silicosis: A lung disease caused by breathing in silica particles. Dusty jobs, like cutting roof tiles, can lead to silicosis if performed without respiratory protection.

Standing seam panels: Architectural metal panels with overlapping vertical legs that get folded together during installation.

Strip in: Apply membrane flashing strips to seal the edges of a metal flashing component.

Trim tile: An accessory piece provided by the manufacturer used to attractively weatherproof various areas of a roof, such as at ridges, hips, and rakes.

Valley: The internal angle formed by the intersection of two sloping roof planes.

Steep-slope roofs (*Figure 6*) are water-shedding, meaning they rely on gravity to drain water off the system and protect the building. These roofs are typically covered with overlapping materials such as shingles, tiles, or metal panels. Leaks on steep-slope roof systems often occur at penetrations, corners, edges, valleys, and other components where the field is interrupted (*Figure 7*).

Steep-slope roof systems usually include an underlayment installed between the roof deck and roof covering as secondary weather protection (*Figure 8*). When conducting repairs on these systems, assume the underlayment is damaged until you visually determine otherwise. To repair underlayment, remove the shingles, tiles, or panels, to expose it. Then remove the fasteners, cut out damaged sections, and replace with matching materials following the manufacturer's instructions.

2.1.0 Asphalt Shingle Roof Systems

Though the specific repair steps for asphalt shingle roof systems (*Figure 9*) vary depending on the type of damage, they each follow the same basic pattern:

1. Remove the affected shingles.
2. Inspect the substrate.
3. Replace any damaged materials (such as underlayment or flashings).
4. Install new shingles according to the original installation method.

NOTE

Asphalt shingle roof systems are covered in more detail in NCCER Module 16201, *Asphalt Shingle Roof Systems*. Refer to this module for more details about roof system components, safety considerations, and installation procedures.

This section describes repair processes at a very general level. Always check applicable documentation, such as manufacturer instructions, for exact measurements and guidelines.

Figure 6 Steep-slope roofs.

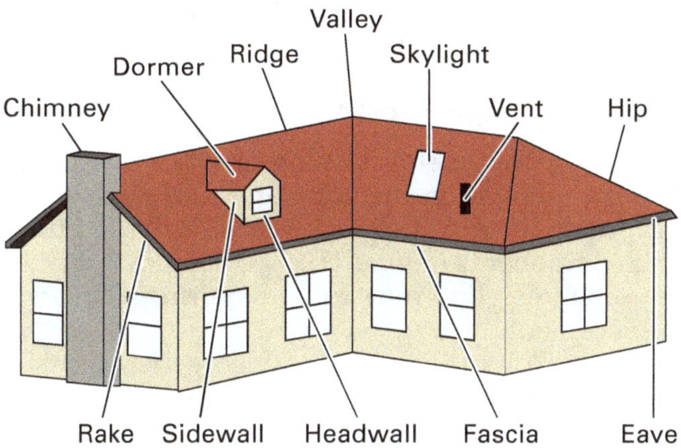

Figure 7 Components of a roof system.

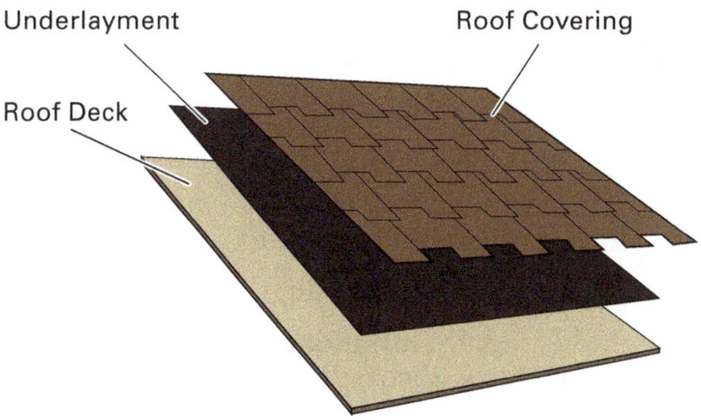

Figure 8 Basic steep-slope roof construction.

2.1.1 Damaged Shingle

Asphalt shingles are a popular steep-slope roofing option due in part to their combination of affordability and durability. Asphalt shingles can last 15–40 years. Individual shingles that have deteriorated or torn can be removed and replaced without significant disruption of the roof system (*Figure 10*).

> **CAUTION**
> When replacing a damaged asphalt shingle, always replace the whole shingle (not just part of it). Avoid flexing or bending shingles too much.

The following steps outline a general procedure for replacing a damaged shingle on an asphalt shingle roof system:

Step 1 Use a **flat bar** tool or flat bladed tool to break the sealant of the damaged shingle and the two shingle courses directly above it.

> **NOTE**
> Even if only one shingle is damaged, several other shingles may need to be removed. The number of shingles that must be removed depends on factors such as the age of the roof.

Step 2 Insert the flat bar tool under the damaged shingle and locate all nails.

Step 3 With your flat bar tool, grasp each nail's shank and apply upward pressure on the shingle and nail head until it pulls out of the deck. Lift each nail in the same fashion. Remove the nails.

Step 4 Repeat this process for the two courses of shingles directly above the damaged shingle.

Step 5 Install new shingles. Follow manufacturer instructions for where to drive nails.

Step 6 Use quarter-sized dabs of roof cement under the new shingles to fill all nail holes and be sure the shingles you have installed are properly adhered. This is called **hand sealing**.

Figure 9 Asphalt shingle roof systems.

Figure 10 Torn shingle in need of replacement.

> **CAUTION**
>
> Avoid using too much sealant on shingles. Always refer to manufacturer instructions and other applicable documentation.

2.1.2 Damaged Penetration Flashing

Penetration flashings on asphalt shingle roof systems are typically metal pieces that fit tightly around roof penetrations, such as pipe collars and soil stack boots.

The following steps outline a general procedure for repairing a damaged penetration flashing on an asphalt shingle roof system:

Step 1 Use a flat bar tool or flat bladed tool to remove the shingles surrounding the damaged penetration flashing.

Step 2 Remove the damaged flashing.

Step 3 Inspect the penetration and deck and clear any obstructions.

Step 4 Measure and fit a new flashing to the penetration.

Step 5 Replace shingles according to the sequence they were originally installed.

Step 6 Seal and cement according to the manufacturer's recommendations.

2.1.3 Missing Vent

If a damaged or loosened vent has blown off an asphalt shingle roof system, some shingles will need to be removed before the new vent is installed.

The following steps outline a general procedure for repairing a missing vent on an asphalt shingle roof system:

Step 1 Use a flat bar tool or flat bladed tool to remove the shingles surrounding the vent hole.

Step 2 Inspect the vent hole and deck and clear any obstructions.

Step 3 Ensure the vent hole is sized correctly for the new vent. Enlarge if it is too small.

Step 4 Install the new vent according to the manufacturer's recommendations.

Step 5 Replace shingles according to the sequence they were originally installed.

Step 6 Seal and cement according to the manufacturer's recommendations.

2.1.4 Drip Edge Replacement

Drip edge is a metal flashing or other overhanging component on the edge of a roof that directs dripping water to help protect underlying building components (*Figure 11*). The metal can rust or become unsightly over time, requiring drip edge replacement.

The following steps outline a general procedure for replacing a rake on an asphalt shingle roof system:

> **NOTE**
>
> This procedure assumes the underlayment is self-adhered.

Step 1 Gently lift the shingles on the lower edge of the roof.

Step 2 Use a flat bar tool to remove the nails holding the drip edge to the substrate.

Step 3 Remove the drip edge.

Step 4 Inspect the underlayment and deck and clear any obstructions.

Step 5 Slide the new drip edge in place, under the underlayment.

Step 6 Nail down the new drip edge while holding up the shingles, per manufacturer recommendations.

Figure 11 Example of a metal drip edge.

2.1.5 Valley Flashing Replacement

A valley is the internal angle formed by the intersection of two sloping roof planes. Because falling water gets directed into valleys, these areas of the roof are especially vulnerable to moisture intrusion. Metal valley flashings provide an additional layer of protection against water (*Figure 12*).

There are two main types of valleys: closed valleys and open valleys. Closed valleys use the primary roof material (such as shingles) all the way to the center of the valley. This creates a seamless look. Open valleys include metal in the center of the valley. The shingles or tiles extend several inches short of the center of the valley, as shown in *Figure 12*.

The following steps outline a general procedure for repairing an open, no-cut metal valley on an asphalt shingle roof system:

> **NOTE**
> The steps for repairing a valley differ depending on if it is open or closed. The following is a basic example.

Step 1 Remove shingles within at least 3' on both sides of the valley.

Step 2 Remove valley flashing.

Step 3 Inspect the underlayment and deck and clear any obstructions.

Step 4 Install new underlayment.

Step 5 Mechanically fasten the new metal valley flashing to the deck.

Step 6 Strip in the outside edges of the valley with two new courses of shingles vertically up the valley.

Step 7 Replace shingles according to the sequence they were originally installed.

Step 8 Seal and cement according to the manufacturer's recommendations.

2.2.0 Clay and Concrete Tile Roof Systems

Clay and concrete roof tiles overlap on a steep-slope roof to shed water downwards (*Figure 13*).

Be mindful of tile damage that can occur during the repair process. Step carefully on the roof and always inspect the system for newly broken tiles before you consider the repair job complete.

> **NOTE**
> Tile roof systems are covered in more detail in NCCER Module 16202, *Clay and Concrete Tile Roof Systems.* Refer to this module for more details about roof system components, safety considerations, and installation procedures.
> This section describes repair processes at a very general level. Always check applicable documentation, such as manufacturer instructions, for exact measurements and guidelines.

Figure 12 Open, no-cut metal valley flashing.

WARNING!

Silica is a common mineral found in tile. Cutting tile creates dust that contains crystalline silica particles that are too small to see with the naked eye (*Figure 14*). Inhaling this silica dust puts workers at risk for developing lung cancer, chronic obstructive pulmonary disease (COPD), and a form of lung disease called **silicosis**. Silica exposure has also been linked to other forms of cancer, as well as kidney disease. Always wear appropriate respiratory protection while cutting or grinding tile. OSHA guidelines for controlling exposure to silica are found in *29 CFR 1926.1153, Table 1.* This table outlines what respiratory protection you need when working with different materials and tools.

2.2.1 Damaged Tile

If a roofing tile is broken, it will likely need to be removed from the system entirely. In some cases, a broken tile is easy to identify, but be on the lookout for less obvious damage in the adjacent tiles as well. A broken tile is shown in *Figure 15.*

The following steps outline a general procedure for replacing a damaged tile:

Step 1 Use a flat bar tool to gently lift the roof tile above the damaged one.

Step 2 If the tiles are bonded with adhesive, use a flat bar tool or a manufacturer-recommended solvent to separate them.

Step 3 Remove the fasteners on the damaged tile. If necessary to access the fastener, tap on the damaged tile with your wedge or a hammer to break it into smaller pieces.

Step 4 Inspect the deck and underlayment and clear any obstructions.

Step 5 Seal any holes in the area with sealant.

Step 6 Install the replacement tile and fasten according to the manufacturer's recommendations.

- For tile installed on battens, reinstallation involves hanging a new tile on that batten.
- For tile installed with foam, cut out the old piece of foam to slip in a new tile before adding new foam.

2.2.2 Damaged Penetration Flashing

Penetration flashings on tile roof systems typically have two parts: a top flashing that sits over a layer of tile, and a deck flashing that sits between the bottom layer of tile and the roof deck.

The following steps outline a general procedure for replacing a damaged boot collar on a clay or concrete tile roof system:

Step 1 Detach and remove the tiles surrounding the penetration. Save any undamaged tiles for reuse. Separate adhered tiles with a flat bar tool or manufacturer-recommended solvent as necessary.

Step 2 Remove the top flashing.

Step 3 Inspect the underlayment and deck and clear any obstructions.

Figure 13 Clay tile roof systems.

Figure 14 Microscopic view of crystalline silica particles on a filter.

Figure 15 Example of a broken roof tile.

Step 4 Install the boot and seal according to the manufacturer's recommendations.

Step 5 Replace remaining tiles according to the sequence they were originally installed and fasten according to the manufacturer's recommendations.

2.2.3 Trim Tile (Hip, Ridge, or Rake) Replacement

A **trim tile** is a clay or concrete tile that sits along intersections of different roof planes in a roof system, giving it a polished appearance. Trim tiles sit in the following locations (refer to *Figure 7*):

- The hip, which is formed by the meeting of two sloping sides of a roof with eaves that meet at a right angle.
- The ridge, or the highest point on a roof represented by a horizontal line where two roof planes intersect.
- The rake, or the sloped edge of a roof.

The following steps outline a general procedure for replacing damaged trim tile on a clay or concrete tile roof system:

Step 1 Remove the damaged tile and fasteners.

- If possible, use an appropriate tool to gently lift the roof tile above the damaged one. Remove any exposed fasteners on this tile so that it will be easier to lift.
- If the tiles are bonded with adhesive, use a flat bar tool or a manufacturer-recommended solvent to separate them.
- If necessary to access the fasteners, gently tap on the damaged tile with a hammer to break it into smaller pieces.

NOTE

If you need to remove and reuse any surrounding tiles, be careful to remove the fasteners without cracking the tile.

Step 2 Inspect any tiles above or below the trim tile and clean off sealant. Inspect the deck and underlayment and clear any obstructions.

Step 3 Seal any holes in the area with sealant.

Step 4 Put down a bead of sealant to keep water out.

Step 5 Install the replacement tile and fasten according to the manufacturer's recommendations.

Step 6 Secure any tiles that were loosened during the repair according to the manufacturer's specifications.

2.2.4 Valley Flashing Replacement

The following steps outline a general procedure for replacing a valley on a clay or concrete tile roof system:

NOTE

The steps for repairing a valley on a tile roof system differ depending on a variety of factors. For example, the steps will be different for open or closed valleys, as well as for batten-hung or interlocking systems. The following procedure is a basic example.

Step 1 Detach and remove tiles within 3' on both sides of the valley.

Step 2 Remove valley flashing.

Step 3 Inspect the underlayment and deck and clear any obstructions.

Step 4 Mechanically fasten the new valley flashing to the deck.

Step 5 Cement, seal, or strip in the outside edges of the valley.

Step 6 Replace remaining tiles according to the sequence they were originally installed and fasten according to the manufacturer's recommendations.

2.3.0 Architectural Metal Panel Roof Systems

Architectural metal panel roof systems are made with overlapping or interlocking metal sheets (*Figure 16*). They are known for being especially long-lasting and durable. However, they may sustain damage such as rusting or warping from temperature changes.

There are two common types of architectural metal panel: standing seam and corrugated. **Standing seam panels** have overlapping vertical legs that get folded together during installation, so they cannot be easily separated for a roof repair. When these panels are damaged, they can be patched for a short-term fix, or cut out and replaced with a customized piece of sheet metal.

Corrugated panels have alternating ridges and grooves that overlap at seams (*Figure 17*). Unlike standing seam panels, corrugated roof panels have exposed mechanical fasteners, making them more prone to leaks. However, they tend to be easier to repair because individual panels can be removed and replaced.

Because corrugated panels have a significant market share and a replicable repair process, the instructions below will apply to corrugated panel roof systems.

Figure 16 Metal roof installation.

> **NOTE**
>
> Architectural metal panel roof systems are covered in more detail in NCCER Module 16205, *Metal Roof Systems*. Refer to this module for more details about roof system components, safety considerations, and installation procedures.
>
> This section describes repair processes at a very general level. Always check applicable documentation, such as manufacturers' instructions, for exact measurements and guidelines.

2.3.1 Damaged Penetration Flashing

Penetration flashings on architectural metal panel roof systems typically consist of a flexible boot that sits over the metal and seals the penetration. Some damaged boot flashings may be repaired with caulk, while others may need to be replaced completely.

The following steps outline a general procedure for replacing a damaged flexible boot flashing on an architectural metal panel roof system:

Step 1 Remove the damaged penetration flashing by detaching mechanical fasteners and sealants.

Step 2 Clean the penetration of residual sealant and other obstructions. Inspect the deck through the penetration hole.

Step 3 Size the replacement flashing so that it fits tightly to the penetration and covers up existing holes from the old flashing fasteners.

Step 4 Fasten and seal according to the manufacturer's recommendations.

2.3.2 Damaged or Missing Hip/Ridge Cap

Hips and ridges are areas where roof planes intersect at an upward angle, in contrast to valleys which intersect downwards.

Metal roof hip and ridge caps are premanufactured components that protect these areas.

The following steps outline a general procedure for repairing damaged or missing hip or ridge caps on an architectural metal panel roof system.

Step 1 If the damaged hip or ridge cap is still attached to the roof system, remove all mechanical fasteners from the damaged piece.

Step 2 Inspect the underlying nailer and underlayment to make sure they are intact. (Replace as needed.)

Step 3 Inspect and clean the joints of adjacent pieces.

Step 4 Size and insert the replacement hip or ridge cap component.

Step 5 Fasten and seal per manufacturer recommendations.

Figure 17 Example of a corrugated metal panel roof system.

2.0.0 Section Review

1. Which of the following tools should be used to break the weather sealant on a damaged asphalt shingle and lift out nails?
   a. A flat bar
   b. A Phillips head screwdriver
   c. A manufacturer-recommended solvent
   d. A hand grinder

2. What are the common components of penetration flashing on tile roof systems?
   a. Fabric reinforcement and liquid flashing materials
   b. Valley flashing and two intersecting roof planes
   c. Top flashing and deck flashing
   d. Underlayment and cap sheet

3. Which of the following statements about architectural metal panels is *true*?
   a. Standing seam panels are typically less durable but more affordable than corrugated panels.
   b. Architectural metal panel roof systems have an average lifespan of 10 years.
   c. Corrugated roof panels can be removed and replaced more easily than standing seam panels.
   d. When damaged, corrugated roof panels should be patched.

SECTION THREE

3.0.0 LOW-SLOPE REPAIR

Objective

List and describe common low-slope repair procedures.

a. Describe common repairs for thermoplastic roof systems.
b. Describe common repairs for EPDM roof systems.
c. Describe common repairs for built-up roof (BUR) systems.
d. Describe common repairs for modified bitumen (mod-bit) roof systems.

Performance Tasks

4. Repair a blister in a BUR or mod-bit roof membrane.
5. Patch a damaged thermoplastic or EPDM roof membrane.
6. Repair a flashing in a low-slope roof system.

Trade Terms

Blindside patch: A membrane patch that is adhered from underneath the field membrane.

Degranulate: To embed or remove the granule surfacing on a mod-bit sheet, exposing the bitumen.

Fishmouth: An opening in the edge of a membrane caused by wrinkling or shifting of sheets during installation.

Lap: The part of a roofing, waterproofing, or flashing component that overlaps or covers any portion of the same or another type of adjacent component.

Perimeter metals: Metal components that seal the membrane at the edge of the roof and create a transition to the building wall.

Seam: The area where two separate, overlapping sheets of roof membrane are joined together.

Seam probe: A small tool that is angled and pointed on the end and used to slide along a seam to test for loose sections.

Spudding: Removing the gravel aggregate surface on a BUR system to expose the roof membrane.

Strip flashing: Membrane flashing strips used for sealing or flashing metal flashing flanges into the roof or waterproofing membrane.

Three-course patch: A repair method commonly used for BUR and modified bitumen flashings. It involves successive applications of a base course of asphalt roof cement, reinforcing fabric, and a second course of asphalt roof cement.

Low-slope roofs (*Figure 18*) are weatherproof instead of water-shedding. They generally rely on a weatherproof membrane covering, or roof membrane, to keep water from entering the structure. Damage to this membrane or to roof components it attaches to can create opportunities for moisture to enter the system.

Most common repair procedures for low-slope roofs involve removing damaged material and creating a weatherproof patch in the membrane. Regardless of the type of damage, one of the first steps is always to cut away the damaged membrane and peel it back to examine for dirt or debris. Always install patches over a clean surface.

Some common types of damage include damaged flashings, perimeters, seams, and field membrane.

3.1.0 Thermoplastic Roof Systems

Thermoplastic roof systems are popular single-ply systems that consist of a field membrane attached to the substrate by fasteners, ballast, adhesives, or induction welding plates (*Figure 19*). Repair materials and patches used on these roof systems should be compatible with the type of thermoplastic membrane (*Figure 20*). For example, if the field membrane is PVC, the repair patch should be compatible with PVC.

> **NOTE**
>
> Thermoplastic roof systems are covered in more detail in NCCER Module 16206, *Thermoplastic Roof Systems*. Refer to this module for more details about roof system components, safety considerations, and installation procedures.
>
> This section describes repair processes at a very general level. Always check applicable documentation, such as manufacturers' instructions, for exact measurements and guidelines.

Figure 18 Example of low-slope roof.

Figure 19 Thermoplastic roof system.

Figure 20 Thermoplastic patch.

3.1.1 Damaged Flashings

Common thermoplastic roof flashings include pipe flashings (*Figure 21*) and curb flashings. Flashings are susceptible to damage such as punctures, tears, and deteriorated membrane.

The following steps outline a general procedure for repairing a damaged flashing on a thermoplastic roof system:

Step 1 Cut out and remove damaged flashing materials.

Step 2 Using a manufacturer-approved method, clean a work area that extends beyond the perimeter of the patch area. Inspect the substrate and clear any obstructions.

Step 3 Cut a target patch with rounded corners from a membrane sheet that is compatible with the existing roof membrane. For example, if the roof membrane is reinforced, the patch should also be reinforced. The patch should extend 4" beyond any part of the damaged flashing material.

Step 4 Apply the patch over the damaged area by hot-air welding (*Figure 22*).

Step 5 After the welded area has cooled, check the seam around the patch for unadhered areas using a seam probe. Seal the edges of the patch using a manufacturer-recommended sealant.

3.1.2 Damaged Perimeters

Perimeter metals seal the membrane at the edge of the roof and create a transition to the building wall (*Figure 23*). Perimeter metals shed water off the roof and provide a surface for membrane stripping to seal the edge. They may also be used to hold gravel/aggregate on the roof.

Damage at this area poses a significant risk for leaks. Often, the membrane near the perimeter metal may also have damage. Stripping plies may tear due to thermal expansion and contraction.

The following steps outline a general procedure for repairing a damaged thermoplastic stripping ply on an embedded metal edge:

Step 1 Cut and remove the damaged stripping ply.

Step 2 Clean a work area that extends beyond the perimeter of the patch area.

Did You Know?

The Yale Whale

The Ingalls Ice Arena, known as "The Yale Whale," was the first building to use a single-ply roof. It was designed by Eero Saarinen in the 1950s. Its roof was made of large timbers hung on cables.

Saarinen wanted to cover the roof in a layer of rubber that would be elastic enough to tolerate the swaying roof, yet retain its watertight characteristics. However, adhering 48" rolls of rubber would not work without sanding down the entire wood deck. The larger rolls were substituted with narrow strips of rubber adhered to strips of metal. The metal strips were then nailed to the deck and disguised as standing seams.

After Saarinen's pioneering work, the use of single-ply membranes increased dramatically. Saarinen is also known for designing the former TWA terminal at Kennedy Airport, Dulles International Airport outside of Washington, D.C., and the St. Louis Arch.

Figure Credit: iStock@sainaniritu

Figure 21 Thermoplastic pipe flashing.

Figure 22 Hot-air welding.

Figure 23 Example of perimeter metal.

Step 3 Replace the stripping ply per manufacturer recommendations. Repair any damaged membrane using a compatible patch material cut with rounded corners. Allow at least 4" of overlap around the damaged area.

Step 4 Attach the stripping ply. Stripping plies may be self-adhered or hot-air welded.

> **NOTE**
> If using a hot-air welder, wait until the area has cooled and check the seams for unadhered areas using a seam probe. Self-adhering patches should be pressed onto the substrate using a hand roller to ensure adequate adhesion.

Step 5 Seal the edges of the patch using a manufacturer-approved process.

3.1.3 Seam Repair

In single-ply roofing, a seam is formed where two overlapping sheets of membrane adhere together. Seams are vulnerable to water intrusion if they come apart. The edge of a sheet may also become folded, raised, or wrinkled, causing a fishmouth. Cold welds are a seam problem where the seam did not fuse properly during installation.

Seam damage is often repaired using a hand welder to fuse the seam back together. If the seam damage is very long, you may need a robotic welder to repair it. The following steps outline a general procedure for seam repair on a thermoplastic roof system:

Step 1 Pull back the material in the unbonded lap until you reach a properly bonded area.

Step 2 Clean a work area that extends beyond the perimeter of the patch area. Clean inside the unbonded lap. Inspect the substrate and clear any obstructions.

Step 3 Cut a patch with rounded corners from a membrane sheet that is compatible with the existing roof membrane. The patch should extend at least 4" beyond any part of the damaged material.

> **NOTE**
> Refer to manufacturer guidelines and other applicable documentation for exact measurements.

Step 4 Apply the patch over the damaged area by hot-air welding.

Step 5 After the welded area has cooled, check the seams around the patch for unadhered areas using a seam probe. Seal the edges of the patch using a manufacturer-approved process.

3.1.4 Field Repair

There are several common forms of field membrane damage on single-ply roof systems, such as a puncture or cut. The following steps outline a general procedure for repairing field damage on a thermoplastic roof:

Step 1 Examine the damaged area and remove any debris / clean the area.

Step 2 Remove the damaged membrane by cutting or slicing.

Step 3 Clean a work area that extends beyond the perimeter of the patch area.

Step 4 Cut a patch with rounded corners from a membrane sheet that is compatible with the existing roof membrane. The patch should extend at least 4" beyond any part of the damaged membrane.

> **NOTE**
> Refer to manufacturer guidelines and other applicable documentation for exact measurements.

Step 5 Apply the patch over the damaged area by hot-air welding (*Figure 24*).

> **NOTE**
> If the surface of the field membrane is too deteriorated or dirty to create a long-lasting waterproof seal with a patch, it is possible to apply the patch from underneath. This is known as a **blindside patch**. Insert and center the patch under the cut-out area and apply by hot-air welding. Note that blindside patches only work for mechanically attached installations.

Step 6 After the welded area has cooled, check the seams around the patch for unadhered areas using a seam probe. Seal the edges of the patch using a manufacturer-approved process.

3.2.0 EPDM Roof Systems

EPDM roof systems are a single-ply membrane that is typically seamed with seam tape (*Figure 25*). EPDM roof systems consist of field membrane attached to the substrate by adhesives, fasteners, or ballast. In EPDM roof repairs, compatible patches are bonded to the membrane with liquid-applied adhesive or adhesive seam tape.

Self-adhering membrane materials are most often used to repair EPDM roof systems. Always defer to the manufacturer's instructions for applying patches.

> **NOTE**
> EPDM roof systems are covered in more detail in NCCER Module 16207, *EPDM Roof Systems.* Refer to this module for more details about roof system components, safety considerations, and installation procedures.
>
> This section describes repair processes at a very general level. Always check applicable documentation, such as manufacturer instructions, for exact measurements and guidelines.

3.2.1 Damaged Flashings

EPDM flashings may deteriorate over time and with exposure to weather. The following steps outline a general procedure for repairing a damaged flashing on an EPDM roof system:

Step 1 Examine the damaged area and remove any debris / clean the area.

Step 2 Remove the damaged membrane by cutting or slicing.

Step 3 Cut a patch with rounded corners from a membrane sheet that is compatible with the existing roof membrane. The patch should extend at least 4" beyond any part of the damaged flashing material.

Figure 24 Hot-air welding a blindside patch.

> **NOTE**
> Refer to manufacturer guidelines and other applicable documentation for exact measurements.

Step 4 Place the patch centered over the damaged area. Use a crayon or marker to mark a perimeter around the patch 1" to 2" outside the edges.

Step 5 Apply primer to the patch area extending to your markings and to the underside of the patch.

Step 6 Check the seams around the patch for unadhered areas. Seal the edges of the patch using a manufacturer-recommended process (*Figure 26*).

3.2.2 Damaged Perimeters

The following steps outline a general procedure for repairing damaged EPDM stripping ply on an embedded metal edge:

Step 1 Cut and remove the damaged stripping ply.

Step 2 Clean a work area that extends beyond the perimeter of the patch area.

Step 3 Cut the membrane and remove the metal flashing. Inspect the area for dirt or debris and clean as needed.

Step 4 Extend field membrane back over the edge.

Step 5 Replace damaged sections of perimeter metals. Refasten loose perimeter metals.

Step 6 Strip in the metal edge.

Step 7 Seal the edges of the patch using a manufacturer-recommended process.

3.2.3 Seam Repair

As with thermoplastic roofing systems, EPDM systems are vulnerable to water damage at areas where seams have opened (*Figure 27*).

The following steps outline a general procedure for seam repair on an EPDM roof system:

Step 1 Pull back the material in the unbonded lap until you reach a properly bonded area.

Step 2 Clean a work area that extends beyond the perimeter of the patch area. Clean inside the unbonded lap. Inspect the substrate and clear any obstructions.

Figure 25 EPDM roof system.

Figure 26 Sealing a patch on an EPDM flashing.

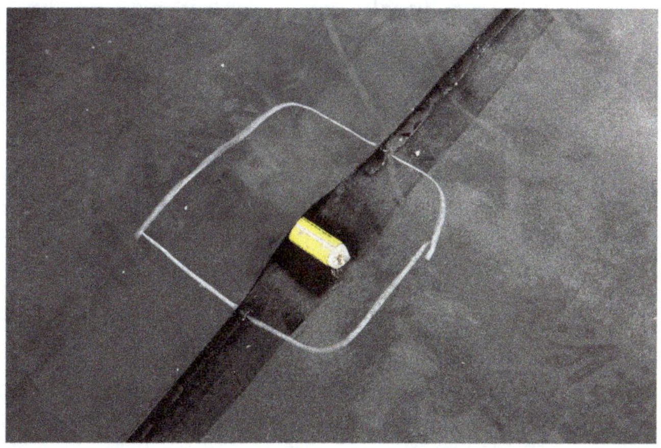

Figure 27 Open EPDM lap.

Step 3 Cut the patch material with rounded corners from a membrane sheet that is compatible with the existing roof membrane. The patch should extend at least 4" beyond any part of the damaged membrane.

> **NOTE**
> Refer to manufacturer guidelines and other applicable documentation for exact measurements.

Step 4 Place patch over the repair area. Use a crayon or marker to mark a perimeter around the patch.

Step 5 Apply primer to the patch area extending 1" to 2" beyond your markings. Apply more primer to the underside of the patch.

Step 6 Center the patch over the repair area, ensuring seam tape edges extend $\frac{1}{4}$" outside the patch perimeter.

Step 7 Check the seams around the patch for unadhered areas. Seal the edges of the patch using a manufacturer-recommended process.

3.2.4 Field Repair

The following steps outline a general procedure for field repair on an EPDM roof system:

Step 1 Examine the damaged area and remove any debris / clean the area.

Step 2 Remove the damaged membrane by cutting or slicing.

Step 3 Clean a work area that extends beyond the perimeter of the patch area.

Step 4 Round the edges of the damaged area.

Step 5 Cut patch material with rounded corners from a self-adhering membrane sheet that is compatible with the existing roof membrane. The patch should extend 4" beyond any part of the damaged membrane.

Step 6 Place patch over the repair area. Use a crayon or marker to mark a perimeter around the patch 1" to 2" outside the edges. This establishes a target area to apply primer.

Step 7 Apply primer to the patch area extending to your markings and to the underside of the patch.

Step 8 Center the patch over the repair area.

Step 9 Check the seams around the patch for unadhered areas. Seal the edges of the patch using a manufacturer-recommended process.

> **CAUTION**
> Blindside patches should not be used on EPDM roof systems because they may leak at the seams.

3.3.0 Built-Up Roof (BUR) Systems

Built-up roof (BUR) systems consist of reinforcing ply sheets layered with bitumen for weatherproofing and adhesion (*Figure 28*). They are surfaced with granules or a coating material for UV protection.

Three-coursing is commonly used to repair BUR and mod-bit flashings. This repair involves successive applications of a base course of asphalt roof cement, reinforcing fabric, and a second course of asphalt roof cement. Where a three-course repair will remain exposed, it should include a surfacing material.

> **NOTE**
> BUR roofs are covered in more detail in NCCER Module 16208, *Built-Up Roof Systems*. Refer to this module for more details about roof system components, safety considerations, and installation procedures.
>
> This section describes repair processes at a very general level. Always check applicable documentation, such as manufacturer instructions, for exact measurements and guidelines.

3.3.1 Damaged Flashings

Most leaks on BUR systems occur at flashings. Flashings may loosen and open up, slide, or split, creating opportunities for moisture to penetrate the roof system.

The following steps outline a general procedure for repairing a damaged wall base flashing on a BUR system:

Step 1 Clean work area that extends beyond the perimeter of the patch area. Inspect the substrate and clear any obstructions. Be sure to clean and inspect both horizontal and vertical surfaces.

Figure 28 BUR membrane.

Step 2 Degranulate the surface of granule-surfaced membranes where the patch will tie in using the appropriate spudding, scraping, and/or heating method for the surface material (*Figure 29*). This helps the patch adhere.

- For asphalt BUR, where the repair area was not degranulated or heated (such as when using a cold-applied adhesive), lightly prime the surface with asphalt primer.

Step 3 Cut out and remove damaged flashing materials.

Step 4 Measure, cut, and fit at least one ply of a compatible mod-bit patch to extend at least 4" beyond the damage.

Step 5 Secure the flashing using a method compatible with the roof system.

Step 6 Close the seam using either hot- or cold-applied bitumen.

Step 7 Apply a three-course patch on top of the repaired area.

Figure 29 Prepare the surface by spudding or degranulating.

3.3.2 Damaged Perimeters

The following steps outline a general procedure for repairing stripping plies on perimeter metal on a BUR roof system (*Figure 30*):

Step 1 Clean a work area that extends beyond the perimeter of the patch area.

Step 2 Clear the surface of gravel or aggregate using a spud bar or scraper where the patch will tie into the existing membrane.

Step 3 Remove unbonded flashing membrane.

Step 4 Lightly prime the surface with asphalt primer.

Step 5 Where plies are removed from the perimeter metal flashing, install the same number of plies of reinforcing material. Extend the bottom ply at least 6" beyond the area to be repaired and each succeeding ply at least 3" beyond the previous ply.

3.3.3 Liquid-Applied Penetration Flashing Repair

Liquid-applied flashings can restore existing penetrations to a watertight state on low-slope roof systems. These flashings consist of fabric reinforcement embedded in multiple coats of liquid flashing material.

The following steps outline a general procedure for liquid-applied penetration pipe flashing repair on a BUR roof system:

Step 1 Clean a work area that extends about 1' in all directions around the penetration.

Step 2 Establish a target square using tape.

Step 3 Place fabric in a square around the penetration, extending about 6" to 8" from the penetration on all sides. Wrap the fabric around the base of the penetration and spread cut fingers, or fanned strips, onto the roof surface below (*Figure 31*). Refer to manufacturer specifications for how far up the penetration to install the fabric.

Step 4 Apply compatible sealant around the base of the penetration. Consult manufacturer recommendations for how far around the penetration to apply the sealant.

Step 5 Clean and prime the penetration according to the manufacturer's specifications.

Step 6 Apply the flashing liquid around the base of the penetration and extend it to the taped boundaries.

Step 7 Embed the fabric into the liquid.

Step 8 Apply a top coat of flashing liquid according to the manufacturer's recommendations.

Step 9 Remove tape before the flashing liquid has cured.

3.3.4 Field Repair

If a BUR field membrane has been punctured, cut, or otherwise damaged, it should be patched to return the system to a watertight condition (*Figure 32*).

The following steps outline a general procedure for field repair on a BUR roof system:

Step 1 Clean a work area that extends beyond the perimeter of the patch area.

Step 2 Degranulate the surface where the patch will tie into the existing membrane using the appropriate spudding, scraping, and/or heating method for the surface material.

- For asphalt BUR, where the repair area was not degranulated or heated (such as when using a cold-applied adhesive), lightly prime the surface with asphalt primer.

Step 3 Where plies are not removed from the membrane, apply a three-course patch.

Step 4 Where plies are removed from the membrane, install the same number of plies of reinforcing material. Extend the bottom ply at least 6" beyond the area to be repaired and each succeeding ply at least 3" beyond the previous ply.

Step 5 Reapply surfacing.

Figure 30 Gravel stop.

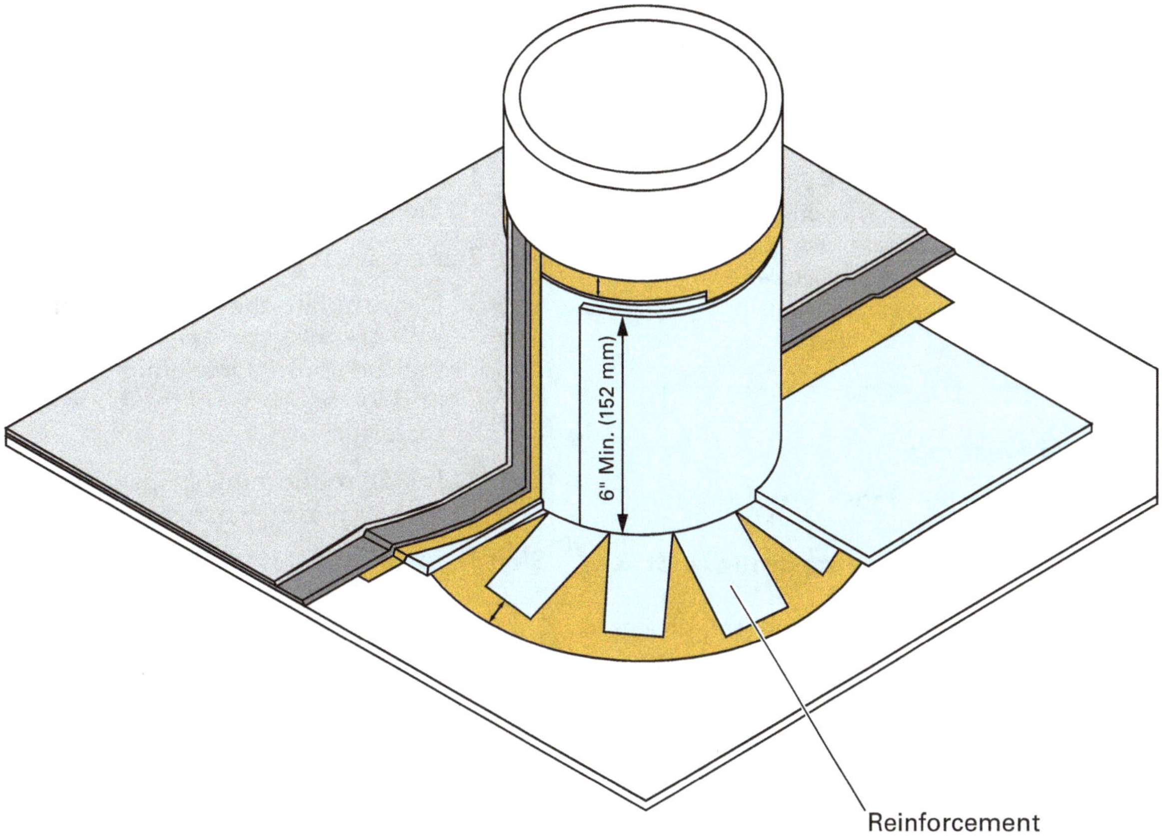

Figure 31 Flashing reinforcement cut with fingers at base of penetration.

3.4.0 Modified Bitumen Roof Systems

Modified bitumen (mod-bit) roof systems are an evolution of BUR (*Figure 33*). They are similarly made up of bitumen, reinforcement, and surfacing material. However, unlike BUR systems that are fabricated in the field, the components for modified bitumen systems are fabricated into sheets at a factory before being installed.

Two common types of modified bitumen products are styrene butadiene styrene (SBS) and atactic polypropylene (APP), which each have slightly different repair procedures. Unlike APP, SBS usually has granules that need to be embedded prior to installing a patch. As always, refer to the manufacturer's instructions when choosing the appropriate patch installation method for a roof system.

Figure 32 Damaged BUR field membrane.

Figure 33 Mod-bit roof.

NOTE

Mod-bit roof systems are covered in more detail in NCCER Module 16209, *Modified Bitumen Roof Systems*. Refer to this module for more details about roof system components, safety considerations, and installation procedures.

This section describes repair processes at a very general level. Always check applicable documentation, such as manufacturer instructions, for exact measurements and guidelines.

3.4.1 Damaged Flashings

Like other roof systems, most leaks on mod-bit roofs occur at flashings. Repairing a mod-bit flashing is very similar to repairing a BUR flashing.

The following steps outline a general procedure for repairing a damaged flashing on a mod-bit roof system:

Step 1 Clean work area that extends beyond the perimeter of the patch area. Inspect the substrate and clear any obstructions. Be sure to clean and inspect both horizontal and vertical surfaces.

Step 2 Degranulate the surface where the patch will tie in using the appropriate embedding, scraping, and/or heating method for the surface material. This helps the patch adhere.

- Where the repair area was not degranulated or heated (such as when using a cold-applied adhesive), lightly prime the surface with asphalt primer.

Step 3 Cut out and remove damaged flashing materials.

Step 4 Measure, cut, and fit at least one ply of a compatible mod-bit patch to extend at least 4" beyond the damage.

Step 5 Secure the flashing using a method compatible with the roof system.

Step 6 Close the seam using either hot- or cold-applied bitumen.

Step 7 Three-course the repair.

3.4.2 Damaged Perimeters

The following steps outline a general procedure for repairing damaged stripping plies on perimeter metal on a mod-bit roof system:

Step 1 Clean a work area that extends beyond the perimeter of the patch area.

Step 2 Remove unbonded flashing membrane.

Step 3 Degranulate the surface where the patch will tie into the existing membrane using the appropriate embedding, scraping, and/or heating method for the surface material.

Step 4 Install reinforcing strips of modified bitumen at perimeter metal end joints.

Step 5 Install strip flashing so that it extends at least 3" laterally and 3" beyond the reinforcing strip onto adjoining roof membrane.

3.4.3 Penetration Repair

The following steps outline a general procedure for liquid-applied pipe flashing repair on a mod-bit roof system (*Figure 27*):

Step 1 Clean a work area that extends about 1' in all directions around the penetration.

Step 2 Place tape in a square around the penetration, sitting about 6" to 8" away from the penetration on all sides.

Step 3 Apply compatible sealant around the base of the penetration.

Step 4 Clean and prime the penetration according to the manufacturer's specifications.

Step 5 Apply the flashing liquid around the base of the penetration and extending to the taped boundaries.

Step 6 Wrap flashing reinforcement around the base of the penetration and spread cut fingers, or fanned strips, onto the roof surface below.

Step 7 Apply a top coat of flashing liquid according to the manufacturer's recommendations.

Step 8 Remove tape before the flashing liquid has cured.

Step 9 Broadcast (or sprinkle) granules into liquid to give it a finished look.

3.4.4 Field Repair

The following steps outline a general procedure for field repair on a mod-bit roof system:

Step 1 Clean a work area that extends beyond the perimeter of the patch area.

Step 2 Degranulate the surface where the patch will tie into the existing membrane using the appropriate embedding, scraping, and/or heating method for the surface material. Alternatively, lightly prime the surface of the membrane with asphalt primer and allow it to dry.

Step 3 Cut a patch of compatible modified bitumen with rounded corners 8" larger in all dimensions than the damaged area.

Step 4 Install the patch over the repair area using an appropriate adhesion method for the material.

Step 5 Close the seam using either hot- or cold-applied bitumen.

Step 6 Three-course the repair.

3.0.0 Section Review

1. Which of the following statements about patches used in thermoplastic seam repair is *true*?
   a. The patch should extend 1" beyond any part of the damaged area.
   b. The patch should have rounded corners.
   c. The patch should always be applied over the damaged area with mechanical fasteners.
   d. The patch should be degranulated before application.

2. Which of the following is a *likely* scenario for an EPDM roof system repair?
   a. The surface material is spudded where the patch will tie into the existing membrane.
   b. The surface material is degranulated where the patch will tie into the existing membrane.
   c. Compatible patches are hot-air welded to the membrane.
   d. Compatible patches are bonded to the membrane with liquid-applied adhesive or adhesive seam tape.

3. How should reinforcing fabric be placed during liquid-applied penetration flashing repair on a BUR roof system?
   a. Around the penetration with fingers that fan out from the base
   b. In a circle with a diameter 2' wider than the penetration
   c. In a square around the penetration, creating a target area
   d. In a square encompassing the entire repair work area

4. What do you do with granules when repairing a mod-bit roof?
   a. Sprinkle them over the primer
   b. Remove or embed them
   c. Sweep them with a broom
   d. Secure them with one-component adhesive

Review Questions

1. Which of the following steps always comes *after* hot-air welding a patch during field repair on a thermoplastic roof system?
   a. Install a blindside patch.
   b. Remove the fishmouth by cutting or slicing.
   c. Pull back the cut membrane to examine for dirt or moisture.
   d. Check the seams around the patch using a seam probe.

2. Which of the following is *not* a typical requirement for a career in roofing repair?
   a. Problem-solving skills
   b. Customer service skills
   c. A master's degree
   d. Several years of on-site job experience

3. Which of the following steps comes *first* in the repair process?
   a. Service technicians dispatched to investigate leak.
   b. Roofing contractor schedules service.
   c. An invoice is created.
   d. Customer discovers leak and notifies roofing company.

4. The necessary tasks for a roofing repair job are outlined in the _____.
   a. work order
   b. conversation with the customer
   c. jobsite documentation
   d. leak investigation

5. Which of the following is *not* a step in the field repair procedure for modified bitumen roof systems?
   a. Cut a patch of compatible material with rounded corners.
   b. Cut fasteners with a hand grinder.
   c. Degranulate or lightly prime the surface where the patch will tie into the existing membrane.
   d. Clean a work area that extends beyond the perimeter of the patch area.

6. What is a seam in single-ply roofing?
   a. The area where two separate, overlapping pieces of roof membrane are joined together
   b. The part of a roofing, waterproofing, or flashing component that overlaps or covers any portion of the same or another type of adjacent component
   c. Components that seal the membrane at the edge of the roof and create a transition to the building wall
   d. Openings or voids in a lapped edge, usually caused by wrinkling or shifting of membrane during installation

7. What should a roofing professional do after encountering a situation *not* covered by the work order?
   a. Use their best judgment to make the necessary additional repairs.
   b. Stop work and contact the building owner.
   c. Tell their supervisor the work order has been completed.
   d. Take photos and contact their supervisor.

8. Which of the following roof systems are known for being especially long-lasting and durable on steep-slope roofs?
   a. Modified bitumen roof systems
   b. Architectural metal panel roof systems
   c. Thermoplastic roof systems
   d. Clay or concrete tile roof systems

9. Which of the following is *not* a common use for a flat bar tool in damaged shingle replacement?
   a. Breaking the weather sealant of the damaged shingle
   b. Filling all nail holes with sealant
   c. Removing nails from the damaged shingle
   d. Removing nails from the two shingles directly above the damaged shingle

10. When should a three-course patch be applied during flashing repair on a BUR roof system?
   a. Before degranulating the surface where the patch will tie into the existing membrane
   b. Immediately after cleaning a work area that extends beyond the perimeter of the patch area
   c. After application of a patch adhered in cold-applied adhesive
   d. Before hot-air welding a patch to the membrane

11. Which of the following is a common penetration flashing for an architectural metal panel roof system?
   a. Panels with alternating ridges and grooves that overlap at seams
   b. A pre-manufactured component that sits at hips or ridges
   c. A flexible boot that sits over the metal and hugs the penetration
   d. Bitumen, reinforcement, and surfacing material that have been fabricated into sheets at a factory

12. During drip edge replacement on an asphalt shingle roof system, which of the following *must* happen before the damaged drip edge is removed?
   a. Use a flat bar tool to remove the nails holding the drip edge to the substrate.
   b. Slide the new drip edge in place, over the underlayment but under the shingles.
   c. Inspect the underlayment and deck.
   d. Nail down the new drip edge while holding up the shingles.

13. What is a trim tile?
   a. A clay or concrete tile that sits along intersections of different roof planes, such as the hip, ridge, or rake.
   b. A clay or concrete tile that sits along the drip edge.
   c. A clay or concrete tile that has been installed to replace a damaged tile.
   d. A clay or concrete tile that sits within 3' of the roof valley on either side.

14. What is a valley?
   a. The meeting of two sloping sides of a roof with eaves that meet at a right angle
   b. The highest point on a roof represented by a horizontal line where two roof planes intersect
   c. The internal angle formed by the intersection of two sloping roof planes
   d. The sloped edge of a roof at or adjacent to the first or last rafter

15. What are perimeter metals?
   a. Components that anchor, seal, and flash the membrane at the edge of the roof
   b. Components that cover skylights
   c. An additional layer of protection against water at steep-slope roof valleys
   d. Flexible boots that sit over architectural metal panels and hug roof penetrations

Trade Terms Introduced in This Module

Blindside patch: A membrane patch that is adhered from underneath the field membrane.

Corrugated panels: Architectural metal panels with alternating ridges and grooves that overlap at seams.

Degranulate: To embed or remove the granule surfacing on a mod-bit sheet, exposing the bitumen.

Fishmouth: An opening in the edge of a membrane caused by wrinkling or shifting of sheets during installation.

Flat bar: A long and narrow hand-held tool with an L-shaped curve at the head. In roofing repair, flat bars can be used for wedging between and separating attached components such as asphalt shingles, or for removing nails.

Hand sealing: The process of adding additional adhesive to ensure new shingles adhere to old shingles, especially in cold weather. Also called *hand tabbing*.

Lap: The part of a roofing, waterproofing, or flashing component that overlaps or covers any portion of the same or another type of adjacent component.

Nailer: A piece or pieces of dimensional lumber and/or plywood secured to a structural deck or walls that provide(s) a receiving medium for the fasteners used to attach tiles or panels.

Perimeter metals: Metal components that seal the membrane at the edge of the roof and create a transition to the building wall.

Roof repair: Roofing work that involves fixing a deficiency, leak, or other damage to a roof system.

Roof service: Roofing work that involves routine maintenance of a roof system to prevent damage in the future and extend the life of the roof system.

Seam: The area where two separate, overlapping sheets of roof membrane are joined together.

Seam probe: A small tool that is angled and pointed on the end and used to slide along a seam to test for loose sections.

Silicosis: A lung disease caused by breathing in silica particles. Dusty jobs, like cutting roof tiles, can lead to silicosis if performed without respiratory protection.

Spudding: Removing the gravel aggregate surface on a BUR system to expose the roof membrane.

Standing seam panels: Architectural metal panels with overlapping vertical legs that get folded together during installation.

Strip flashing: Membrane flashing strips used for sealing or flashing metal flashing flanges into the roof or waterproofing membrane.

Strip in: Apply membrane flashing strips to seal the edges of a metal flashing component.

Three-course patch: A repair method commonly used for BUR and modified bitumen flashings. It involves successive applications of a base course of asphalt roof cement, reinforcing fabric, and a second course of asphalt roof cement.

Trim tile: An accessory piece provided by the manufacturer used to attractively weatherproof various areas of a roof, such as at ridges, hips, and rakes.

Valley: The internal angle formed by the intersection of two sloping roof planes.

Work order: A document that outlines the necessary tasks for a contracted job by showing the roofer where to look and what to do. On a roof repair job, the work order will outline the specific repair tasks the roofing company is expected to complete.

Additional Resources

This module presents thorough resources for task training. The following reference material is suggested for further study.

Careers in Roofing, **www.careersinroofing.com**.

Code of Federal Regulations, **ecfr.gov**.

National Roofing Contractors Association (NRCA), **www.nrca.net**.

NCCER Module 16201, *Asphalt Shingle Roof Systems*.

NCCER Module 16202, *Clay and Concrete Tile Roof Systems*.

NCCER Module 16205, *Metal Roof Systems*.

NCCER Module 16206, *Thermoplastic Roof Systems*.

NCCER Module 16207, *EPDM Roof Systems*.

NCCER Module 16208, *Built-Up Roof Systems*.

NCCER Module 16209, *Modified Bitumen Roof Systems*.

NRCA Pocket Guide to Safety, NRCA. 2015. Rosemont, IL: National Roofing Contractors Association.

NRCA Repair Manual for Low-slope Membrane Roof Systems, NRCA. 2018. Rosemont, IL: National Roofing Contractors Association.

The NRCA Roofing Manual: Architectural Metal Flashing and Condensation and Air Leakage Control, NRCA. 2018. Rosemont, IL: National Roofing Contractors Association.

The NRCA Roofing Manual: Membrane Roof Systems, NRCA. 2019. Rosemont, IL: National Roofing Contractors Association.

The NRCA Roofing Manual: Metal Panel and SPF Roof Systems, NRCA. 2016. Rosemont, IL: National Roofing Contractors Association.

The NRCA Roofing Manual: Steep-slope Roof Systems, NRCA. 2017. Rosemont, IL: National Roofing Contractors Association.

The NRCA Safety Manual, NRCA. 2012. Rosemont, IL: National Roofing Contractors Association.

Figure Credits

iStock@tsvibrav, Module Opener

Deer Park Roofing, Inc., Cincinnati, OH, Figure 2

Lee Company, 2017 ACCA Contractor of the Year, Figure 3

Courtesy of Jobba Trade Technologies, Figure 4

Eagle Roofing Products, Figure 5 (A)

Courtesy of GAF, Figures 5 (D, F), 21, 31

Courtesy of Latite Roofing and Sheet Metal LLC, Figures 5 (E), 16

Patrick D Murphy Co., Inc., Architects, Figures 5 (H), 28

iStock@Rajesh Pandit, Figure 6

Courtesy of the National Roofing Contractors Assoc., Figures 8, 11–12, 20, 22–27, 29–30, 32

iStock@cmannphoto, Figure 10

U.S. Geological Survey, Figure 14

iStock@StockSolutions, Figure 17

iStock@Svetlana123, Figure 18

iStock@kokodrill, Figure 33

Section Review Answer Key

SECTION 1.0.0

Answer	Section Reference	Objective
1. b	1.1.1	1a

SECTION 2.0.0

Answer	Section Reference	Objective
1. a	2.1.1	2a
2. c	2.2.2	2b
3. c	2.3.0	2c

SECTION 3.0.0

Answer	Section Reference	Objective
1. b	3.1.3	3a
2. d	3.2.0	3b
3. a	3.3.3	3c
4. b	3.4.0	3d

User Update

NCCER makes every effort to keep its textbooks up-to-date and free of technical errors. We appreciate your help in this process. If you find an error, a typographical mistake, or an inaccuracy in NCCER's curricula, please submit a User Update form by visiting **https://www.nccer.org/olf**. You can also scan the QR code using the camera on your phone or mobile device to access the form.

Glossary

Adhesion test: A test performed on an existing roof to verify that a substrate can successfully bond to a liquid coating. Also called a *peel test*.

Alignment marks: Marked or chalked lines extending the length of membrane and used to align the lap of the adjacent sheet accurately and consistently.

Apex tile: An accessory piece that helps waterproof systems by covering the area where hips and/or ridges converge from different directions.

Application rate: The average quantity (mass, volume, or thickness) of material applied per unit area, typically specified in gallons per square or liters per square meter.

Architectural metal roofs: A metal panel system constructed of 26-gauge or thinner metals used in steep-slope applications that require a continuous or closely spaced substrate underneath the system for support.

Atactic polypropylene (APP): A plastic-like polymer used to modify bitumen.

Backnailing: Nailing the concealed portion of a roofing component so the fasteners are covered by the next ply, or course, and are not exposed to the weather in the finished roof system.

Ballast: A material, such as smooth river stone, crushed stone, standard precast concrete pavers or interlocking, beveled, doweled, or contoured fit lightweight concrete pavers, that weighs down a membrane to hold it in place.

Batten-hung: A style of clay and concrete tile that has a batten lug, typically on the backs of the tile pieces to help hang the tile off batten systems.

Bib: Flashing that covers and weatherproofs the area where a new nail is used to replace individual slates with a new supply.

Bill of lading: A detailed list of items shipped, delivered by the carrier with the shipment.

Bitumen: An asphalt-based product that provides waterproofing and adhesion.

Bleed-out: A continuous line of bitumen at the edge of a seam where it has flowed out of the lap and is visible on the outside, indicating that the proper amount of bitumen was used.

Blindside patch: A membrane patch that is adhered from underneath the field membrane.

Bundle: The measurement unit that mills use to sell shakes and shingles.

Butt edge: The bottom or exposed end of a shake or shingle.

Cant strip: A long, triangular object placed in 90-degree bends before applying the membrane to reduce the stress on the membrane; An angled strip installed at the roof's eaves that raises the butt end of the starter course to match same angle as the above slate courses. Also referred to as *cant*.

Certified Roofing Torch Applicator (CERTA) Program: A training program by the National Roofing Contractors Association (NRCA) that teaches critical safety hazards, considerations, and practices for all torching work.

Corrugated panels: Architectural metal panels with alternating ridges and grooves that overlap at seams.

Degranulate: To embed or remove the granule surfacing on a mod-bit sheet, exposing the bitumen.

Dry film thickness: The thickness, expressed in mils, of an applied and fully cured liquid coating.

Eave closures: Accessory pieces used to close off and flash the open space at a roof's perimeter edge. Also called *eave risers* for low-profile tile and *bird stops* for medium- to high-profile tile.

Edge grain: A log cut perpendicular to the rings of the tree.

Elastomeric: Able to be stretched at least twice its original size and return to its original shape.

EPDM: Acronym for *ethylene propylene diene monomer*; a common type of thermoset used in single-ply roof systems.

Exposure: The amount of tile surface exposed to the weather and not overlapped by the next course or surrounding tiles.

Fasteners: Any of a wide variety of mechanical securement devices, including nails, staples, screws, cleats, clips, and bolts.

Fiberglass base: An asphalt shingle base made with glass spun into fibers and combined using plastic polymers.

Fishmouth: An opening in the edge of a membrane caused by wrinkling or shifting of sheets during installation.

Fixed points: Points of attachment on steep-slope metal roof systems that secure the system and represent an obstacle to the natural expansion and contraction of metal. A fixed point for a panel is usually located at a roof's ridges or hips, or at the eaves or valleys.

Flash point (FP): The temperature at which a material can ignite.

Flashing-grade coatings: Liquid coatings specifically designed for areas where the primary roof covering is interrupted or terminated, such as protrusions and damaged areas. These are generally applied before base and finishing coats and can be applied to vertical surfaces since they have a thicker consistency.

Flashings: Components used to weatherproof or seal edges of a roof system at perimeters, penetrations, walls, expansion joints, valleys, drains, and other places where the roof covering is interrupted or terminated. For example, membrane base flashing covers the edge of the field membrane, and cap flashings or counterflashings shield the upper edges of the base flashing.

Flat bar: A long and narrow hand-held tool with an L-shaped curve at the head. In roofing repair, flat bars can be used for wedging between and separating attached components such as asphalt shingles, or for removing nails.

Flat grain: A log cut parallel to the tree's rings.

Fleece-backing: A nonwoven polyester material adhered to the underside of a thermoplastic sheet for reinforcement.

Flux: A chemical substance that helps remove and prevents the formation of oxides on metal surfaces as they are heated for soldering, and acts as a vehicle to help solder flow more readily.

Galvanic corrosion: A form of corrosion (rusting) resulting from the creation of an electric current flowing between dissimilar metals in contact with each other.

Grade: Classification of materials by quality or specific type.

Hand sealing: The process of adding additional adhesive to ensure new shingles adhere to old shingles, especially in cold weather. Also called *hand tabbing.*

Heartwood: Wood located at the center of a log.

Hook blade: A blade that has sharp, rounded edges used to cut materials, such as underlayment and asphalt shingles, with great precision.

Hot-air welding: A method of melting and fusing together the overlapping edges of separate sheets of roof membrane by the application of heat (in the form of hot air) and pressure. Also called *heat welding.*

Interlayments: Underlayments installed between each course of shakes or shingles.

Interlocking: A style of clay and concrete tile that has integrated connecting mechanisms on each tile piece that help link tile together during installation.

Laps: Areas where membrane sheets or flashings overlap the membrane below. Laps are sealed to create seams.

Liquid-applied roof membranes: Continuous roof membranes constructed in place with a liquid coating and a reinforcing material. The liquid coating is available as a one- or two-component product and is typically applied in two coats. Depending on the chemistry, a catalyst or hardener may be added to induce the curing process. The reinforcement is typically polyester or fiberglass mat set into a still-wet base coat.

Magnet sweeper: A push-broom tool that uses magnets to attract and collect metal waste at a jobsite.

Mechanical fasteners: Attachment methods that secure objects together through fastening. Examples include nails and screws.

Metal shingle roofs: Steep-slope metal roof systems installed in overlapping courses like traditional shingles.

Mil: A unit of measurement equal to one-thousandth of an inch, or 0.001" (0.0254 mm).

Mole run: A long wrinkle in a roof membrane.

Monolithic: Formed or composed of a single material; seamless.

Nailer: A piece or pieces of dimensional lumber and/or plywood secured to a structural deck or walls that provide(s) a receiving medium for the fasteners used to attach tiles or panels.

Overspray: Deposits of sprayed material that fall outside the intended area.

Perimeter metals: Metal components that seal the membrane at the edge of the roof and create a transition to the building wall.

Ply: A layer of reinforcement in a built-up or mod-bit roof system.

Point loading: The act of spacing out roofing materials in different areas on the roof so that all the weight of the roofing materials is not in one place.

Polymer-modified: A term used to describe bitumen that has been modified with polymers (such as APP or SBS) to add plastic or rubber qualities desirable for roofing.

Probing: Checking for voids or open areas in a seam using a seam probe hand tool.

Profiles: The ratio of a tile's height to its width; the lower the profile, the flatter the tile.

PVC: Polyvinyl chloride, a common type of thermoplastic used in roofing systems.

Redundant: Having multiple layers for extra protection.

Roof coatings: Liquids containing essential weatherproofing properties that are applied to roofs to protect, restore, prolong, or enhance performance.

Roof repair: Roofing work that involves fixing a deficiency, leak, or other damage to a roof system.

Roof service: Roofing work that involves routine maintenance of a roof system to prevent damage in the future and extend the life of the roof system.

Roofing hammer: A hammer that allows users to orient and nail shingles using just one tool.

Scrim: Reinforcement fibers in modified-bitumen sheets.

Seam: The area where two separate, overlapping sheets of roof membrane are joined together.

Seam probe: A small tool that is angled and pointed on the end and used to slide along a seam to test for loose sections.

Seam tape: Self-adhering cured or uncured synthetic-rubber tape used for creating seams between membrane sheets.

Self-adhered membrane: A membrane with factory-applied adhesive that is able to bond itself without the use of an additional adhesive.

Selvage edge: The edge of a mod-bit sheet with no granules, allowing it to be heat welded to the next sheet.

Silicosis: A lung disease caused by breathing in silica particles. Dusty jobs, like cutting roof tiles, can lead to silicosis if performed without respiratory protection.

Solder: A metal alloy often containing tin and lead that is melted using a hot iron to seal seams and join metals.

Soldering: A non-fusion process used to bond metal objects together with solder. See *solder*.

Solvent-based: Containing a petroleum-based solvent as a liquifying agent, leading to a higher content of organic compounds (see *volatile organic compounds*).

Spudding: Removing a gravel surface to expose the roof membrane.

Square: The amount of slate it takes to cover 100 square feet of roofing surface when installed with the appropriate head lap.

Standing seam panels: Architectural metal panels with overlapping vertical legs that get folded together during installation.

Standing-seam roofs: Metal roofs with a panel seam made by turning up the edges of adjacent metal panels and interlocking them in a variety of ways. Common to both architectural and structural metal roofs.

Starter strip shingles: Roll roofing or shingle strips applied along the downslope eave line before the first course of roof covering. These are intended to fill spaces between cutouts and joints of the first course.

Strip flashing: Membrane flashing strips used for sealing or flashing metal flashing flanges into the roof or waterproofing membrane.

Strip in: Apply membrane flashing strips to seal the edges of a metal flashing component.

Structural metal roofs: A metal panel system constructed of 24-gauge or thicker metals used primarily in low-slope applications but also for steep-slope applications.

Styrene butadiene styrene (SBS): A rubber-like polymer used to modify bitumen.

Surfacing material: Protects the membrane from ultraviolet (UV) degradation and weather erosion. A variety of materials are used for surfacing, such as a mineral aggregate, a liquid-applied coating, or a sheet material.

Taper: Increasing width uniformly to gradually widen or narrow.

Tapered edge boards: Strips used to modify the angle where the roofing membrane meets an eave; uniformly props up tile courses starting from the eave.

Tapered edge strip: A rigid board shaped with built-in slope and used in penetration flashings to provide slope toward drains.

Thermography: The creation of digital images using infrared cameras to detect the amount of heat energy emitted.

Thermoplastic: A roofing material designed to be heated and fused to itself. When cooled, it remains flexible, waterproof, and strong.

Thermoset: A class of polymers that when cured using heat, chemical, or other means, change into a substantially infusible and insoluble material.

Three-course patch: A repair method commonly used for BUR and modified bitumen flashings. It involves successive applications of a base course of asphalt roof cement, reinforcing fabric, and a second course of asphalt roof cement.

Tile adhesives: Attachment methods that adhere, or stick, objects together. For example, foam or mortar.

Tin snips: Useful tools for cutting and trimming thin but tough materials, such as sheet metal and shingles.

T-joint: An area where multiple field membrane sheets overlap.

TPO: Thermoplastic polyolefin, a common type of thermoplastic used in roofing systems.

Trim tile: An accessory piece provided by the manufacturer used to attractively weatherproof various areas of a roof, such as at ridges, hips, and rakes.

Ultraviolet radiation (UV) resistance: The ability of shingles to resist ultraviolet radiation from the sun degrading their quality over time.

Undercourse: A course of shakes or shingles installed beneath the covering course.

Valley: The internal angle formed by the intersection of two sloping roof planes.

Valley centerline: The center of the valley; no fasteners should be used within 6" (15 cm) of this point.

Viscosity: The consistency of a liquid. Liquids with high viscosity are thicker, while liquids with low viscosity are thinner.

Volatile organic compounds (VOCs): Compounds and chemicals that easily evaporate at room temperature. They are found in products such as solvents, paints, and adhesives.

Weighted roller: A heavy object pushed over adhered or self-adhered membrane to smooth it out and ensure sufficient adhesion to the substrate.

Wet film thickness: The thickness, expressed in mils, of a liquid coating as applied (wet).

Wood billet: Wood blocks from which shakes or shingles are made.

Wood shakes: Roof covering typically split from a wood billet and then resawed or shaped by a mill.

Wood shingles: Sawn products with uniform butt thickness relative to the specific length used as a roof covering.

Work order: A document that outlines the necessary tasks for a contracted job by showing the roofer where to look and what to do. On a roof repair job, the work order will outline the specific repair tasks the roofing company is expected to complete.

www.ingramcontent.com/pod-product-compliance
Lightning Source LLC
LaVergne TN
LVHW080023210926
848481LV00011B/3099